Value of Information for Healthcare Decision-Making

Value of Information for Healthcare Decision-Making introduces the concept of Value of Information (VOI's) use in health policy decision-making to determine the sensitivity of decisions to assumptions, and to prioritise and design future research. These methods, and their use in cost-effectiveness analysis, are increasingly acknowledged by health technology assessment authorities as vital.

Key Features:

- Provides a comprehensive overview of VOI
- Simplifies VOI
- Showcases state-of-the-art techniques for computing VOI
- Includes an R statistical software package
- Provides results when using VOI methods
- Uses a realistic decision model to illustrate key concepts

The primary audience for this book is health economic modellers and researchers, in industry, government, or academia, who wish to perform VOI analysis in health economic evaluations. It is relevant for postgraduate researchers and students in health economics or medical statistics who are required to learn the principles of VOI or undertake VOI analyses in their projects. The overall goal is to improve the understanding of these methods and make them easier to use.

Dr. Anna Heath is a Scientist at The Hospital for Sick Children, with affiliations at the University of Toronto and University College London. Her research aims to develop innovative statistical methods to design, prioritise and analyse clinical research within a Bayesian framework, with a focus on Value of Information methods.

Dr. Natalia Kunst is a Senior Research Fellow at the Centre for Health Economics at the University of York, with additional affiliations at the Yale University Schools of Public Health and Medicine and Harvard Medical School & Harvard Pilgrim Health Care Institute. Dr. Kunst is a decision sciences and health economics researcher with interests in uncertainty and evidence in decision-analytic modelling and health economic evaluations, value of information analysis, and health disparities.

Dr. Christopher Jackson is a Senior Statistician at the Medical Research Council Biostatistics Unit at the University of Cambridge. He works on developing statistical methods with applications to population health, in particular, methods for combining different sources of data, Bayesian methods, survival and multistate models, and developing statistical software.

Chapman & Hall/CRC Biostatistics Series

Recently Published Titles

Real World Evidence in a Patient-Centric Digital Era
Edited by Kelly H. Zou, Lobna A. Salem, Amrit Ray

Data Science, AI, and Machine Learning in Pharma
Harry Yang

Model-Assisted Bayesian Designs for Dose Finding and Optimization
Methods and Applications
Ying Yuan, Ruitao Lin and J. Jack Lee

Digital Therapeutics: Strategic, Scientific, Developmental, and Regulatory Aspects
Oleksandr Sverdlov, Joris van Dam

Quantitative Methods for Precision Medicine
Pharmacogenomics in Action
Rongling Wu

Drug Development for Rare Diseases
Edited by Bo Yang, Yang Song and Yijie Zhou

Case Studies in Bayesian Methods for Biopharmaceutical CMC
Edited by Paul Faya and Tony Pourmohamad

Statistical Analytics for Health Data Science with SAS and R
Jeffrey Wilson, Ding-Geng Chen and Karl E. Peace

Design and Analysis of Pragmatic Trials
Song Zhang, Chul Ahn and Hong Zhu

ROC Analysis for Classification and Prediction in Practice
Christos Nakas, Leonidas Bantis and Constantine Gatsonis

Controlled Epidemiological Studies
Marie Reilly

Statistical Methods in Health Disparity Research
J. Sunil Rao, Ph.D.

Case Studies in Innovative Clinical Trials
Edited by Binbing Yu and Kristine Broglio

Value of Information for Healthcare Decision-Making
Edited by Anna Heath, Natalia Kunst, and Christopher Jackson

For more information about this series, please visit: https://www.routledge.com/Chapman--Hall-CRC-Biostatistics-Series/book-series/CHBIOSTATIS

Value of Information for Healthcare Decision-Making

Edited by
Anna Heath, Natalia Kunst, and
Christopher Jackson

CRC Press is an imprint of the
Taylor & Francis Group, an **informa** business
A CHAPMAN & HALL BOOK

First edition published 2024
by CRC Press
62385 Executive Center Drive, Suite 320, Boca Raton, FL 33431, U.S.A.

and by CRC Press
4 Park Square, Milton Park, Abingdon, Oxon, OX14 4RN

CRC Press is an imprint of Taylor & Francis Group, LLC

ISBN: 978-0-367-74101-3 (hbk)
ISBN: 978-0-367-74111-2 (pbk)
ISBN: 978-1-003-15610-9 (ebk)

DOI: 10.1201/9781003156109

Typeset in Nimbus font
by KnowledgeWorks Global Ltd.

Publisher's note: This book has been prepared from camera-ready copy provided by the authors.

To Sam Evans. [AH]

To my mentors, Cary P. Gross, Karen M. Kuntz, A. David Paltiel and Ann C. Wu. Your kindness, knowledge and unselfish investments and belief in me have enriched my life beyond measure. [NK]

In memory of Richard Nixon, a statistician who made valuable contributions to decision science in medicine. [CJ]

Contents

Acknowledgements

This book was first conceived as a project for the Collaborative Network on Value of Information (ConVOI), a network of researchers aiming to improve the use of VOI methods in practice. While many of the members of ConVOI are named authors on chapters or sections of this book, their informal contributions are seen throughout and we would like to thank them all for their contributions to the field of VOI and the development of this resource. These include (in alphabetical order) Fernanado Alarid-Escudero, Gianluca Baio, Alan Brennan, Doug Coyle, Michael Fairley, David Glynn, Jeremy Goldhaber-Fiebert, Sabine Grimm, Hawre Jalal, Erik Koffijberg, Nicholas Menzies, Claire Rothery, Mark Strong, Howard Thom, Haitham Tuffaha, Nicky Welton and Edward C. F. Wilson. In addition to these collaborators and friends, the final version of this book has been informed by comments, discussion and suggestions from Petros Pechlivanoglou, Myriam Hunink, Eline Krijkamp, Janneke Grutters, Annisa Siu and Richard Nixon. We would also like to thank Sherry Thomas, Rob Calver and Vaishali Singh at Taylor & Francis Group for their support and patience during the development of this book, especially during the extended delays due to the COVID-19 pandemic.

Abbreviations

CEA Cost-effectiveness analysis

CEAC Cost-effectiveness acceptability curve

CEAF Cost-effectiveness acceptability frontier

CHEERS Consolidated Health Economic Evaluation Reporting Standards

CI Credible interval (or confidence interval)

COSS Curve of optimal sample size

ELC Expected loss curve

ENBS Expected net benefit of sampling

EVPI Expected value of perfect information

EVPPI Expected value of partial perfect information

EVSI Expected value of sample information

GAM Generalised additive model

GA Gaussian approximation (method for calculating EVSI)

GP Gaussian process

HRQoL Health-related quality of life

HTA Health technology assessment

ICER Incremental cost-effectiveness ratio

INLA Integrated nested Laplace approximation

IS Importance sampling (method for calculating EVSI)

JAGS Just Another Gibbs Sampler (software for Bayesian statistical modelling)

MARS Multivariate adaptive regression splines

AIC Akaike's information criterion

MC Monte Carlo

MCMC Markov chain Monte Carlo

MM Moment matching (method for calculating EVSI)

NB Net benefit

NHB Net health benefit

NMB Net monetary benefit

PA Probabilistic analysis

PSA Probabilistic sensitivity analysis

QALY Quality-adjusted life year

RCT Randomised controlled trial

RB Regression-based (method for calculating EVSI)

RWD Real-world data

SAVI Sheffield Accelerated Value of Information

SD Standard deviation

SE Standard error

VOI Value of Information

WTP Willingness-to-pay threshold

Preface

"Further Research Is Needed"

A stereotypical conclusion for a piece of scientific research is that *"further research is needed."*

But is this always correct? Is the investment in gathering more evidence always the best use of resources? Clearly, it is not always the case. Further research is pointless if the scientific question is settled. In that case, research would never change our decision. However, if there is uncertainty – then additional research may be worthwhile.

How do we determine whether collecting further evidence is worthwhile? Some important questions to ask are: Could the extra knowledge gained make a difference to any policy or practice? What are the expected benefits of better-informed decision-making? How do these benefits compare to the costs associated with the additional research?

This book is about mathematical methods to answer questions like these. These methods are collectively known as *Value of Information* analysis, which we sometimes abbreviate as *VOI.*

Value of Information methods are based on *Bayesian decision theory.* This theory is about situations where

- We have to decide between different courses of action, and identify the action that leads to the greatest overall benefit, while accounting for both good and bad outcomes (benefits and costs).
- The evidence informing the decision is incomplete, thus we are uncertain about the expected outcomes of the actions.
- We can specify a mathematical model: a simplification of reality which describes how benefits and costs arise under different courses of action.
- We can describe uncertainty about the truth by using probability statements about quantities in this model.

Once all these ingredients are specified, we can deduce the best course of action, and assess the uncertainty around that decision. Consequently, we can determine whether further research might reduce our decision uncertainty and lead to better informed decision-making. In practice, however, the mathematics required can make this process complicated, which has perhaps been the biggest barrier to the development and application of VOI. This book aims to make this process easier, hence to make the use of VOI methods more appealing to practitioners.

This book mainly describes the use of VOI in health and medicine, as part of *health economic evaluations*, a field where VOI methods have been developed extensively. However, many of the ideas we present can apply to any situation where decisions are made based on mathematical models with uncertain parameters.

Health Economic Evaluations

To set the context for the book, we briefly describe what health economic evaluations are and how Value of Information analysis fits in – though this is explained in more formal detail in Chapter 1.

Economics deals with situations where resources are finite, and assumes that people want to make the best use of those resources. Healthcare providers (whether financed through the state or through private insurance) frequently face policy decisions, such as whether a new drug should be reimbursed, which are subject to resource constraints. This requires considering the impact of different decisions on different "stakeholders." These include not only the patients who would receive the treatment but other people in the healthcare system, including patients who would be affected by resources being diverted to the new treatment.

Health economic evaluation is a tool to help with these decisions. There are agencies in several countries[1] who have developed formal frameworks for conducting health economic evaluations to inform decisions about treatment reimbursement. The most common framework is based around recommending the decision that would lead to the highest expected *net benefit* for the population. This is defined as a function of the expected health effects and expected costs, over some period of time, compared between different courses of action (treatments or other healthcare interventions). Health effects are most often expressed in terms of *quality-adjusted life years* (*QALYs*), a combined measure of expected survival and expected health. Costs include both the

1. For example, the Canadian Agency for Drugs and Technologies in Health (CADTH), the National Institute for Health and Care Excellence (NICE) in the United Kingdom, the Norwegian Medicines Agency (NoMA) and the Pharmaceutical Benefits Advisory Committee (PBAC) in Australia.

direct costs of delivering the intervention, and changes in costs elsewhere in the system if the intervention is paid for, often valued through the amount that a provider is *willing to pay* for a unit of health effects.

These health effects and costs should be estimated based on the *best available evidence* about the disease and the interventions considered, relevant to the decision being made. This can be a complex process, which often involves a mathematical *model* – a simplification of reality to analyse the consequences of decisions. This model is usually a *function* of various quantities describing various health states and pathways, for example, the expected survival of patients, disease progression, effectiveness of treatments and the costs of healthcare. These quantities are known as model *parameters*.

Model parameters are typically estimated based on observing what has happened in the past to patients similar to those of interest, for example, through clinical trials, or routinely collected, observational data. Parameters are naturally *uncertain* to some extent, because we can never observe the whole population of interest. This is why the discipline of *statistics* is important to health economic evaluations. A closely associated idea is *probability*. Probability can be used not only to describe variability in outcomes between individuals, but also to describe uncertainty about our knowledge of the model parameters. This gives us a way of describing mathematically how much uncertainty there is in a decision, and a way of prioritising the collection of further research to reduce uncertainty. This is Bayesian decision theory and Value of Information analysis.

Value of Information methods answer questions like:

- Is further research potentially beneficial?
- What are the drivers of the current uncertainty in our decision? In other words, for which uncertain quantities is it most worthwhile to conduct further research?
- What would be the expected benefits from conducting a specific research study of a particular design or sample size?
- Do the potential benefits of that study outweigh the costs of conducting it?

The essential idea is to compare the decision based on current evidence with the decisions that might be made if more information was available, and estimate the expected improvements (combining health effects and costs) that would result from this additional information and a better informed decision.

A Brief History of Value of Information Analysis

The first description of the Value of Information concept that we are aware of is in the book by Raiffa and Schlaifer (1961), which was written during a period of rapid developments in statistical theory, and Bayesian statistics in particular, following the work of researchers such as Fisher, Neyman, Jeffreys and Lindley. As with many statistical concepts, the ideas are simple to state as mathematical abstractions – but their routine use in practice is harder.

VOI methods became better-known in healthcare contexts many years later, around the end of the century (see, e.g., Claxton 1999a). This was motivated by the rapid development of health economics as a field of research, due to an increasing awareness of the constraints on healthcare systems. During the same period, the importance of evidence-based decision-making in health became better appreciated. This inspired many developments in research methods, for example, in randomised controlled trials, systematic reviews of evidence, and meta-analysis of published research. Statistical methods are central to these fields, due to the need to carefully consider uncertainty and variability.

In the health economic perspective on evidence-based healthcare, the role of decision-making, and the explicit consideration of costs and effects, is made central. A lot of work was done around the turn of the century to develop a quantitative framework for health economic evaluation, encompassing decision theory and the probabilistic (or Bayesian) view of uncertainty, and describing the relation of this framework to other common statistical concepts (such as hypothesis testing, Claxton (1999b)).

These foundational developments have in turn inspired practical developments, which have made VOI methods gradually easier to use. While the mathematical framework is elegant, performing the computations has traditionally been hard. However, the past 20 years have seen an explosion of research on computational algorithms to calculate VOI more efficiently. The core computational tool is *Monte Carlo simulation*, or generating large samples of random numbers governed by probability distributions. This has been facilitated by computers becoming more powerful, and software becoming more user-friendly. Meanwhile, as the underlying principles of evidence-based research, decision theory and probability have become better known, they have become more embedded in practice, with the publication of various "good practice" guidelines for health economic evaluations by healthcare policy and academic institutions (Briggs et al. 2012; Caro et al. 2012; CADTH: Canadian Agency for Drugs and Technologies in Health 2017). Some of these have focused on emerging good practices in VOI analysis (Fenwick et al. 2020; Rothery et al. 2020).

With this book, we hope to bring together relevant information that has previously been scattered around academic papers, books and methods guidelines, to explain VOI methods and show how to use them efficiently in practice for decision-making in health. Thus, while we acknowledge that VOI methods have been used in many other fields, e.g., environmental risk (Yokota and Thompson 2004), engineering (Davis, Kisiel and Duckstein 1972; Bratvold, Bickel and Lohne 2009; Eidsvik, Mukerji and Bhattacharjya 2015) and management (Repo 1989), this book will not explicitly address how VOI can be used in these settings, where our experience is limited!

Aims and Broad Overview of the Book

This book aims to describe the state of the art of VOI methods, and to explain how to use them in health economic evaluations. We show you that it is possible to use VOI methods to improve decision-making and inform research planning in a range of useful situations.

In Chapter 1, we explain the fundamentals of health economic modelling, Bayesian decision theory and uncertainty quantification required to understand Value of Information analysis, and introduce a running case study in Chapter 2.

The core of the book is two chapters (Chapters 3 and 4) that explain the main tools of VOI analysis. We explain the different computational methods that can be employed, and give worked examples that show how to use them in practice (Chapters 3–5). We then describe a series of examples where VOI analyses have been used in real policy-making situations (Chapter 6).

The final part of the book (Chapters 7–12) aims to showcase VOI methodology as an active field of research. This is formed of several chapters, contributed by a range of different authors, describing variants and special applications of VOI ideas.

Who Should Benefit from the Book? What Is and Isn't Assumed

The main intended audience of the book is practitioners of health economic modelling. The methods, however, are applicable to any situation where decisions are made based on models with uncertain parameters. So the book may also help people working in other decision-making situations. We explain the basics of health economic modelling required to understand VOI analysis, though the book isn't intended as a guide to health economics or how to do health economic modelling. Similarly, this book is not about how to conduct statistical analyses of data, or how to review evidence to inform health economic models. We do, however, explain the ideas of uncertainty quantification that are necessary to perform VOI analysis, and give examples of how these ideas can be used in typical situations.

The technical level is roughly that of a masters' level or postgraduate scientific textbook. We assume you are comfortable with mathematical notation, and understand the idea of probability and basic statistics (e.g., expectations and variances).

In general, we do not assume knowledge of Bayesian methods or decision theory, and the basics of these are described in Chapter 2. However, for readers to implement some of the methods for calculating the expected value of sample information that are explained in Chapter 4, knowledge of Bayesian methods and general statistical skills are required – see Table 4.2 for a list of these skills.

For the parts of Chapters 3 and 4 where we demonstrate the use of the `voi` `R` package to calculate measures of the expected value of information, knowledge of `R` is assumed. In particular, the ability to write `R` functions is required to implement some of the methods in Chapter 4.

Complementary Resources

For more information about health economics in general, and health economic evaluation, see, for example, Drummond et al. (2015). Briggs, Sculpher and Claxton (2006) give a practical tutorial on decision modelling, and Willan and Briggs (2006) cover the statistical analysis of cost-effectiveness data. A series of papers on good research practices in modelling to inform health technology assessment were also published by the International Society for Pharmacoeconomics and Outcomes Research (ISPOR) and the Society for Medical Decision Making (SMDM) (Caro et al. 2012).

See Baio (2012) for an overview of Bayesian methods in health economics, which also covers the principles that underlie VOI analysis and some of the simpler methods for VOI computation. A related software package with tools for applied cost-effectiveness analysis is presented in Baio, Berardi and Heath (2017). For resources on evidence synthesis and meta-analysis, see Welton et al. (2012) and Dias et al. (2018), and see Spiegelhalter, Abrams and Myles (2004) for a broader perspective on statistical methods in healthcare evaluation, and Bayesian methods in particular. Finally, see Parmigiani and Inoue (2009) for a comprehensive introduction to Bayesian decision theory.

Detailed Overview of the Book

- In Chapter 1, we will give a rapid overview of the process of health economic modelling and evaluation, and the mathematical and conceptual principles that underpin the methods of VOI analysis. We describe the ingredients of a health economic evaluation, including health outcomes such as survival and health-related quality of life, and healthcare costs. We describe the decision-theoretic approach that is used to combine these ingredients to inform policy-making, and explain the idea of decision-analytic models and how they are informed by evidence.

A key notion introduced here is *uncertainty* in models, and how this can be described by probability distributions for model parameters. This is the essence of Bayesian decision theory and VOI methods. We explain how *probabilistic analysis* of economic models is implemented by using Monte Carlo simulation from the joint distribution of model parameters and associated model outcomes (expected costs, effectiveness and net benefits). This is the key procedure that enables VOI measures to be computed in practice. We also explain the relation between VOI and other measures of uncertainty such as the "probability of cost-effectiveness."

- Chapter 2 introduces a case study of a chemotherapy treatment, that will be used throughout the book to demonstrate VOI methods. While it is artificial, it is designed to look like a typical model used in health economic evaluations. This is a discrete-time Markov model for transitions between clinical states, where the transition probabilities, health utilities, state costs and treatment effects on these can be estimated from simple summaries of data.

- Chapter 3 explains the expected value of perfect information (EVPI) and the expected value of partial perfect (or perfect parameter) information (EVPPI). The EVPI describes the expected benefits of eliminating all decision uncertainty by learning the exact values of all the parameters in a model. The EVPPI describes the expected benefits of learning about the exact values of specific input parameters or sets of parameters. Hence, it provides the expected value of reducing uncertainty in a model by eliminating uncertainty from specific input parameters. These values can be compared to the costs of research about the parameters. If the costs of research are greater, that research can be judged to be not worthwhile.

  We explain in detail the different methods that are available to compute EVPPI in practice. While these are easy to use in general, they sometimes need some care. The chapter ends with a worked example of how EVPPI is computed, presented and interpreted in the case study.

- Chapter 4 explains the expected value of sample information (EVSI). This describes the expected value of a proposed study of a finite sample size, which will yield "imperfect" information about one or more model parameters. This is slightly more conceptually and computationally challenging, since it requires an extra ingredient of specifying a statistical model for how these data might be generated. Similar to EVPPI, there are several alternative methods for computing EVSI, and we explain the most widely applicable of these methods. The use of these methods is demonstrated in various realistic situations where we might want to obtain the kind of data that are of interest in health economic evaluations. We focus in particular on the presentation and communication of results.

- Chapter 5 focuses on the use of VOI in real decision-making contexts, and in particular how VOI analyses are presented. Firstly, we present a set of standardised guidelines for the reporting of VOI analyses. Secondly, we present a series of graphics and measures which can help the process of communicating the results of VOI analyses in the context of real-world decision-making about reimbursement of health technologies.

- Chapter 6 presents a series of VOI "success stories": examples where VOI methods have been used in the context of real health economic policy-making situations.

The final part of the book showcases various topics of current research in VOI methodology. An international network has recently been founded to promote and develop VOI methods in health economic evaluations, the Collaborative Network for Value of Information (ConVOI): `http://convoi-group.org`. Recent ConVOI publications have included Heath et al. (2020), Kunst et al. (2020), Heath et al. (2022) and Kunst et al. (2023a). This part is inspired by the interests of members of this network. The material in this part involves a mixture of statistical theory, economic theory and descriptions of applied tools for specialised varieties of VOI analyses.

- Chapter 7 focuses on the use of VOI methods in problems where, instead of explicit decision-making, mathematical models are built to simply estimate a quantity of interest. VOI techniques are still useful here for sensitivity analysis, research prioritisation and study design. While this work was motivated by examples in public health, the ideas are general to all mathematical models of the same form.

- Chapter 8 discusses how VOI methods relate to the idea of *structural uncertainty* – the notion that models themselves are imperfect, but VOI methods can be useful to assess sensitivity of models to their assumptions and prioritise model improvements.

- Chapter 9 is a presentation of how VOI methods can be implemented without building a comprehensive health economic model. This is known as the *minimal modelling* approach, and enables VOI methods to be used efficiently and conveniently in certain contexts.

- Chapter 10 describes the extension of VOI ideas to *portfolio optimisation*: where a decision maker responsible for research funding must decide which out of a set of proposed research studies should be funded, given a fixed budget.

- Chapter 11 discusses the assessment of the expected VOI in situations of *non-optimal decisions*. This acknowledges that in practice, the treatments or health technologies that get used are not always those with the highest net benefit.

- Chapter 12 further discusses the assessment of the expected VOI in specific kinds of non-optimal decisions, where the implementation of the optimal decision is *imperfect*.

R Package and Demonstration Code

- The VOI analyses that we demonstrate in the book will be implemented using the `R` software. We have written an `R` package called `voi`, which implements the most practically useful methods for computation of the EVPPI, EVPI and EVSI. This is freely available from the standard CRAN repository of `R` packages, and full information about it is available at `https://chjackson.github.io/voi/`.
- All code demonstrated in the worked examples is made freely available, and is collected and linked from `https://chjackson.github.io/voi/articles/book.html`.
- Section 3.3.8 also describes some web-based applications that could be used to calculate many of these measures, and discusses their features and limitations.

Contributors

Fernando Alarid-Escudero
Assistant Professor of Health Policy
Stanford University School of Medicine
Stanford, California, USA

Alan Brennan
Professor of Health Economics and Decision Modelling
Population Health, School of Medicine and Population Health
University of Sheffield
Sheffield, UK

Doug Coyle
Professor
School of Epidemiology and Public Health
Faculty of Medicine
University of Ottawa
Ottawa, ON, Canada

Michael Fairley
Senior Applied Scientist
Uber
San Francisco, California, USA

David Glynn
Research Fellow
Centre for Health Economics
University of York
York, UK

Jeremy Goldhaber-Fiebert
Professor of Health Policy
Stanford University School of Medicine
Stanford, California, USA

Sabine Grimm
Health Economist
Department of Epidemiology and Medical Technology Assessment (KEMTA)
Maastricht Health Economics and Technology Assessment (Maastricht HETA) Center
Maastricht University Medical Center
Maastricht, The Netherlands

Anna Heath
Scientist
Child Health Evaluative Sciences
The Hospital for Sick Children

Assistant Professor
Division of Biostatistics
Dalla Lana School of Public Health
University of Toronto
Toronto, ON, Canada

Honorary Research Fellow
Department of Statistical Science
University College London
London, UK

Chris Jackson
Senior Statistician
MRC Biostatistics Unit
University of Cambridge
Cambridge, UK

Hawre Jalal
Associate Professor
School of Epidemiology and Public Health
Faculty of Medicine
University of Ottawa
Ottawa, ON, Canada

Hendrik Koffijberg
Full Professor
Health Technology & Services Research
University of Twente
Twente, The Netherlands

Natalia Kunst
Senior Research Fellow
Centre for Health Economics
University of York
York, UK

Claire Rothery
Senior Research Fellow
Centre for Health Economics
University of York
York, UK

Mark Strong
Professor of Public Health
Population Health, School of Medicine and Population Health
University of Sheffield
Sheffield, UK

Howard Thom
Senior Lecturer in Health Economics
Bristol Medical School (PHS)
University of Bristol
Bristol, UK

Haitham Tuffaha
Professor
Centre for the Business and Economics of Health
University of Queensland
Brisbane, Australia

Nicky Welton
Professor
Bristol Medical School (PHS)
University of Bristol
Bristol, UK

Edward C. F. Wilson
Professor of Health Economics and Health Policy
Peninsula Technology Assessment Group
University of Exeter
Exeter, UK

Part I

Theory and Calculation of Value of Information

1

Health Economic Modelling

Natalia Kunst, Christopher Jackson, and Anna Heath

Healthcare policy makers frequently face decisions about how to allocate scarce resources to maximise health. Although these decisions must be made, they are generally based on imperfect information, often resulting in uncertainty. To guide policy makers, the consequences of different decisions are often estimated using a model-based cost-effectiveness analysis (CEA), or other types of economic evaluation. This allows evaluation of the expected health effects and costs of each decision option, as well as the *uncertainty* surrounding these expected values and the decision. Understanding the consequences of making a suboptimal decision helps inform the potential need for gathering more evidence to support decision making. A value of information (VOI) analysis plays a central role in this assessment.

This chapter introduces the theoretical background for healthcare policy making, and the mathematical and conceptual principles that underpin the methods of VOI analysis. We start with a brief introduction to CEA and the different components and measures that are involved. We thereafter outline the main types of decision-analytic models, and provide a brief overview of the typical kinds of evidence that are used to inform CEAs. Finally, we introduce the idea of *uncertainty* in model-based CEAs. Specifically, we describe *probabilistic analysis*: the framework of representing uncertainty by probability distributions for model input parameters. This forms the foundation of VOI analysis. We explain how such probability distributions can be defined based on data, and how probabilistic analysis can be implemented by using Monte Carlo simulation from the joint distribution of model input parameters to propagate uncertainty to model outcomes such as expected costs, health outcomes and net benefits. Finally, we describe different measures of decision uncertainty that can be obtained from a probabilistic analysis, which motivates the idea of VOI.

DOI: 10.1201/9781003156109-1

1.1 Economic Evaluation in Health and Medicine

Decisions are constantly being made in health and medicine. These decisions concern the broad spectrum of healthcare, including implementation of different policies, services, diagnostics, treatments and medications. Choosing one decision option instead of another involves tradeoffs (Hunink et al. 2014). In health and medicine, all alternative decision options are associated with specific health outcomes and use of resources. While new interventions may provide better health outcomes, they often require greater resource use (Drummond et al. 2015). One of the decision makers' aims is to maximise health outcomes given scarce resources. Thus, decision makers face the need to consider a decision's consequences for both expected health effects and resource use. These consequences might include the effects outside of the healthcare sector (e.g., productivity loss) (Neumann et al. 2016). An economic evaluation plays an essential role in achieving this aim, and is often used to guide decision and policy makers in their choices.

1.1.1 Theoretical Foundations

An *economic evaluation* is an analysis that compares costs and outcomes of alternative decision options (e.g., treatment regimens, diagnostic strategies and preventive interventions), and aims to identify the decision option that optimises the use of scarce resources to maximise the utility associated with the desired outcomes (Drummond et al. 2015; Neumann et al. 2016). Economic evaluation is based on expected utility theory, which was reviewed in detail in Savage (1954), Raiffa (1968), Lindley (1956), Berger (2013), Bernardo and Smith (2009) and Parmigiani and Inoue (2009). Economic evaluation has two main characteristics:

1. It links use of resources to a set of resulting health outcomes and costs.
2. It is associated with choosing between different possible decision alternatives.

There are a number of economic evaluation types, but only those evaluations that consider both costs and health outcomes, and compare at least two alternatives, are considered a full economic evaluation, with all other options being considered partial evaluations (Drummond et al. 2015). A number of techniques for economic evaluations can be distinguished, mostly depending on the applied approaches taken to measure health outcomes (Drummond et al. 2015). Here, a *cost-effectiveness analysis* (*CEA*) (including cost-utility analysis (CUA)) represents the most widely used economic evaluation technique in health and medicine.

CEA denotes an economic evaluation that compares two or more decision options, in terms of their costs and effects, where effects are expressed in natural health units (e.g., life years or screen-detected cases) or health-related quality of life (HRQoL) (Hunink et al. 2014; Drummond et al. 2015). It aims to maximise population health with limited resources (Hunink et al. 2014), and can be grounded in welfare-economic theory (Basu 2020). Given that most decision makers have budgetary or resource constraints, choosing one decision option over another is associated with opportunity costs representing the value given up when making the decision (i.e., the value placed on the rejected decision option) (Buchanan 2008; Drummond et al. 2015). A CEA aims to estimate this opportunity cost and to compare it with the value of the benefits achievable with the decision option being evaluated (Drummond et al. 2015). Because most healthcare decision makers operate within a defined budget, the opportunity cost of choosing a new intervention represents the health forgone resulting from no longer available resources. These no longer available resources were previously used to provide effective healthcare to patients with different healthcare needs. Thus, determining if a new intervention is cost-effective requires examining whether the expected additional health gained exceeds the health forgone due to resources being diverted (Drummond et al. 2015). Most CEAs reflect these opportunity costs (i.e., health forgone) using a cost-effectiveness threshold, which can be defined as an estimate of the cost per health forgone (Neumann et al. 2014; Suhrcke et al., accessed 2023). The CEA is a broader concept that includes CUA. While CEA denotes evaluations that express health effects in either natural health units or HRQoL, CUA describes the subset of CEAs that express health effects in HRQoL.

1.1.2 Components of Cost-Effectiveness Analysis

Each CEA should start by identifying all decision options that are relevant in clinical practice. Importantly, the CEA should not be limited to those decision options for which data are available, but rather all those that are feasible and relevant should be considered (Sculpher, Drummond and Buxton 1997; Caro et al. 2012; Neumann et al. 2016). Once all relevant decision options have been identified, the desired health outcomes must be specified, and the resource use associated with each decision option identified.

1.1.2.1 Health Outcomes

Health outcomes in CEAs can be expressed in various natural units, both single and multidimensional (Hunink et al. 2014; Drummond et al. 2015).

- Measures of single health outcomes include, for example, observations of survival or mortality, screen-detected cases or cases averted.

- Multidimensional measures of health outcome account for both clinical outcome and quality of life, and can be categorised into disease-specific and generic measures of HRQoL (Drummond et al. 2015).

Although disease-specific measures may perform better in capturing small changes in HRQoL because their instruments measure aspects relevant to a particular disease or condition, there are several advantages of using generic measures that are applicable across various diseases or conditions (Michel et al. 2016). More specifically, the use of generic measures in CEAs allows decision makers to compare health outcomes across different health states and diseases. Consequently, decision makers can compare health gained to the health forgone in other areas covered by the same budget due to resources being diverted. Furthermore, the use of generic measures enables medical decision-making to be consistent between different diseases and populations (Drummond et al. 2015).

Over time, the quality-adjusted life year (QALY) has become the most commonly used generic measure of health outcomes applied in CEA in health and medicine (Neumann et al. 2016; Brazier et al. 2017). QALYs account for both the HRQoL and the length of life (see Figure 1.1). The HRQoL is incorporated into QALYs by assigning different utility values to various health states. These values are expressed on a scale between 0 (analogous to death) and 1 (a state of perfect health), and reflect an individual's preferences for a specific state of health, in line with expected-utility theory (Drummond et al. 2015; Hunink et al. 2014).

A number of techniques can be distinguished to measure HRQoL. These techniques can be broadly grouped into direct and indirect methods. Two main theoretical traditions for direct methods are expected utility theory and psychophysics (Neumann et al. 2016). The key techniques to assess the utility of a given health state, following expected utility theory, include the *standard gamble* and the *time tradeoff* (Hunink et al. 2014). The standard gamble method follows the axioms of expected utility theory, and involves providing the respondent with a choice between a certain option and a risky option. The time tradeoff method involves determining the length of time that the respondent would be willing to give up for a more desirable health state. For the methods that follow psychophysics, rating scale methods, including category scaling and visual analog scales, are well-known approaches. These methods involve assigning a number to each considered health state, typically using a scale from 0 to 100, where 0 represents the least desirable health state, and 100 represents the most desirable. An alternative approach to measure HRQoL is the use of an "indirect" method, involving "multi-attribute" utility measures. This approach predicts utility measures from descriptive assessments of health states, and consists of two main components: a classification system of health status and a preference-based scoring system (Hunink et al. 2014).

As outlined above, the utility values used to estimate QALYs in CEAs are often obtained either directly using various utility measuring techniques or

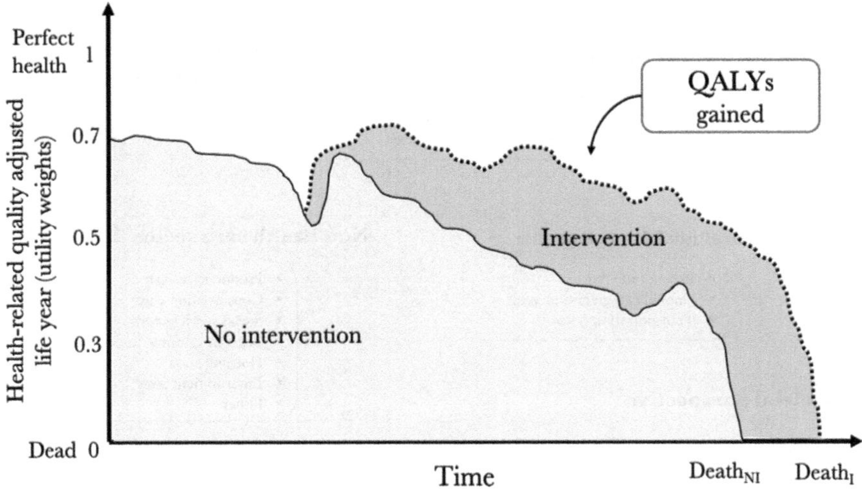

FIGURE 1.1
Quality-adjusted life years with and without intervention. The figure was adapted from Gold, Siegel and Russell (1996). DeathNI denotes the time of death without intervention (i.e., "No intervention") and DeathI denotes the time of death with intervention (i.e., "Intervention").

indirectly using so-called multiattribute utility measures. Examples of the most commonly used preference models include the five-dimensional EuroQoL (EQ-5D), the Health Utilities Index (HUI) and the six-dimensional Short Form (SF-6D) (Donlan 1997; Brazier, Roberts and Deverill 2002; Brazier et al. 2010). A CEA with health effects expressed in QALYs is sometimes referred to as a CUA (Hunink et al. 2014).

Costs

For the purpose of CEA, resource use associated with all considered decision options should be carefully identified, quantified and valued, and incorporated into a decision-making framework (e.g., decision-analytic model) (Neumann et al. 2016). This framework allows the estimation of the incremental, or decremental, costs resulting from a new intervention. Healthcare interventions may have an impact on use of resources both in the healthcare sector as well as in sectors outside of healthcare.

Cost components that may be included in a CEA are presented in Figure 1.2. As proposed in the recommendations of the Second Panel on Cost-Effectiveness in Health and Medicine (Neumann et al. 2016), resource use and the associated costs can be categorised into consumption and consequences associated with the considered decision options. Both costs of consumption and costs of consequences can take place within and outside of the healthcare sector. The type of resources that should be included in a CEA depends on

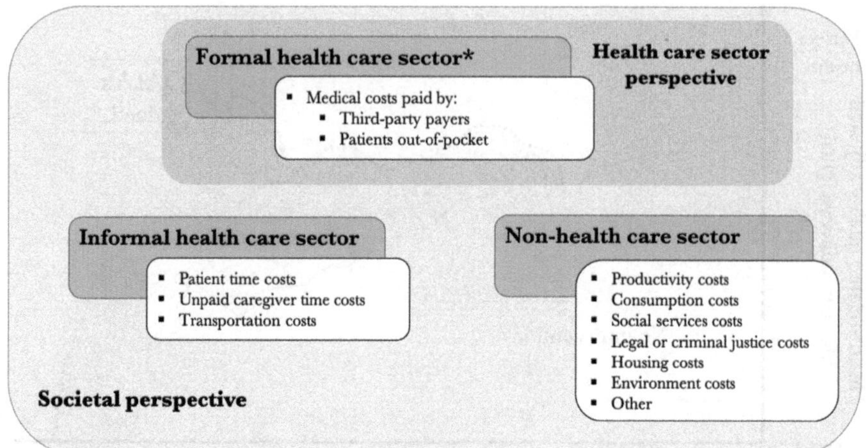

FIGURE 1.2
Cost components that may be included in a cost-effectiveness analysis. The figure was based on the Impact Inventory proposed by Neumann et al. (2016). *Considers both current costs and future costs that are related or unrelated to the health condition of interest.

the perspective of the analysis. CEAs are most commonly performed from a societal, healthcare sector, or healthcare payer perspective (Kim et al. 2020). While the societal perspective is the widest point of view, as this perspective considers resources used in both healthcare and non-healthcare sectors, the healthcare sector perspective considers resources used in the healthcare sector, and the healthcare payer perspective considers only those resources that are relevant to a particular payer and are, therefore, directly relevant for the objective of maximising the health of the population (Neumann et al. 2016). Most importantly, the perspective should be chosen based on the context of the decision that is to be informed.

1.2 Decision Criterion: Measures

As previously stated, health economic modelling involves estimating the health effects and costs for each decision d under consideration, denoted by the pair (e_d, c_d). These can be defined as the *total* costs and effects over a population that would be affected by the decision.

However, for the purpose of modelling, they are usually defined as *expected* costs and effects for an individual i in this population: as $e_d = \text{E}(e_{di}), c_d =$

$E(c_{di})$. The expectation is defined with respect to the *distribution* of actual costs and effects experienced by different individuals.

These definitions are practically equivalent for decision-making, since the expected value (or "mean") is estimated by the total divided by the size of the population. However, working with expectations is more convenient for modelling. The population size is difficult to define in practice, and statistical models for expected outcomes are generally constructed in terms of parameters that can be interpreted as population average effects.

1.2.1 Comparing Costs and Effects

When evaluating two or more mutually exclusive alternative decision options, the aim is to identify the decision option that leads to the highest health effects at acceptable costs (Drummond et al. 2015).

In some cases, the decision is clear (Hunink et al. 2014; Drummond et al. 2015). This includes circumstances in which one decision option is associated with higher health effects and lower costs compared with the other alternative decision options (i.e., it is cost-saving). This type of decision option is clearly the optimal choice as it "*dominates*" all the other alternatives.

Other examples of clear decisions include circumstances where one decision option results in lower health effects and higher costs than one or more of the other considered alternatives. In this case, the decision option is clearly not cost-effective as it is "*dominated by*" the other alternative(s).

If a decision option leads to higher health effects, but at increased costs compared with the other alternatives, the decision about its cost-effectiveness depends on the *willingness-to-pay threshold* (sometimes abbreviated to "willingness-to-pay," or WTP). The WTP threshold accounts for health forgone due to other healthcare activities being displaced to accommodate the additional costs of the decision option considered (Claxton et al. 2015). In other words, if the objective of the decision maker is to maximise the health of the population, the threshold must represent the value forgone as a consequence of the additional cost – this value is known as the "opportunity cost."

In practice, to determine the optimal decision option, the health and cost outcomes must be combined, using the willingness-to-pay threshold, into a quantity that represents the overall value of each of the decision options. We now discuss two related approaches to this problem: (a) the *net benefit*, the more general approach and the basis of Value of Information (VOI) analysis, and (b) the *incremental cost-effectiveness ratio*, a more conventional, but more technically awkward, approach.

1.2.2 Net Benefit

In health economic evaluations, the overall measure of value of a decision is defined by the *net benefit* (NB) (Briggs, Sculpher and Claxton 2006; Stinnett and Mullahy 1998; Claxton 1999b). NB can be expressed on the cost scale as the *net monetary benefit* (NMB) or on the effect scale as the *net health benefit* (NHB). These lead to equivalent decision-making in practice, and are defined as

$$\begin{aligned} \text{NMB}_d &= k \times e_d - c_d \\ \text{NHB}_d &= e_d - \frac{c_d}{k} \end{aligned}$$

recalling that c_d and e_d are the expected costs and effects of decision option d for an individual in the population of interest.

k indicates the cost that a decision maker is willing to pay for a unit of health effects (WTP). This quantity is often also denoted by λ. NMB_d or NHB_d is estimated for each considered decision option $d = 1, \ldots, D$. The decision option providing the highest NB estimate is considered the optimal and cost-effective option at a particular WTP threshold (Hunink et al. 2014).

At this point, it is important to note that the ranking of the decision options is based on the willingness-to-pay threshold k, which may be defined as an interval or as a fixed number. As the net benefit is linear in k, it is easy to calculate the expected NMB or NHB for many different values of k, and therefore perform sensitivity analysis to the assumed willingness-to-pay threshold.

1.2.3 Incremental Cost-Effectiveness Ratio

A related quantity that measures the economic value of decision options is the *incremental cost-effectiveness ratio* (ICER, Hunink et al. 2014; Drummond et al. 2015; Neumann et al. 2016). This is defined as:

$$\text{ICER}_{d,\ d-1} = \frac{c_d - c_{d-1}}{e_d - e_{d-1}} = \frac{\Delta c_{d,d-1}}{\Delta e_{d,d-1}},$$

where c_d and e_d are the population average costs and effects of decision option $d = 2, \ldots, D$, c_{d-1} and e_{d-1} are the population average costs and effects of the next least costly decision option, while $\Delta c_{d,d-1}$ and $\Delta e_{d,d-1}$ are the population average incremental costs and incremental effects gained.

The quantities in this formula can be illustrated in an *incremental cost-effectiveness plane* (Figure 1.3), with the x-axis representing incremental health effects $\Delta e_{d,d-1}$ and the y-axis representing incremental costs $\Delta c_{d,d-1}$.

- If $\text{ICER}_{d,d-1}$ is located in quadrant II (i.e., reduced effects ($\Delta e_{d,d-1} < 0$) and increased costs ($\Delta c_{d,d-1} > 0$), the decision option is *dominated.* Thus, it is deemed not cost-effective and is excluded from the evaluation.

- If $\text{ICER}_{d,d-1}$ is located in quadrant IV (i.e., increased effects $[\Delta e_{d,d-1} > 0]$ and reduced costs $[\Delta c_{d,d-1} < 0]$), the evaluated decision option is dominant and is considered effective and cost-saving.

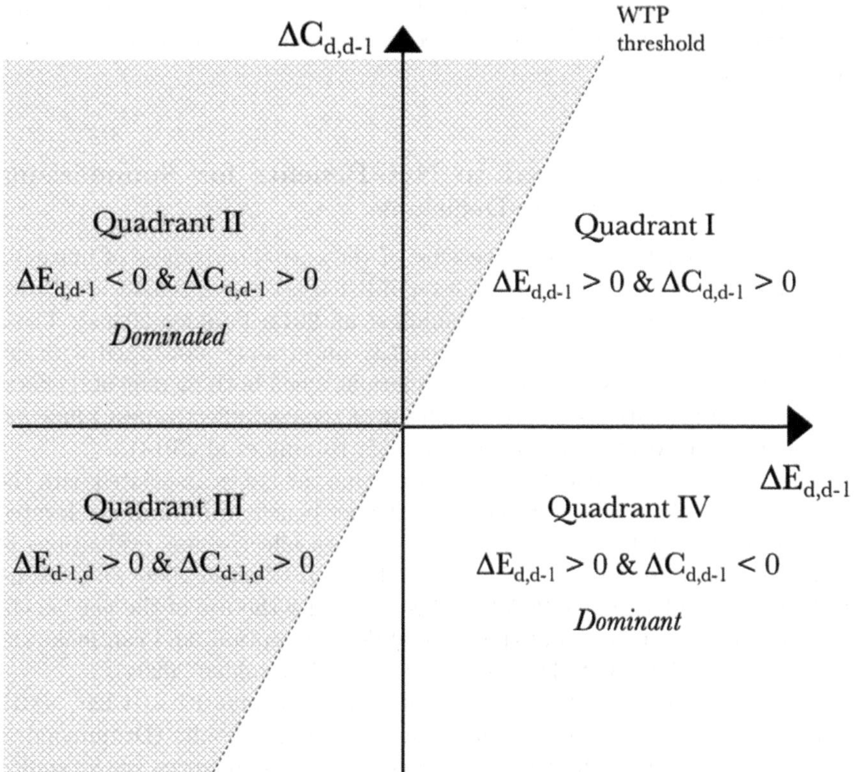

FIGURE 1.3
Incremental cost-effectiveness plane. The grey, dotted area indicates cost-ineffectiveness given a specific willingness-to-pay (WTP) threshold. Specifically, if the ICER estimate is located in this grey, dotted area, the decision option is considered not cost-effective compared with the comparator, given the assumed WTP threshold. The figure was based on Hunink et al. (2014). Δc, incremental costs; Δe, incremental effects; WTP, willingness-to-pay.

- Most often $\text{ICER}_{d,d-1}$ is located in quadrant I (i.e., increased effects [$\Delta e_{d,d-1} > 0$] and increased costs [$\Delta c_{d,d-1} > 0$]), requiring an assessment of whether the increased costs ($\Delta c_{d,d-1} > 0$) are justified by the increased effects ($\Delta e_{d,d-1} > 0$) using the WTP threshold. Here, the decision option d would be considered cost-effective if the $\text{ICER}_{d,d-1}$ lies below the WTP threshold.
- In quadrant III, it is good practice to define the ICER as $\text{ICER}_{d-1,d}$, where the evaluated decision option d is the next least costly, and the alternative is the more costly and more effective decision option (i.e., increased effects $\Delta e_{d-1,d} > 0$ and increased costs ($\Delta c_{d-1,d} > 0$]). In these circumstances,

the WTP threshold line crosses $\text{ICER}_{d,d-1}$ (Figure 2.3). If $\text{ICER}_{d-1,d}$ lies below the WTP threshold line, the decision option $d-1$ would be considered cost-effective.

1.2.4 ICERs Compared to Net Benefits for Summarising Health Economic Decisions

Although the ICER is a useful measure of the cost-effectiveness of mutually exclusive alternatives, the interpretation of ICER estimates is associated with several issues (Briggs et al. 2012; Hunink et al. 2014; Paulden 2020a). These issues include how to appropriately quantify uncertainty associated with decisions based on ICERs, especially in situations when both incremental effects and incremental costs are in any quadrant of the cost-effectiveness plane except quadrant I (Wakker and Klaassen 1995; Hunink et al. 2014).

The use of the net benefit can help overcome the issues associated with the ICER. Specifically, it is useful to present the net benefit when ICER is located in quadrants other than quadrant I, and in cost-effectiveness evaluations of more than two alternatives (Briggs, Sculpher and Claxton 2006; Drummond et al. 2015; Paulden 2020a, 2020b). This is because the use of the net benefit simplifies the identification of the optimal decision option, and can make the results easier to interpret (Drummond et al. 2015; Paulden 2020a).

It is recommended that one or more of these three quantities (ICER, NMB, NHB) are presented when reporting cost-effectiveness results (Drummond et al. 2015; Neumann et al. 2016). All three of these summaries are grounded in constrained optimisation and apply the same decision rule (Neumann et al. 2016). That is, regardless of which framework is reported, the decision resulting from the analysis will be the same. To illustrate this, suppose we are simply comparing two decisions, $d = 1$ and $d = 0$, say. The *incremental net benefit* (for a WTP threshold of k) is defined as

$$\text{INB} = \text{NMB}_1 - \text{NMB}_0 = k \times (e_1 - e_0) - (c_1 - c_0)$$

(using monetary benefit, but the principle is the same for health benefit). Whatever quadrant the ICER lies in, if treatment 1 is preferred by the ICER, then the incremental net benefit is positive. In quadrant I, for example, if the ICER is less than the WTP threshold k, it is easy to see that $\text{INB} > 0$. Thus the decision is being made by maximising the net benefit, a principle which extends to situations with more than two decision options d.

From the point of view of this book, we focus on net benefits since they form the basis for Value of Information analysis, as will be seen in subsequent chapters.

1.3 Estimating Expected Costs and Effects

The core challenge that this book is concerned with is that the expected costs and effects e_d and c_d are *uncertain* in practice. That is, we do not know their exact values for sure. Instead, we have to *estimate* them, based on observed or published evidence gained from limited observation. In the next few sections, we will explain how this is done.

Then once the evidence has been gathered and analysed, a common practice is to base decision-making on net benefits or ICERs calculated using *point estimates* of e_d and c_d. However, this fails to account for the uncertainty inherent in most evidence that informs health economic models. In Section 1.6, we will describe a more formal framework for decision-making based on *Bayesian* decision theory, which accounts for uncertainty by using *probability* distributions for model parameters. This forms the basis of Value of Information methods.

In some circumstances, individual-level data on health effects and costs of decision options can be taken directly from one clinical research study. These data can then be analysed directly to estimate the expected costs and health benefits and calculate the ICERs, NMB or NHB (O'Brien et al. 1994). However, most often, a single clinical research study does not consider all relevant interventions, or has a limited time horizon, or may not measure all relevant outcomes (or be powered to do so). Moreover, patient-level data from clinical studies are usually not publicly available (Sculpher et al. 2006). In addition, a CEA based on one clinical research study ignores all other potentially relevant evidence, and would only be applicable if the study population represents the real-world clinical population to which the CEA's results would be applied in practice, which is rarely the case given typical inclusion and exclusion criteria.

Consequently, health effects and costs are more commonly estimated in a more general framework known as a *decision-analytic model* (Briggs, Sculpher and Claxton 2006; Greenberg et al. 2010; Neumann et al. 2016).

1.4 Decision-Analytic Models

A decision-analytic model is a mathematical framework that represents how the consequences of different actions may arise. In healthcare decision-making, decision-analytic models define and quantify how states of health or clinical events are experienced by individuals following the decision of interest (Caro et al. 2012; Roberts et al. 2012; Drummond et al. 2015). Models make use of a wide range of evidence, and predict a set of outcomes of interest that are needed to perform a CEA (i.e., health effects and costs) (Ades et al. 2006).

Models are generally defined as functions of quantities known as *parameters.* These are quantities that can be informed or estimated more easily than the overall expected health effects and costs. Clinical parameters may include, for example, relative effectiveness of interventions, risks of various clinical events (e.g., risk of distant cancer recurrence), disease prevalence and incidence, or the accuracy of diagnostic tests. Non-clinical parameters may include, among others, the effect of interventions on HRQoL, resource use and associated costs.

We will now extend the definition of the expected costs and effects c_d and e_d, to consider these as *functions of parameters*: $c_d(\boldsymbol{\theta})$, and $e_d(\boldsymbol{\theta})$. The symbol $\boldsymbol{\theta}$ represents a vector that comprises all the parameters of interest.

A health economic model defines the functions that relate the parameters to the expected costs and effects. A number of health economic model types can be distinguished, which differ by how they define this function.

1.4.1 Types of Decision Models

Selection of the appropriate model type should be based on the decision problem and the level of detail and complexity required. When selecting the model type, the analyst should consider the appropriate time horizon (short term or long term) and time measurement (continuous or discrete), the unit of representation (simulation of heterogeneous individuals or of a homogeneous cohort) and interactions between individuals and model components (Roberts et al. 2012).

The primary types of decision models used in health economic evaluations are decision trees, state-transition models, discrete event simulations, agent-based models and dynamic transmission models (Caro et al. 2012; Karnon et al. 2012; Roberts et al. 2012; Siebert et al. 2012). The selection of the type of model is dependent on the attributes of the model and the decision problem. Decision trees and state-transition models constitute the two most commonly used modelling types. A decision tree is suitable for decision problems with a short-term time horizon and with limited complexity. State-transition models allow evaluation of more complex decision problems, that require simulation of the disease prognosis and the effects of alternative decision options over a long-term time horizon. The example model used to illustrate VOI methods in this book combines a decision tree with a Markov state-transition model (Chapter 2).

More complex model structures are needed for decision problems where an individual's risks, behaviours or other features depend on a combination of their own characteristics, current situation and/or past history, or on what happens to other individuals (e.g., transmission of infectious diseases). More complex models may also be needed to represent how resources are limited or constrained, for example admissions to hospital, clinic appointments, number of consulting rooms or donor organs. Microsimulation models, discrete event simulations, agent-based models or dynamic transmission models may be suitable in these cases.

1.5 Evidence in Cost-Effectiveness Analysis

Model-based CEAs require gathering evidence from a range of sources to inform model input parameters and estimate the desired model outcomes. Appropriate evidence is needed to design and structure a decision-analytic model, as well as to inform its clinical and non-clinical parameters (Drummond et al. 2015). Structuring the model requires appropriate evidence to simulate, for example, the risks of diseases and effects of treatments.

There are a number of types of research design that can be used to gather relevant evidence. A classification of research design types is presented in Figure 1.4. These can be categorised as experimental studies and observational studies (Grimes and Schulz 2002a).

In experimental studies, the study subjects are assigned to either an exposure group (i.e., intervention) or a control group (i.e., current clinical care). The study subjects can be assigned randomly (as in a randomised controlled trial (RCT)) or by some other allocation scheme (as in a non-randomised controlled trial). Alternatives to randomised experiments include quasi-experimental methods such as interrupted time-series, difference-in-differences, propensity score matching, difference-in-differences or synthetic control methods (Valentelyte, Keegan and Sorensen 2022). These methods use non-experimental data, and might be applied when randomisation is infeasible or unethical.

In observational studies, study subjects and their outcomes of interest are observed without researchers making any intervention on the subjects' exposure. If the observational study has no comparison or control group, the study is called descriptive. Descriptive studies are often the first step in analysing a new phenomenon in health and medicine (e.g., new area, event, disease, condition), and they are used to describe the phenomenon: its patterns, frequency, determinants and features. The main types of descriptive studies include surveillance studies, cross-sectional (descriptive) studies, ecological correlational studies, case series and case reports (Grimes and Schulz 2002b).

Three common types of observational study designs that can have comparison or control groups include the cohort study, the case-control study and the cross-sectional (analytical) study (Grimes and Schulz 2002a). A cohort study tracks individuals from the time of their exposure to the occurrence of the outcome of interest. A case-control study, on the other hand, tracks individuals from the time of outcome back to the time of exposure. A cross-sectional study considers both exposure and outcome at the same time.

1.5.1 Evidence to Inform Decision-Analytic Models

1.5.1.1 Randomised Controlled Trials

RCTs are considered the "gold standard" for assessing the efficacy of interventions because, if designed and conducted properly, this study type removes the risk of bias and confounding (Akobeng 2005). However, RCTs are generally performed in selected populations, under highly controlled and idealised settings, with short time horizons (Garrison Jr et al. 2007). These settings limit the generalisability of RCT results to larger populations in actual clinical practice.

1.5.1.2 Real-World Evidence

Because evidence used in a model-based CEA should reflect clinical practice, various sources of real-world data (RWD) are essential to appropriately populate decision-analytic models (Garrison Jr et al. 2007). RWD is often broadly defined as healthcare data collected from a variety of sources outside of the conventional RCT framework (Garrison Jr et al. 2007; Sherman et al. 2016; Hubbard, accessed 2023). These data can provide information on a wide range of outcomes including, but not limited to, clinical outcomes, economic outcomes, HRQoL outcomes and patient-reported outcomes (Makady et al. 2017). RWD can be collected from a number of sources. The main types of RWD sources include: administrative claims, electronic health records, medical record reviews, patient-generated data and patient registries (Garrison Jr et al. 2007; Nabhan, Klink and Prasad 2019). Real-world evidence is evidence derived from RWD through the application of appropriate statistical methods (Garrison Jr et al. 2007). RWD can be perceived as raw material that is uninformative without being processed. Real-world evidence, on the other hand, is an informative and shaped source of knowledge processed from RWD. Although RWD have a great potential to provide valuable evidence on a variety of health outcomes, they are also associated with a number of challenges. Specifically, the quality of RWD may be uncertain and the non-interventional design of RWD collection complicates the process of making causal inferences. Consequently, special efforts are required to ascertain the quality of the data source, to minimise bias and confounding, and to generate correct and reliable conclusions by using adequate, often more sophisticated, analytical methods (Sherman et al. 2016; Hernan and Robins 2020).

1.5.1.3 Selecting Relevant Evidence

Decision-analytic models may include a large number of input parameters of various kinds, including clinical parameters, economic parameters and HRQoL parameters. Some type of literature review is recommended to populate input parameters with identified relevant evidence (Drummond et al. 2015). Model-based CEA should follow principles of evidence-based

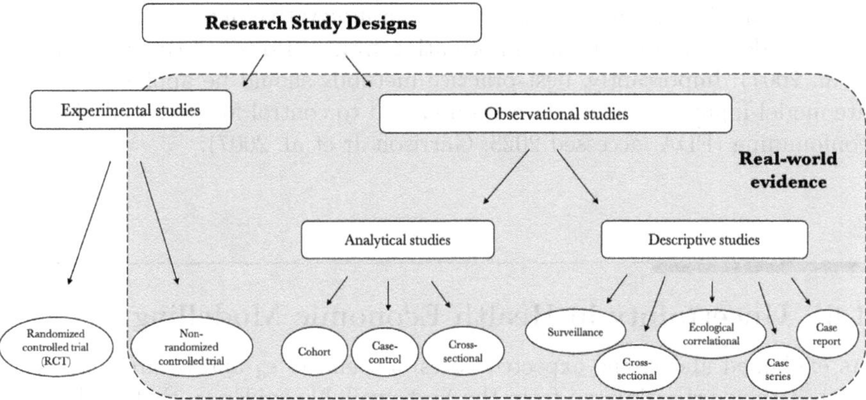

FIGURE 1.4
Types of research studies. The presented classification is based on Grimes and Schulz (2002a), Grimes and Schulz (2002b), Garrison Jr et al. (2007), Glover et al. (2006).

medicine, which indicate that all available evidence should be appropriately identified, synthesised, analysed and interpreted before it is used to inform decision-making in health and medicine (Sackett et al. 1996). The identified evidence is often assessed using a hierarchy of evidence (Walden University Library, accessed 2020). This hierarchy ranks evidence by its strength based on the quality of its source and the measured outcomes (National Cancer Institute, accessed 2020). The evidence hierarchy ranks the evidence sources from strongest to weakest as follows: (1) systematic reviews and meta analyses, (2) RCTs, (3) cohort studies, (4) case-control studies, (5) cross-sectional studies, (6) descriptive studies and (7) expert opinions (Figure 1.4) (Guyatt et al. 1995; Greenhalgh 1997; Glover et al. 2006).

While the hierarchy of evidence is predominantly used to practice evidence-based medicine in making decisions about clinical effectiveness of interventions, model-based CEAs may incorporate many parameters that are not related to clinical effectiveness. In identifying the most relevant evidence for clinical parameters other than comparative effects (e.g., disease prevalence, risk of disease recurrence), resource use, costs, and HRQoL, the role of the traditional evidence hierarchy is limited (Drummond et al. 2015). For these types of parameters, RWD are more appropriate than RCTs, which are often conducted in highly controlled conditions, for deriving relevant evidence (Garrison Jr et al. 2007). Furthermore, it has been shown that in some circumstances, RWD may also be useful to derive evidence on clinical effectiveness (Berger et al. 2009), but the use of RWD is associated with a number of challenges and requires the application of adequate methods. It should be noted that RCTs remain the gold standard to derive this type of evidence (Garrison Jr et al. 2007). As each research design has its advantages and disadvantages and may represent an appropriate evidence source for different model inputs, a model-based CEA can be described as an integrative process of seeking relevant evidence and combining information from various sources (Garrison Jr et al. 2007). Importantly, best-practice methods should be applied to estimate model inputs to ensure data validity and to control for potential bias and confounding (FDA, accessed 2023; Garrison Jr et al. 2007).

1.6 Uncertainty in Health Economic Modelling

As explained above, the expected costs and effects e_d and c_d are *uncertain.* Even if they are estimated from the best available evidence, that evidence is never perfect. Research studies observe finite samples of people. Individuals vary, and people studied in the past will never be completely identical to the decision population. However, models represent assumptions about reality in a way that can be helpful for decision-making, and we have well-understood tools for *quantifying* uncertainty in models. This can give a useful representation of the *strength* of the evidence that informs the decision, and help to determine if or where better evidence should be obtained.

The standard framework for quantifying uncertainty in health economic modelling is based on two core principles:

1. Quantities that are relevant to a decision can be represented as *parameters* in models.
2. Uncertainty about knowledge of those quantities can be represented as *probability distributions* for these parameters.

These are the principles behind *Bayesian* methods – which also give us a way of *learning* or updating our knowledge given observed data (using Bayes' theorem, which we will discuss in Section 1.6.3).

In the next few sections, we will explain probabilistic health economic modelling in detail. Firstly, we explain what is meant by "uncertainty" in this context, compared with some similar concepts (Section 1.6.1). We will explain how the decision-making framework described in Section 1.2 is modified to deal with uncertainty (Section 1.6.2), how probability distributions are defined (Section 1.6.3), how probabilistic analyses are computed (Section 1.6.4) and how the results of these analyses can be summarised and illustrated (Section 1.6.5).

1.6.1 Uncertainty and Similar Concepts

In this book, when we talk about "uncertainty," we will generally be referring to uncertainty in knowledge, which is sometimes called "epistemic" uncertainty. Uncertainty has been described in many different ways in literature about health economic modelling, and in other scientific fields – therefore, to clarify our perspective, we now briefly discuss some related concepts.

1.6.1.1 "Uncertainty" versus "Variability"

Recall that in Section 1.2 we said that costs and effects may differ for every individual. Even if individuals have the same characteristics or risk factors, they may experience different events or clinical outcomes. For example, even when individuals have a similar risk of hospitalisation under an intervention, one individual may stay in hospital for one week, the second individual may stay in hospital for three days, and the third may not be hospitalised at all. In practice, it is not possible to explain how and why every individual differs.

In the language of modelling, we say that individuals share the same *parameters*, but there is random *variability* in their outcomes given those parameters. Many different terms have been used to describe this notion (e.g., stochastic or first-order uncertainty, Briggs et al. (2012)). For the decision-making situations discussed in this book, rather than outcomes for an individual, we are interested in *summaries* of costs and effects for a *population* – which we conceptualise as parameters or functions of parameters in models. Therefore, when we talk about uncertainty, we will generally be referring to imperfect knowledge of these parameters – which has also been called "second-order" uncertainty (Briggs et al. 2012).

Note also that while most of our examples of decision-making are based on totals or averages of costs and effects over a population, sometimes health *inequalities* are of interest for decision-making, e.g., the extent to which health outcomes vary between people. The variance of an outcome can also be conceived as a model parameter, which will be uncertain in practice.

Some decision-analytic models (microsimulations, discrete event simulations or agent-based models) are designed to describe individual variability explicitly, by simulating outcomes for a large sample of individuals. In these, we suppose that the interest is in a summary of the simulated individuals (e.g., total or average costs and effects, or the variance). Uncertainty about these summaries will mainly be influenced by the uncertainty about the parameters driving the simulation.

1.6.1.2 Heterogeneity

Heterogeneity refers to differences in outcomes that can be explained by diversity in the characteristics of individuals in a population. Sometimes a decision-maker will be interested in whether the optimal decision is different for different subgroups of the population. In this situation, individuals with similar characteristics can be grouped into subpopulations, and evaluated separately. Separate CEAs can be used to guide decision-making for each of the subpopulations. If the decision maker simply wants to make a single decision for the overall population, but acknowledging the contribution of heterogeneity to the total outcome in the population, then the subgroup analyses could be aggregated. See, e.g., Sculpher (2008) for more discussion of these issues. Note in particular that more data are required in general to distinguish outcomes from different subgroups, compared to a single analysis for the full population.

1.6.1.3 Structural versus Parameter Uncertainty

The standard framework for uncertainty in decision modelling assumes that all uncertainties can be represented as parameters in a specific model, and that all uncertainties can be represented by probability distributions for those parameters. However, there may be more than one way to structure and parameterise a model and express uncertainty. More generally, while models may be useful, they are, by definition, approximations of a more complex reality and the state of our knowledge. This is the concept of "structural" uncertainty, which is often contrasted with "parameter" uncertainty. Generally, in this book, we will deal with uncertainties that can be parameterised in a model. Some structural uncertainties can be parameterised using techniques such as expert elicitation or model averaging, see, e.g., Jackson et al. (2011). In Section 1.6.3 and Chapter 8, we will briefly discuss situations where parameterisation is challenging.

A related concept sometimes described (see, e.g., Briggs et al. 2012) is *methodological* uncertainty, a loosely defined term referring to the principles and methods used to tackle similar kinds of decision problems. In general, these kinds of uncertainties are less suitable for formal, probabilistic analyses. As recommended in Briggs et al. (2012), these principles should be agreed upon by stakeholders and made transparent (sometimes called a "reference case"), and the impact of any plausible alternative approaches discussed if necessary.

1.6.2 Bayesian Decision-Theoretic Approach to Health Economic Modelling

We now explain how decision-making is conducted when accounting for uncertainty. The decision rule (Section 1.2.2) is modified to acknowledge that the net benefits are uncertain. Instead of maximising the "known" net benefit, the optimal decision now is the one that maximises the *expected* net benefit, with respect to the uncertainty. The expected net benefit is expressed as either a monetary or health benefit, respectively:

$$
\begin{aligned}
\mathrm{E}_{\boldsymbol{\theta}}(\mathrm{NMB}_d) &= k \times \mathrm{E}_{\boldsymbol{\theta}}[e_d(\theta)] - \mathrm{E}_{\boldsymbol{\theta}}[c_d(\boldsymbol{\theta})] \\
\mathrm{E}_{\boldsymbol{\theta}}(\mathrm{NHB}_d) &= \mathrm{E}_{\boldsymbol{\theta}}[e_d(\boldsymbol{\theta})] - \frac{\mathrm{E}_{\boldsymbol{\theta}}[c_d(\boldsymbol{\theta})]}{k}
\end{aligned}
$$

The expectation is denoted $\mathrm{E}_{\boldsymbol{\theta}}$ here, since it refers to the *uncertainty* about parameters $\boldsymbol{\theta}$. Note that this does not refer to the expectation over *individual variability* – that expectation is inherent in the definition of the effects and costs e_d and c_d. We also make explicit here that e_d and c_d are defined as functions of these uncertain $\boldsymbol{\theta}$.

Maximising the expected net benefit is the principle behind *Bayesian decision theory* (see Chapter 7 for a more general definition). See, e.g., Bernardo and Smith (2009) or Savage (1954) for the theory behind this, and Arrow and Lind (1970) for a similar theory of rational decision-making from an economic perspective.

To use this framework in practice, three important steps are required:

1. Defining the joint distribution that reflects our knowledge about the plausible values of $\boldsymbol{\theta}$ – discussed in Section 1.6.3.
2. Performing the computations – discussed in Section 1.6.4.
3. Summarising appropriate results – discussed in Section 1.6.5.

In health economic modelling, this process is often referred to as *probabilistic sensitivity analysis* (PSA). We prefer to call it *probabilistic analysis* (PA), however – we briefly explain why. "Sensitivity analysis" generally means checking whether a "base-case" result is reliable, by examining alternative assumptions. In health economic modelling, a common practice is to ignore uncertainty in e_d and c_d. PSA then represents a check of whether ignoring uncertainty is reasonable to do. But our view is that the "base-case" approach should instead be to account for uncertainty probabilistically. Then if necessary, alternative model assumptions could still be assessed with reference to this base case.

This raises the question: why it is important to routinely consider uncertainty? In general decision making situations, a decision must be made, whatever the level of uncertainty (Claxton 1999b) – even "doing nothing" implies making a decision to continue with current practice. We discuss two reasons for considering uncertainty.

The first is that uncertainty can affect the optimal decision. This is a subtle point. If we were to ignore uncertainty, we would obtain a single "point estimate" $\hat{\boldsymbol{\theta}}$ of the parameters $\boldsymbol{\theta}$, and make the decision based on $\text{NB}_d(\hat{\boldsymbol{\theta}})$ (as in Section 1.2). However, if there is uncertainty, then the rational decision should instead be based on $\text{E}_{\boldsymbol{\theta}}(\text{NB}_d(\boldsymbol{\theta}))$, and this is not necessarily the same as $\text{NB}_d(\hat{\boldsymbol{\theta}})$. The equality holds if we define $\hat{\boldsymbol{\theta}} = \text{E}_{\boldsymbol{\theta}}(\boldsymbol{\theta})$ and $\text{NB}_d(\boldsymbol{\theta})$ is a linear function of $\boldsymbol{\theta}$, but not in more general cases. Therefore, ignoring uncertainty generally gives a biased estimate of the key quantity required for decision-making. Even if this bias might be small in practice, the size of the bias is not known before conducting the analysis. Therefore we think it is sensible to routinely use probabilistic analysis for decision-making where feasible, and base the decision on the expected net benefits.

The second reason is that quantifying uncertainty allows us to determine the potential value of collecting more information. This is the main subject of this book. If there is sufficient uncertainty, the optimal decision may change when we collect more information. If we quantify uncertainty, we can quantify the chance that the decision might change given better information, and the expected net benefits that would be gained from this. These issues are discussed in more detail later.

Quantifying uncertainty probabilistically can be difficult, but in most cases there is some information to guide this choice. See Section 1.6.3 for more about quantifying different kinds of uncertainty, and Chapter 8 for a related discussion of VOI in the context of "structural" uncertainties that are hard to parameterise. For the occasional parameter for which probabilistic uncertainty quantification is too challenging, decision makers still need to know the impact of different assumptions about this parameter. Then it is recommended to at least do *deterministic sensitivity analysis*, that is, simply presenting different results under different assumptions. For this kind of sensitivity analysis to helpfully guide decision-making, the alternative assumptions investigated should be plausible (Hunink et al. 2014). If this is done for a particular parameter, then probability distributions should still be used where possible to represent uncertainty about other parameters in the same model, giving a procedure termed "probabilistic one-way sensitivity analysis" (McCabe et al. 2020).

1.6.3 Obtaining Distributions for Parameters in Models

The distributions for parameters in a probabilistic decision model should represent what is currently known about those parameters – that is, they should give high probability to values that are plausible, and low probability to values that are implausible. There are various ways this can be done, depending on what information is available. Broadly we can distinguish three situations:

- If we can conduct our own analysis of data (individual-level data, or through meta-analysis of published summaries) to estimate the parameters, we can derive a probabilistic quantification of uncertainty from the statistical model that we have fitted.

- If the parameters are informed by a single published summary (e.g., estimates and standard errors), we can derive an appropriate distribution representing this information.

- If there is no formally published information about a parameter, we may be able to make reasonable judgements about what values it can plausibly take.

While it is beyond the scope of this book to explain how to do statistical analyses, we will explain some general principles that are important for uncertainty quantification. See also Briggs, Sculpher and Claxton (2006) and Briggs et al. (2012) for similar advice.

Note that the aim is to derive the *joint* distribution of the parameter vector $\boldsymbol{\theta}$. Any *correlations* between different parameters that comprise $\boldsymbol{\theta}$ (in our knowledge or judgements about their values) should be represented, e.g., the effect of a treatment on the risk of an event may depend on the baseline risk.

1.6.3.1 Bayesian Analysis of Data

Bayesian statistics is an approach to analysing data, which is based on quantifying parameter uncertainty by probability.

Bayesian statistical models represent two sources of "randomness." As with most statistical methods, they represent *sampling variability.* As described in Section 1.6.1, this refers to how different individuals drawn from a population differ from each other. A *sampling distribution* $p(y \mid \boldsymbol{\theta})$ is defined for an observable outcome y, which depends on a set of model *parameters* $\boldsymbol{\theta}$. For example, in a linear regression model, the sampling distribution of an outcome y_i for individuals $i = 1, \ldots, n$ is a normal distribution, with mean defined as a linear function of their predictor x_i: $\mu = \alpha + \beta x_i$, and a variance σ^2. The parameters are $\boldsymbol{\theta} = (\alpha, \beta, \sigma)$. Or if our outcome was the number y of people whose pain is cured by a drug, out of n who are given the drug, and we assumed each person had an equal probability π of cure, our sampling distribution would be a Binomial distribution with an uncertain parameter $\boldsymbol{\theta} = \pi$.

The key ingredient that distinguishes Bayesian methods is that they also describe *uncertainty about knowledge* – "epistemic" uncertainty, as described in Section 1.6.1, by defining probability distributions for the parameters $\boldsymbol{\theta}$. In a Bayesian analysis of data, we define a *prior* distribution $p(\boldsymbol{\theta})$ representing any information we have about the parameters other than the information in the observed data. For example, if the outcome in our linear regression model was blood pressure, the priors for the regression intercept and slope α and β should give low probabilities to biologically implausible predictions for mean blood pressure. Then the *posterior* distribution $p(\boldsymbol{\theta} \mid \mathbf{y})$ is defined by combining the prior with the information from the data $\mathbf{y}$ (e.g., comprising $y_1, \ldots, y_n$ in the linear regression). This is obtained through *Bayes' theorem,* which describes how the *conditional* probability distribution of one quantity $\boldsymbol{\theta}$, given another quantity $\mathbf{y}$, is obtained from the distributions of $\boldsymbol{\theta}$ and $\mathbf{y}$:

$$p(\boldsymbol{\theta} \mid \mathbf{y}) \propto p(\boldsymbol{\theta}) p(\mathbf{y} \mid \boldsymbol{\theta})$$

Most practically applicable methods for obtaining and summarising $p(\boldsymbol{\theta} \mid \mathbf{y})$ are based on *random sampling,* rather than directly evaluating this equation.[1]

In a Bayesian analysis, the posterior distribution $p(\boldsymbol{\theta} \mid \boldsymbol{y})$ describes the uncertainty about our knowledge of the parameters, after taking into account both the data and our background/contextual knowledge. Therefore, if we are using those parameters as inputs to a probabilistic decision model, the required distribution for them is the posterior. In that case, the model outputs (e.g., expected costs and effects) also have an (implicit) posterior distribution

1. The constant of proportionality in Bayes' theorem does not depend on $\boldsymbol{\theta}$, and does not need to be computed in most methods for Bayesian modelling.

– this is usually computed through Monte Carlo techniques, as described in Section 1.6.4.

If there is no observed data $\mathbf{y}$, and our knowledge is simply expressed by the prior, then we would use the prior to express uncertainty in our decision model (in other words, the posterior is the same as the prior, because there would be no data to update our background/contextual knowledge).

After doing a Bayesian analysis, we can define the *predictive distribution* for future data that have not yet been observed, and we do this by integrating over the posterior distribution for the parameters.

$$p(\tilde{\mathbf{y}} \mid \mathbf{y}) = \int p(\tilde{\mathbf{y}} \mid \boldsymbol{\theta}) p(\boldsymbol{\theta} \mid \mathbf{y}) d\boldsymbol{\theta}$$

This concept is essential to the *expected value of sample information*, a method to estimate the expected benefits to decision-making from collecting future data, which is described in detail in Chapter 4.

See, e.g., Baio (2012) and Baio, Berardi and Heath (2017) for more in-depth discussion of Bayesian analysis in the context of health economic modelling.

1.6.3.2 Frequentist Analysis of Data

Statistical analyses to inform health economic models need not necessarily be Bayesian. "Standard" statistical analyses follow an alternative approach to inference, typically referred to as *frequentist*, which conceives uncertainty in terms of *repeated sampling* of a dataset from an unspecified large population. The observed data $\mathbf{y}$ are only one of many datasets we could have sampled, if we were able to repeat the "experiment" (that produced the data) under the same conditions. The statistical model is then specified through the sampling distribution alone. An *estimator* for a parameter is then defined – that is, a function $\boldsymbol{\theta}(\mathbf{y})$ that is applied to data $\mathbf{y}$ to produce an estimate of $\boldsymbol{\theta}$. Commonly the estimator is derived from the model, as the *maximum likelihood* estimator.

In a frequentist analysis, a typical measure of parameter uncertainty is the *standard error*. This is (an estimate of) the standard deviation of the estimator $\boldsymbol{\theta}(\mathbf{y})$ with respect to repeated draws of $\mathbf{y}$ from the sampling distribution. The analogous quantity in a Bayesian analysis is the *standard deviation* of the posterior distribution, which describes a probability judgement about the range of plausible values of the true parameter value $\boldsymbol{\theta}$. While, strictly, the standard error cannot be interpreted in the same way, it can be used to derive plausible probability distributions for parameter uncertainty.

The key theoretical result behind deriving a distribution in this way is the "Bernstein–von Mises theorem": as the sample size increases, the posterior distribution for $\boldsymbol{\theta}$ in a Bayesian analysis approaches a *multivariate normal* distribution, with a mean defined by the maximum likelihood estimate, and a covariance defined by the sampling covariance matrix of the maximum likelihood estimator. In practice, this means that if the sample size of data (that inform a particular parameter) is large, and the prior for that parameter is weak (i.e., there is negligible background information about it), then frequentist analyses will give similar answers to Bayesian analyses. The standard error will be similar to the posterior standard deviation, and the (frequentist) 95% confidence interval will be similar to the 95% Bayesian credible interval.[2] For a given sample size, the assumption of normality of the posterior will be more reliable if the parameters are transformed to an *unrestricted* scale before equating the standard error with the standard deviation, e.g., positive-valued parameters such as odds ratios, hazard ratios or variances should be log-transformed.

Similarly, using *bootstrapping* to obtain the sampling distribution of an estimator (i.e., repeatedly resampling from observed data, and recomputing the parameter estimates) gives a distribution that can be used as an approximation to the posterior from a vague prior.

Caution is required, however, if the sample size of the data is small – whatever framework (Bayesian or frequentist) is being used to analyse it. The data may be unrepresentative by chance, and inferences may be sensitive to assumptions, including the sampling distribution in either framework, and the prior in a Bayesian analysis. With small samples, a strength of Bayesian inference is that background information can be included in the prior. For parameters of scientific interest (e.g., expected clinical outcomes such as survival, or blood pressure) we can usually judge the order of magnitude that the parameter value will take, and derive a reasonable prior from this.

1.6.3.3 Deriving Distributions from Published Estimates

Parameters in health economic models are commonly informed by a single published estimate. If this estimate comes with a measure of uncertainty, such as a standard error, or an interval estimate (confidence or credible interval), then the theory described above can be used to derive a reasonable distribution representing this published knowledge, whether or not it was originally obtained from a Bayesian analysis. A generally applicable procedure is to:

1. Transform the point estimate and interval estimates to an unrestricted scale if necessary (e.g., odds ratios to log odds ratios, hazard ratios to log hazard ratios).
2. Define a normal distribution on the unrestricted scale, with mean defined by the estimate, and standard deviation defined by the standard error. Or if an 95% interval estimate is published, we can assume the width of this interval is about 4 standard deviations

2. Note that while these two kinds of interval estimates are interpreted similarly in practice, they have subtly different meanings: a 95% confidence interval is an interval that would cover the (fixed) true value θ 95% of the time if repeatedly computed from different samples of data, while the credible interval is an interval that we believe (the random) θ lies in with 95% probability.

(the constant 4 here is more precisely $2\Phi^{-1}(0.975)$, where $\Phi()$ is the standard normal quantile function.[3])

1.6.3.4 Deriving Distributions from Judgement

Suppose there is no specific published data about a parameter and its uncertainty. We can often make judgements about what values of the parameter are plausible. As discussed in Briggs et al. (2012), if an uncertain parameter is potentially influential to a decision, it is unwise to ignore uncertainty just because it is difficult to quantify.

For example, this situation can arise if there are concerns about bias in the evidence base, e.g., due to differences between the study population and decision population, differences in how patients are treated, poorly conducted or poorly reported studies, or contradictory estimates provided by multiple studies. Issues like these are sometimes described as a form of "structural uncertainty" – where we can build a model and inform it with data, but we are uncertain about some of the assumptions involved. In these cases, we can sometimes extend the model to include extra parameters that quantify the extent of bias, and place distributions on these parameters to express judgements about how much bias is plausible. Chapter 8 gives an example, and discusses the role of VOI analysis for prioritising model improvements in such cases.

Structured expert elicitation can provide a formal way of obtaining distributions representing judgements about parameters. See O'Hagan et al. (2006) for a general introduction to elicitation, and Bojke et al. (2021) for detailed guidance in the context of healthcare decision-making. The main goal is to judge a typical value for the parameter, along with a range of values that are plausible. Beyond this, the specific choices of distributional form tend to be somewhat arbitrary, and generally chosen for mathematical convenience (e.g., Beta distributions for probabilities, Gamma or log-normal distributions for positive-valued parameters, and normal distributions for unrestricted parameters). Some useful tools for the elicitation process are given by the Sheffield Elicitation Framework (SHELF) (`https://shelf.sites.sheffield.ac.uk/`).

1.6.3.5 Example: Eliciting a Probability

Suppose we need to judge the probability p that a patient will respond to a drug. We do not have direct data, but we have background information about similar situations. From this information, suppose we can judge that the success probability p is 40%, and we are 95% certain that this probability is between 20% and 60%. We interpret these numbers as the 50%, 2.5% and 97.5% quantiles of the distribution (the p% quantile is the value q of X such

3. In R, for example, this can be computed as `2*qnorm(0.975)`.

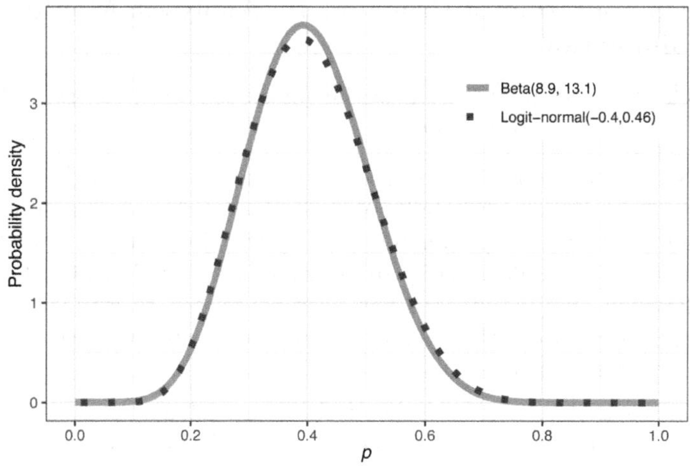

FIGURE 1.5
Comparison of two uncertainty distributions for a probability parameter p, derived via two alternative methods after eliciting a point estimate of 0.4, and a 95% credible interval of 0.2 to 0.6.

that $Pr(X < q) = p\%$). There are many possible approaches to get a distribution that is consistent with this information. We give two examples here: there is negligible difference between the resulting distributions (Figure 1.5), so the choice between them is somewhat arbitrary.

- One approach is to use a Beta distribution for p. The SHELF package for R provides a numerical method to find the Beta distribution that best fits this elicited information (note that an exact fit is not generally possible).

```
> SHELF::fitdist(vals=c(0.2, 0.4, 0.6),
                 probs=c(0.025, 0.5, 0.975),
                 lower=0, upper=1)$Beta
    shape1   shape2
1 8.862633 13.15532
```

  which gives a Beta(8.9, 13.1) distribution. See this package's documentation for more details.

- We might also transform the probability to the logit scale, hence define a normal distribution for $\log(p/(1-p))$. The point estimate of 0.4 and credible interval of (0.2, 0.6) translate to -0.4 $(-1.4, 0.4)$ on the logit scale. Hence, we could define a normal distribution with a mean of -0.4, and a standard deviation σ given by the width of the 95% credible interval (1.8) divided by 4, $\sigma = 0.46$.

1.6.4 Computation of Probabilistic Analyses: Monte Carlo Simulation

The standard way to perform a probabilistic analysis in practice is by using *Monte Carlo simulation*, as follows:

1. Use computer code to draw a set of S random samples $\boldsymbol{\theta}^{(1)}, \ldots, \boldsymbol{\theta}^{(S)}$ from the (joint) probability distribution(s) that represent uncertainty about the parameters. If explicit distributions for $\boldsymbol{\theta}$ are defined, we would directly draw from those distributions. Or if a Bayesian analysis of data has been conducted to estimate $\boldsymbol{\theta}$, we would draw from the posterior of $\boldsymbol{\theta}$, commonly by Markov Chain Monte Carlo methods.
2. For each $\boldsymbol{\theta}^{(s)}$, evaluate the decision model at these parameter values, obtaining the expected costs, effects and net benefits $c_d(\boldsymbol{\theta}^{(s)}), e_d(\boldsymbol{\theta}^{(s)}), \text{NB}_d(\boldsymbol{\theta}^{(s)})$.

The resulting sample represents a draw from the uncertainty distribution of costs, effects and net benefits. The expected values $\text{E}(\text{NB}_d(\boldsymbol{\theta}))$ required for decision-making are simply estimated by the empirical mean of the sampled net benefits for each decision option: $\frac{1}{S}\sum_{s=1}^{S} \text{NB}_d\left(\boldsymbol{\theta}^{(s)}\right)$. This sample can be summarised in various ways to quantify decision uncertainty, as described in Section 1.6.5.

In common practice, around $S = 1000$ to 10000 samples are drawn. A more thoughtful approach to choosing the number of samples, however, would be to decide how much *precision* is needed for the results of interest (e.g., expected costs and effects). For example, we might want to present expected net monetary benefits that are accurate to the nearest \$100. We would then ensure that S is high enough to ensure this margin of error. Crudely, we could keep drawing more samples until the desired results do not appear to change beyond this amount. A more quantitative approach is to compute the *Monte Carlo error*: the standard error of the sample mean $\frac{1}{S}\sum_{s=1}^{S} \text{NB}_d\left(\boldsymbol{\theta}^{(s)}\right)$, that is, the error in the estimate of the mean that results from having a limited number of samples. If the samples are independent, this is simply the empirical standard deviation of the sample, divided by the square root of the sample size: $\text{SD}\left\{\text{NB}_d(\boldsymbol{\theta}^{(1)}), \ldots, \text{NB}_d(\boldsymbol{\theta}^{(s)})\right\}/\sqrt{S}$. This error can be reduced to as small a value as desired by drawing a greater number of samples S.

A note of caution is required here for *microsimulation* and related models that work by simulating a large sample of individuals. To perform Monte Carlo simulation for probabilistic analysis of those models, two levels of "looping" are required. For each iteration s in the procedure above, we:

- Draw random parameter values $\boldsymbol{\theta}^{(s)}$,
- Generate outcomes for a large number of individuals, each given the same parameter values $\boldsymbol{\theta}^{(s)}$,

- Summarise this sample of individuals to generate the net benefits $\text{NB}_d(\boldsymbol{\theta}^{(s)})$ or other decision-making quantities.

This can be computationally intensive. "Emulators" (or "metamodels") are sometimes used to efficiently quantify uncertainty in these situations (see, e.g., Carvalho et al. 2022, for a recent review).

1.6.5 Measures of Decision Uncertainty Computed from Probabilistic Analyses

After a probabilistic analysis, various quantities can be computed. The expected values of net benefit and related quantities $\text{E}_{\boldsymbol{\theta}}(\text{NB}_d(\boldsymbol{\theta}))$ are computed for decision-making, as we have described. This can be presented together with measures of decision uncertainty. A simple measure of decision uncertainty is the *probability of cost-effectiveness* for each decision option d. This is defined as the probability that decision d is the one with the highest net benefit, formally,

$$Pr\{\text{NB}_d(\boldsymbol{\theta}) > \text{NB}_i(\boldsymbol{\theta}) \text{ for all } i \neq d\}.$$

For example, suppose there are two decision options, e.g., a new treatment $d = 2$ and a standard of care $d = 1$. The probability that the new treatment is cost-effective is the probability that its *incremental net benefit*, $\text{NB}_2(\boldsymbol{\theta}) - \text{NB}_1(\boldsymbol{\theta})$, is positive.

Note that the net benefit is a function of the willingness-to-pay threshold k. Thus a common illustration of decision uncertainty is the *cost-effectiveness acceptability curve* (CEAC) – a graph of the probability of cost-effectiveness as a function of k.

An example is given in Figure 1.6 for a situation with two decision options. For lower values of the willingesss-to-pay there is lower decision uncertainty, as we are confident that option II is cost-effective. The uncertainty about the choice becomes higher at higher WTP thresholds.

There is an important but subtle point here – the *optimal decision option* is the one with the highest *expected* net benefit, but this is not necessarily the same as the one with the highest probability of cost-effectiveness. In Figure 1.6, the optimal decision option for each WTP is illustrated with a square around the corresponding CEAC point. For thresholds of between \$280,000 and \$310,000 per QALY gained, option I is optimal, but it has a slightly lower probability of cost-effectiveness than option II.

The curve defined by the probability of cost-effectiveness for the optimal decision, as a function of the WTP, is known as the *cost-effectiveness acceptability frontier* (CEAF). If CEACs are being illustrated, it is essential to also show the corresponding CEAF, since this shows the optimal decision as well as the uncertainty surrounding that decision (Barton, Briggs and Fenwick 2008).

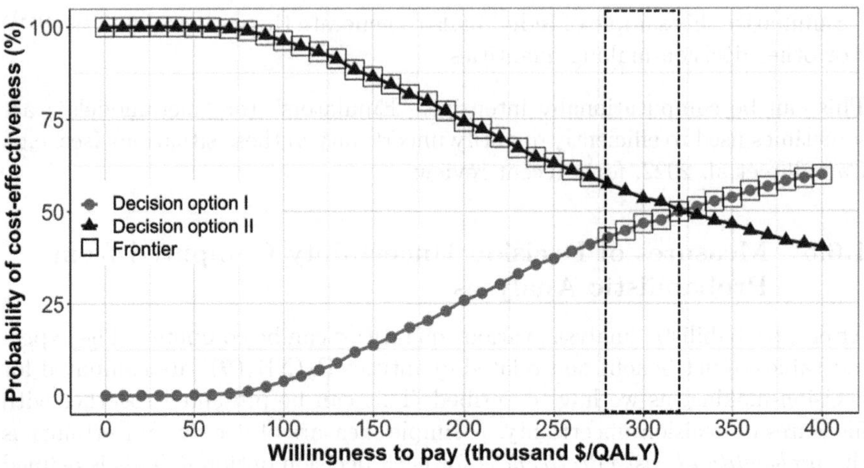

FIGURE 1.6
Cost-effectiveness acceptability curves and cost-effectiveness acceptability frontier.

The consequences of uncertainty can be communicated by the expected loss curve (ELC) (Alarid-Escudero et al. 2019a). An ELC depicts the expected difference between the NMB of each specific decision option and the maximum NMB achieved for the same parameter values, as a function of the WTP threshold. Thus, the ELC reflects the expected average loss in monetary terms that would result from choosing a specific decision option instead of the one that has the highest NB. Furthermore, the ELC can be used to rank the decision options based on their expected loss and to identify the optimal decision option.

Note here that:

- The CEAC describes the probability that the decision is the right one.
- The ELC describes the potential loss from making a wrong decision.

However, a decision maker is interested in both of these things. Large potential losses do not matter if they are unlikely, and a wrong decision does not matter if the consequences of the error are negligible.

This leads us on to the topic of the rest of the book: *Value of Information* analysis. The expected value of perfect information (Chapter 3) is the overall expected loss due to a wrong decision, which encapsulates both the chance of error and the consequences of error. This forms a starting point to measuring the potential *reductions* in this loss from more information of various kinds – which we will describe in detail in Chapters 3 and 4. Probabilistic analysis of uncertainty, and computation by Monte Carlo, forms the basis of VOI analysis. See also Section 5.2 for further discussion of ways to communicate decision uncertainty.

2

A Case Study: A Novel Chemotherapy Treatment

Anna Heath, Natalia Kunst, Nicky J. Welton, and Doug Coyle

This chapter describes a stylised example of a model-based health economic evaluation. We will use this model as an example throughout this book to present, discuss and implement the key concepts for a Value of Information (VOI) analysis. While it is not a real health economic evaluation, it is designed around the decision tree and Markov model structures that are commonly used in practice. Note that while we are using these particular model structures to illustrate the value of information calculations presented in this book, VOI methods can be used irrespective of the complexity of the underlying health economic model.

We have implemented our model in the statistical programming language `R` (R Core Team 2014), coded using the current best practice guidelines for model development in `R` for health economic evaluations (Alarid-Escudero et al. 2019b). While decision-analytic models can be implemented in a range of software (Jalal et al. 2017), VOI calculations require advanced statistical modelling capabilities, which are available, for example, in `R`. Thus, the decision-analytic model and VOI calculations are integrated within the same software package, minimising errors that could arise from translation between packages (Baio and Heath 2017). Further advantages of using full programming languages such as `R` for decision modelling are discussed by, e.g., Incerti et al. (2019) and Baio and Heath (2017).

The accompanying code for this model is available on GitHub in the `convoigroup/Chemotherapy_Book` repository, and is also provided with the `voi` `R` package that we present in Chapters 3 and 4.

DOI: 10.1201/9781003156109-2

2.1 Example: A Novel Chemotherapy Treatment to Reduce Treatment-Related Side Effects

In our stylised example, decision makers are determining the optimal chemotherapy treatment from $D = 2$ potential alternatives. One of the available treatments is currently used in standard clinical practice, and is referred to as the standard care ($d = 1$). A new chemotherapy treatment has been developed, which we will refer to as the novel treatment ($d = 2$), and it is compared to the standard care. To simplify the model structure, we assume that the standard care and the novel treatment do not differ in their effectiveness (i.e., that the remission rates are the same across the two treatment options). However, they have different side effect profiles and costs, which we will model to determine the optimal treatment. Broadly, this implies that our decision-analytic model is considering whether the decreased side effects related to the novel treatment are worth its increased cost, compared to the standard care. We use quality-adjusted life years (QALYs) as our measure of health effects. The costs, including the cost of treating side effects and the one off cost of treatment, are measured in British pounds (£).

2.1.1 Description of Model Structure

Our decision-analytic model combines two commonly encountered model structures, a decision tree and a Markov model. The decision tree determines the expected number of individuals in a treated population who would experience side effects for each of the two treatment options. For the cohort of individuals that do experience treatment-related side effects, irrespective of the chemotherapy treatment they received, their costs and health effects related to the treatment of side effects are estimated using the four state Markov model depicted in Figure 2.1.

All individuals who experience side effects are initially assumed to manage the impact of their side effects at home. However, if their condition deteriorates, they will require more intensive care that is administered in hospital. If the individual's side effects are sufficiently bad to be treated in hospital, then we assume that there is a chance they may die from complications associated with their side effects. We make the simplifying assumption that individuals are not at a direct risk of dying when they are receiving care for their side effects at home. Irrespective of whether the individual experienced side effects at home or were hospitalised, they can then recover from their side effects and we assume that they cannot experience side effects again once they have recovered. All patients who have recovered from their side effects have a risk of death associated with their underlying cancer and baseline risk of mortality. Note that this Markov model does not directly model the natural history

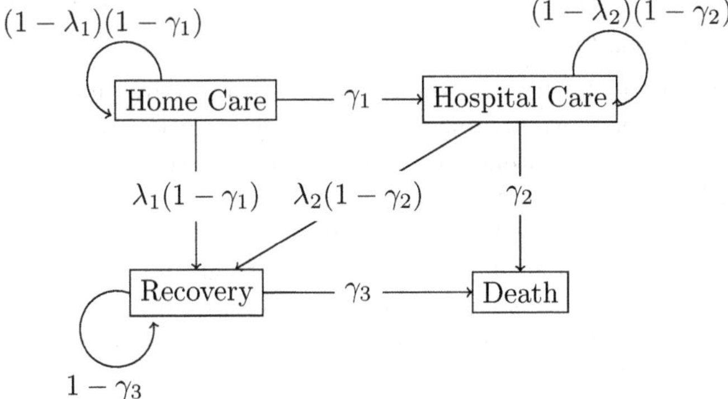

FIGURE 2.1
The Markov model structure for our case study and used to compute the costs and health effects for patients who experience side effects.

of the underlying cancer, implying that individuals who have recovered from their side effects may or may not be in remission for their underlying cancer.

Our Markov model to understand the costs and health related quality of life associated with the treatment side effects has weekly cycles and a time horizon of 1 year. Subsequently, it is assumed that no patients experience further side effects, i.e., they have either recovered or died, and those patients that remain alive after the first year, including those who did not experience side effects at all, continue to be at risk of death from cancer and other causes. The lifetime costs and effects for these patients who survived to 1 year is measured over a lifetime time horizon of 40 years.

To estimate costs and effects, we assume that once an individual recovers from their side effects, they incur no further costs and experience the same quality of life as patients who did not experience side effects. We assume that hospital care is more expensive and results in a lower health related quality of life (utility) than home care. Finally, we assume that patients who die no longer incur healthcare related costs but also have a health related utility of 0. We do not consider discounting.

2.1.2 Description of Model Parameters

Initially, the decision tree determines the proportion of individuals who experience side effects after receiving either the standard care or the novel treatment. We denote the probability of side effects after receiving treatment as π_d, for the standard care and novel treatment ($d = 1, 2$), respectively. As is typical in decision-analytic modelling, we assume that evidence is available to estimate the baseline risk of side effects for the standard of care and the odds ratio of side effects for the novel treatment compared to the standard care. Thus, we denote ρ as the odds ratio and specify that $\text{logit}(\pi_2) = \text{logit}(\pi_1) + \log(\rho)$, where $\text{logit}(x) = \log\left(\frac{x}{1-x}\right)$.

To compute the health economic outcomes for individuals who do not experience side effects, we assume that they only incur the cost of their treatment, and that they experience a quality of life of q throughout the entire 1 year model horizon for the Markov model. This quality of life weight is uncertain and subject to a probabilistic analysis but the drug costs are known with certainty. We assume that the standard care has a single one-off cost of £120 in the first year of treatment, and the novel treatment costs £1975.

Our Markov model is parameterised in terms of five key probabilities, marked in Figure 2.1:

- γ_1, the probability of developing severe side effects that require hospitalisation in the following week for those managing mild side effects at home,
- γ_2, the probability of dying in the following week for someone requiring hospital treatment for severe side effects,
- γ_3, the probability of dying in the following week for someone who has recovered from their side effects (regardless of whether the individual received treatment at home or in the hospital before recovering),
- λ_1, the probability of recovering in the following week for someone who continues to manage their side effects at home,
- λ_2, the probability of recovering in a given week for someone treated in hospital who does not die.

These five key probabilities completely define the transition probabilities in our Markov model as follows:

- Transition probability from Home Care to Hospital Care: γ_1
- Transition probability from Home Care to Recovery: $\lambda_1(1-\gamma_1)$
- Probability of remaining in Home Care: $(1-\lambda_1)(1-\gamma_1)$
- Transition probability from Hospital Care to Death: γ_2
- Transition probability from Hospital Care to Recovery: $\lambda_2(1-\gamma_2)$
- Probability of remaining in Hospital Care: $(1-\lambda_2)(1-\gamma_2)$.
- Transition probability from Recovery to Death: λ_3.

As stated above, once individuals have recovered from their side effects, they experience a quality of life (utility) weight of q, the same as for those who did not experience side effects, and a side-effect related cost of 0. This is assumed to remain constant throughout the 40 year time horizon. Individuals who die have a quality of life weight of 0 after death. We also assume that they require a one-time cost of terminal care, which we denote c_D. For individuals who are treated at home or in hospital, we assign quality of life weights, denoted q_{HC} and q_H, respectively. These individuals also incur healthcare system costs, denoted c_{HC} and c_H, respectively.

2.2 Probabilistic Analysis for Our Model

Unless noted otherwise, all model parameters defined in the previous section are subject to probabilistic analysis, where the relevant distributions are assumed to be obtained from the literature and primary data sources. We assume that the distributions represent our genuine state of belief or uncertainty around the estimates, which assumes that the reported studies are representative of the situation that we want to model. In other situations, we might want to account for uncertainty due to hidden biases, e.g., a mismatch between aspects of study (such as patient characteristics, or treatments given) and the decision situation. This would require extending the model to include parameters that describe those biases, perhaps with uncertainty distributions obtained by expert elicitation (see Section 1.6.3 and Chapter 8). In general, it is crucial to remember that using Value of Information methods to quantify the consequences of uncertainty requires that all uncertainties are represented as model parameters (Section 1.6).

2.2.1 Evidence for Model Parameters

2.2.1.1 Probability of Side Effects for the Standard Care, π_1

We suppose this is informed by data from a previous study, where 111 individuals received the standard care and 52 of these individuals experienced side effects. To determine the probability distribution that represents our uncertainty in π_1, we use a Bayesian analysis. Firstly, we set a uniform prior between 0 and 1 for π_1, indicating that we have minimal prior information. A uniform prior on π_1 is equivalent to a Beta prior with parameters $(1, 1)$. The data are then assumed to follow a binomial distribution, which is conjugate to the Beta distribution. This means that the posterior distribution of π_1 can be calculated analytically and is equal to $\text{Beta}(1+52, 1+111-52)$. Note that the mean of this distribution is equal to $\frac{1+52}{1+111-52+1+52} = \frac{53}{113}$, which is approximately equal to the proportion of patients who experience side effects. The uncertainty about π_1 (posterior variance) is related to the number of patients we observed in order to estimate π_1, 111 in this case.

2.2.1.2 Odds Ratio for Side Effects ρ

We suppose that individual-level data on the odds ratio are not available, as in the common situation where the study comparing the therapies is conducted by a pharmaceutical company, and summary statistics are published but the individual data are confidential. We suppose that the information to inform ρ is derived from the published study comparing the novel chemotherapy to standard care. In this study, the odds ratio is estimated as 0.54 and the standard error for the log odds ratio is 0.3. Thus, we represent uncertainty in the estimate of ρ using a normal distribution for the log odds ratio $\log(\rho)$ with a mean -0.616 and standard deviation 0.3. From these two distributions, we can obtain a sample from the distribution of π_2, by simulating S values from the distributions of π_1 and ρ (π_1^s and ρ^s for $s = 1, \ldots, S$) before computing $\text{logit}(\pi_2^s) = \text{logit}(\pi_1^s) + \log(\rho^s)$. Thus, the simulated values for π_1 and π_2 are correlated.

2.2.1.3 Transition Probability Parameters γ_1, γ_2

The distributions for the key model probabilities γ_1, the transition probability from home care to hospital care, and γ_2, the transition probability from hospital care to death, are also estimated from a primary data source. From the same primary data source, we observe that from the 52 individuals who experienced side effects, 43 were given treatment in hospital and eight died from their side effects. To estimate γ_1 and γ_2, we first note that the primary data sources give us information about the proportion of patients who are treated in hospital and die in 1 year. Thus, we define

- Γ_1, the probability that a patient is hospitalised for side effects in the year after receiving treatment.
- Γ_2, the probability that a patient dies in the year after receiving treatment.

We assume that the transition probabilities γ_1 and γ_2 are constant across the time horizon and can then estimate the weekly transition probabilities as

$$\gamma_i = 1 - (1 - \Gamma_i)^{\frac{1}{52}},$$

for $i = 1, 2$. This formula is only able to approximate γ_2 as we should adjust for the fact that individuals transition to hospital care, and are therefore at risk of dying, over time and that individuals can die having recovered. Thus, the uncertainty distributions for γ_1 and γ_2 are defined by specifying distributions for Γ_1 and Γ_2. Similar to π_1, we use Bayesian modelling with uniform priors to specify the distributions for Γ_1 and Γ_2. The distributions for the transition probabilities are then obtained through simulation.

2.2.1.4 Transition Probability to Death Given Recovery, γ_3

The risk of dying once the individual has recovered is estimated from the literature as it is the baseline risk of death for patients who are diagnosed with this type of cancer. We make the simplifying assumption that the risk of death is constant over the 40-year time horizon we are considering for the baseline risk of death. Thus, we extract information on the constant rate of death r and associated uncertainty from the literature. We suppose there is an estimate of r based on a 95% chance of death at 20 years, representing a yearly death rate of $\frac{0.95}{20} = 0.0475$. This is estimated from population level data meaning it is associated with minimal uncertainty with a variance of 0.001 (standard error of 0.0316). To determine the transition probability for the Markov model, we then transfer the rate r into weekly transition probabilities by calculating:

$$\gamma_3 = 1 - e^{-\frac{r}{52}}.$$

2.2.1.5 Transition Probabilities to Recovery, λ_1,λ_2

The two remaining transition probabilities define the probability of recovery in a given year for a person who is treated at home and does not transition to hospital care (λ_1) and for a person who is treated in hospital and does not die (λ_2). Information to estimate these parameters is not available from our primary data. Thus, they are estimated from the literature where the parameter estimates are 0.21 and 0.03, respectively for λ_1 and λ_2, and the standard errors are 0.03 and 0.0065. This demonstrates that hospitalised patients have a relatively long recovery time, compared to those treated at home. As λ_1 and λ_2 are constrained between 0 and 1, we use Beta distributions to model our probabilistic uncertainty in the parameter values.

2.2.1.6 Quality of Life and Costs

Finally, we need to specify uncertainty distributions for the quality of life weights and costs associated with the states in our Markov model. We also assume that primary data are not available to estimate these parameters and extract published estimates and standard errors for the parameters from literature.

The estimates and standard errors for each of these six parameters are provided in Table 2.1. To specify the shape of the PA distributions, we use log-normal distributions for costs and Beta distributions for the quality of life weights (Briggs, Sculpher and Claxton 2006). Log-normal distributions are commonly used as the PA distribution for cost parameters as costs are positive and are heavily skewed to the right as some individuals experience very high cost of care (Briggs, Sculpher and Claxton 2006). This means that in primary data collection on the costs, it is possible that very high cost values were not observed and there is a potential that the mean cost in the population could be substantially higher than the mean cost observed in the sample. A Gamma distribution would be another appropriate alternative PA distribution that exhibits this behaviour. Similarly, quality of life weights are often modelled with Beta distributions, as they typically fall between 0 and 1, similar to probabilities. An alternative approach would be to place a distribution around

TABLE 2.1
The probability distributions for the model inputs for the chemotherapy model. This table reports the definition, the distribution family and the published estimate and standard error (which are transformed into parameters of the distributions in Section 2.2.2).

Parameter	Definition	Distribution	Estimate	Standard Error
$\log(\rho)$	Log odds ratio of side effects	Normal	log(0.54)	0.3
r	Rate of death for individuals who have recovered or not experienced side effects	Gamma	0.0475	0.0316
λ_1	Probability of recovery in a given week for someone treated at home who does not transition to hospital care	Beta	0.21	0.03
λ_2	Probability of recovery in a given week for someone treated in hospital who does not die	Beta	0.03	0.0065
q	Quality of life for recovered patients	Beta	0.98	0.0283
q_{HC}	Quality of life for home care patients	Beta	0.7	0.141
q_H	Quality of life for hospitalised patients	Beta	0.03	0.173
c_{death}	One-off cost of death	Log-normal	1710	27.57
c_{HC}	Yearly cost of treatment at home	Log-normal	830	12.25
c_H	Yearly cost of treatment in hospital	Log-normal	2400	43.36

the disutility of the health state (1 – utility), which can be expressed as either a log-normal or Gamma distribution. This would allow the possibility of health states will utilities less than 0, i.e., a health state worse than death.

2.2.2 Converting Published Estimates to Distribution Parameters

Note that the published estimate and standard error for a quantity in a model often do not correspond directly to the parameters of the distribution we want to use to represent uncertainty about that quantity. The published estimates and standard errors must then be transformed to the software's parameters, noting that different software often uses different parameterisations of the same distribution, e.g., the Gamma or Weibull.

We explain how this is done for three parameters in our example, using the notation μ for the point estimate (central tendency) and ν for the standard error in each case.

1. The parameters for the Gamma distribution in the `R` parameterisation in terms of shape α_{Ga} and rate β_{Ga}, can be calculated as:

$$\beta_{Ga} = \mu/\nu^2 \qquad \alpha_{Ga} = (\beta_{Ga})\,\mu.$$

2. The parameters for the Beta distribution in the `R` parameterisation, α_{Beta} and β_{Beta}, can be estimated as:

$$\alpha_{\text{Beta}} = \mu^2\left(\frac{1-\mu}{\nu^2} - \frac{1}{\mu}\right) \qquad \beta_{\text{Beta}} = \alpha_{\text{Beta}}\left(\frac{1}{\mu - 1}\right).$$

3. The log-normal distribution in `R` is defined in terms of the mean $\mu_{\log}$ and standard deviation $sd_{\log}$ on the log scale, while we assume our point estimate μ and standard error ν (in Table 2.1) are the mean and standard deviation on the natural scale. $\mu_{\log}$ and $sd_{\log}$ can be estimated from μ and ν as follows:

$$\mu_{\log} = \log(\mu) - \frac{1}{2}\log\left(1 + \left(\frac{\nu}{\mu}\right)^2\right) \qquad sd_{\log} = \sqrt{\log\left(1 + \left(\frac{\nu}{\mu}\right)^2\right)}.$$

Note that in some settings, the published manuscript will not include the standard deviation of the odds ratio but rather an estimate and confidence interval for the log-odds ratio. In this setting, the standard deviation can be estimated from the confidence interval and a normal distribution for the log-odds ratio can be used directly.

VOI depends on how parameter uncertainty is quantified

VOI analyses calculate the value of eliminating or reducing uncertainty in the parameters of a decision model. Thus, the VOI results presented for this case study later in the book are entirely dependent on the distributions, defined in this section, used to represent this uncertainty for the probabilistic analysis of the model. Alternative choices of distributions for the parameters would lead to different VOI results. More specifically, the main characteristics of a distribution that would be expected to influence VOI are the *location* of the distribution (e.g., its mean or median) and the amount of *uncertainty* in it, for example, as governed by the standard error for parameters in Table 2.1, or the sample size of the data informing the parameters in Table 2.2.

Furthermore, the VOI depends on the model structure, by assuming the uncertainties specified for the model parameters are the only sources of uncertainty in the decision. If there are further uncertainties to be suspected, the model would need to be extended if we wanted to capture the value of reducing them – see Chapter 8.

TABLE 2.2
The prior specification and data used to determine the probability distributions for the model inputs informed by primary data sources for the chemotherapy model. These distributions are derived as the Beta posteriors under the given Beta priors and data.

Parameter	Definition	Prior Distribution	Prior Parameters	Data
π_0	Probability of side effects under standard care	Beta	(1, 1)	N: 111; side effects: 52
Γ_1	1-year probability of hospitalisation, given the patient had side effects	Beta	(1, 1)	Side effects: 52; hospitalisations: 43
Γ_2	1-year probability of death	Beta	(1, 1)	Hospitalisations: 43; death: 8

2.3 Model Structure in R

Throughout this book, we will be using the `voi` package in `R` to undertake VOI calculations. We have also implemented the Chemotherapy decision-analytic model in `R`, hence the entire modelling and analysis workflow can be carried out together.

In accordance with best practice guidelines for coding decision-analytic models in `R` (Alarid-Escudero et al. 2019b), we structure the model in an `R` project, using folders that contain scripts with different functionalities. The following section explains these folders and their contents for the chemotherapy model.

2.3.1 Folder `01_data_raw`

This folder should contain the raw datasets that are used to build the decision-analytic model. To improve consistency, these raw datasets should be the datasets generated from the primary study and should not be manipulated before being saved into this folder.

In this example, we provide a `.csv` file that contains the data on side effects, hospitalisation and deaths from the primary data source.

2.3.2 Folder `02_data`

This folder should contain the scripts required to generate and specify the inputs required to run the decision-analytic model. Thus, this folder contains scripts that manipulate the raw data so it can be used to define the model inputs. It also contains scripts that specify key model inputs that may have been derived from the literature. Finally, it could contain scripts that specify model inputs that denote structural assumptions, i.e., the time horizon of the model or the willingness-to-pay threshold.

In this example, this folder contains two scripts:

1. `01_data_inputs.R`, which loads the file from `01_data_raw` and analyses the data to count the number of individuals with side effects, hospital treatment and death.
2. `02_assumptions_inputs.R`, which defines the other model inputs. Firstly, it defines the means and standard errors for the parameters that were derived from the literature. It then defines several key structural inputs such as the Markov model time horizon of 52 weeks and the full model time horizon of 40 years.

2.3.3 Folder `03_R`

This folder contains all of the scripts that are used to define the decision-analytic models. Best practice is to define the model in a modular fashion using `R` functions, for example, so it can be easily re-run with different inputs, e.g., for the PA.

In this example, our model is based on functions defined in two `R` scripts:

1. `01_misc_functions.R`, containing the functions that determine the parameters for Beta, Gamma and log-normal PA distributions from the mean and standard error. These transformations are not model-specific, and so using a separate file allows us to reuse these functions for other decision-analytic models.
2. `02_model_functions.R`, that contains the four functions required to run the decision-analytic model.

Each of these four model functions performs a separate purpose.

1. `generate_psa_parameters`. This function is used to perform the simulations of parameters from their PA distributions. This function takes a single argument `n` which designates the number of simulations that will be used for the PA. The PA distributions generated using this function are based on the model inputs that are defined in the `02_data` folder, and must be loaded into `R` before running the `generate_psa_parameters` function. The advantage of separating the parameter specification from the PA simulation is that if the model inputs can be updated, e.g., if the literature evolves, the analysis can be easily rerun.
2. `calculate_state_occupancy_markov_model`. This function takes model parameters as arguments, and uses matrix algebra within `R` to calculate the "trace matrices" for the Markov model, i.e., the expected proportion of the population occupying each state at each time. By rerunning this function for each simulated set of parameter values, we can compute the expected proportion of the population in each state at each time point. These values can be combined with the costs and quality of life weights for each state to calculate the economic outcomes associated with experiencing side effects.
3. `calculate_costs_effects` performs this economic analysis. This function also combines the outputs from the Markov model with the decision tree, to calculate the overall costs and effectiveness measures for each treatment. The `calculate_costs_effects` function takes the model parameters as inputs to enable us to recompute the economic outcomes for each simulated parameter set. The output of the `calculate_costs_effects` function can be used to undertake economic analyses such as computing the ICER. However, to

perform a VOI analysis, we must compute the net monetary (or net health) benefit.

4. Thus, the final model function, `calculate_net_benefit`, takes the economic output from the `calculate_costs_effects` function, and calculates the net benefit for a given willingness-to-pay threshold value.

2.3.4 Folder `04_analysis`

This folder should contain the scripts to run the model and perform the analyses. These analysis scripts will use the inputs and the functions defined in the previous two sections.

In this example, the baseline cost-effectiveness and probabilistic analysis are run using two scripts. Firstly, the script `01_model_run.R` utilises the model functions and inputs to run a PA and estimate the distribution of the costs, effects and net monetary benefit for this example. This script is called by all other analysis scripts as it performs the PA. In this example, the PA is relatively quick and so the `01_model_run.R` can be run whenever we are working with the model. However, in other examples, it may also be advisable to save the results of the PA analysis to avoid rerunning the analysis each time your wish to work with the model.

We also provide a script `02_baseline_model_output.R`, which manipulates the outcome of the model to present the results seen in Section 2.4. In addition to the baseline cost-effectiveness and probabilistic analysis, this folder also contains scripts to undertake the VOI analyses and present their results.

This file structure also includes several folders to save the output of the analyses to help with reporting. In particular, separate folders are used to save the model output (`05_output`), including summary results and potentially the full results if the model is expensive to run, figures (`06_figs`), tables (`07_tables`) and any reports generated from the analysis. It is also useful to provide documentation in this folder to support others to use your code (or even support your future self when you have to edit the model).

2.3.5 Using the Chemotherapy Model in `R`

We have designed the chemotherapy model and the example code throughout this book to be implementable directly in `R` by the user. We provide two alternative ways to install the model and make it usable by the reader.

1. As a *project* in RStudio. Projects are used to simplify workflow by associating a particular piece of work with a particular working directory. The following instructions are correct for the RStudio

"Cherry Blossom" release (2023) but may need to be adapted if RStudio evolves.

(a) In RStudio, go to `File`, then `New Project`, then `Version Control`, then `Git`.
(b) In `Repository URL` put `https://github.com/convoigroup/Chemotherapy_Book`.[1]
(c) If you wish, change `Project directory name`, otherwise the project will be named `Chemotherapy_Book` by default.
(d) Select "Create project as subdirectory of..." to wherever you want to store the project.

Your project should now be open, and all filepaths within this book should work correctly to run the model and the analyses. Then in a future working session, reopen the project by using `File`, `Open Project` in RStudio and navigate to wherever you have stored the `Chemotherapy_Book R` project file.

2. By installing the `voi` package, which includes the chemotherapy model material. First install the `voi` package using

```
install.packages("voi")
```

(this only needs to be done once). The chemotherapy material will be found in the following directory:

```
chemo_dir <- system.file("Chemotherapy_Book",package="voi")
```

Then either change your working directory using

```
setwd(chemo_dir)
```

or copy the material to the working directory of your choice.

Otherwise, if you do not want to work with RStudio or `voi`, and are comfortable working with Git and `R`, then you must clone the `Chemotherapy_Book` GitHub repository and either set your `R` working directory to the top level of this repository, or change the paths to the data and source code files in the repository before trying to load them.

2.4 Baseline Cost-Effectiveness and Probabilistic Analysis

Before proceeding with the VOI analyses in the subsequent chapters, we briefly discuss the baseline cost-effectiveness analysis for the chemotherapy model.

1. Assuming this URL remains valid. We intend to keep the `convoigroup` repositories on GitHub.

TABLE 2.3
The expected costs and effects (QALY) over 40 years for the standard care and novel treatment, considering that side effects affect individuals for a maximum of 1 year, in the chemotherapy example. The incremental cost-effectiveness ratio (ICER) is also reported.

Treatment	Costs	Effects	ICER
Standard care	1,451.15	13.452	-
Novel treatment	3,284.60	13.545	19,719

The expected costs and effects for the standard care and the novel treatment are displayed in Table 2.3.

The novel treatment results in higher healthcare expenditure than the standard care. This implies that the reduction in expenditure due to treating fewer side effects is not sufficient to outweigh the higher cost of treatment. However, the novel treatment also results in greater effectiveness as the number of patients experiencing side effects is reduced. This results in an ICER of £19,719/QALY gained. Thus, if a decision maker is willing to pay more than £19,719/QALY gained then the novel chemotherapy treatment is preferred, while the standard treatment is preferred if they are not. While cost-effectiveness thresholds vary between jurisdictions, £19,719/QALY gained is lower than most commonly used thresholds such as £20,000 in the UK (Claxton et al. 2015b), \$100,000–\$150,000 in the US (Neumann et al. 2014), or \$50,000 in Canada (Sapsford 2009). Thus, the novel treatment would be considered cost-effective in these jurisdictions.

Figure 2.2 displays the cost-effectiveness acceptability curves and frontier for willingness-to-pay values between £0/QALY and £50,000/QALY. The probability that the novel treatment is cost-effective increases as the willingness-to-pay increases, increasing quickly for small values of the willingness-to-pay. However, even for larger values of the willingness-to-pay, there is still substantial uncertainty about the cost-effectiveness of the novel treatment. Thus, there may be value in future research to reduce this uncertainty, which we will explore throughout this book using VOI measures.

Note that the cost-effectiveness acceptability frontier (marked with black squares on Figure 2.2) indicates that, in this example, the treatment with the highest probability of cost-effectiveness is always the most cost-effective treatment across the range of willingness-to-pay thresholds we consider. This is not always the case as maximising expected net benefit does not consider the probability of cost-effectiveness (as discussed in Section 1.6.5). Thus, even in a two decision alternative example, a treatment may be cost-effective with a probability of cost-effectiveness of below 50%.

The expected net benefit for both treatments changes with willingness-to-pay threshold. However, we report the net benefit for the threshold of

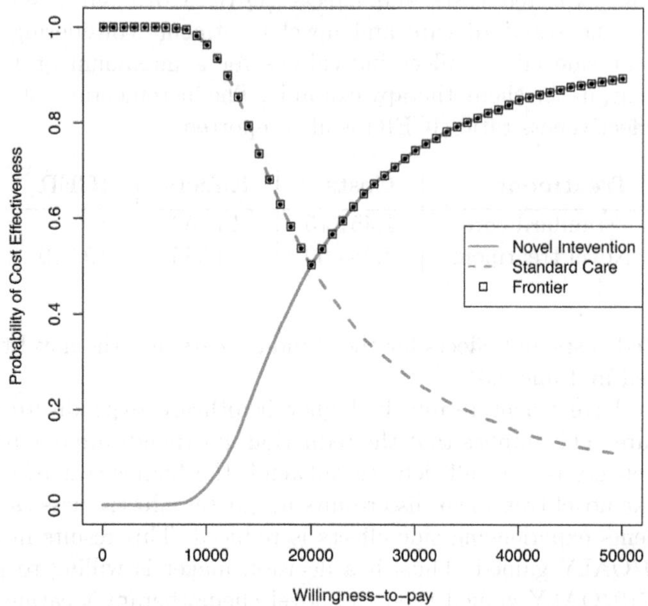

FIGURE 2.2
The cost-effectiveness acceptability curve and frontier for the chemotherapy example.

£20,000/QALY, commonly used in the UK. For the standard care, the expected net monetary benefit is £267,581 and the expected net health benefit is 13.379 QALYs. For the novel treatment, the expected net monetary benefit is £267,607 and the expected net health benefit is 13.380 QALYs. Thus, as seen from the analysis of the ICER, the novel treatment is preferred at a willingness-to-pay of £20,000/QALY, yielding an expected incremental net monetary benefit of £26 (95% interval: −£1,759; £2,022). The 95% interval for the incremental net monetary benefit crosses 0, indicating that there is uncertainty surrounding the optimal treatment when the willingness-to-pay threshold is £20,000/QALY.

Based on this cost-effectiveness and probabilistic analysis, the decision maker can now ask whether there is sufficient evidence to confidently conclude that the novel treatment is the most cost-effective intervention. Additionally, the decision maker may be interested in understanding what are the key assumptions that may make it challenging to determine the optimal chemotherapy treatment in this population. Thus, the following two chapters will explore how VOI methods can be used to answer these questions and provide a comprehensive understanding of the drivers of decision uncertainty and suggest avenues for future research that could support decision-making.

3

The Expected Value of Perfect or Partial Perfect Information

Christopher Jackson, Hawre Jalal, Anna Heath, Natalia Kunst, Howard Thom, Nicky J. Welton, Haitham Tuffaha, and Edward C. F. Wilson

Value of Information (VOI) methods are used to calculate the value of reducing or eliminating uncertainty in the parameters of a decision model (Wilson 2015; Jackson et al. 2022). They are used to answer the following kinds of questions:

- *Sensitivity analysis:* determining which parameters have the highest contribution to decision uncertainty.
- *Research prioritisation:* determining where further research to reduce this uncertainty should be focused.
- *Study design:* predicting the value of specific research that is intended to reduce uncertainty.

In this book, we consider two classes of VOI measures: those that predict the value of eliminating uncertainty (obtaining *perfect* information) and those that predict the value of reducing uncertainty through a proposed research study (i.e., collecting *sample* information).

In this chapter, we focus on determining the expected value of perfect information either through eliminating all parameter uncertainty, or eliminating uncertainty in a specific parameter or set of model parameters. The value of perfect information acts as an *upper bound* for the value of research, either for any research study or a research study focused on the model parameters of interest. It is also typically easier to conceptualise and compute than the value of sample information. Thus, measures of the expected value of perfect information have a crucial role to play in decision-making and research prioritisation, by identifying whether the decision from the model is sensitive to uncertainty, and which parameters are driving that sensitivity.

After a brief introduction to the theory of VOI for decision-making (Section 3.1), we define the EVPI (expected value of perfect information), and

DOI: 10.1201/9781003156109-3

explain how it is easy to calculate (Section 3.2). Section 3.3 explains the expected value of partial perfect information (EVPPI) and the diverse ways that have been used to calculate it. We introduce an R package, `voi`, for doing general VOI calculations, and demonstrate how to implement the most useful methods for calculating EVPPI in it. Finally, Section 3.4 gives a worked example of a EVPPI calculation in a realistic decision model, showing the choices made about what calculations to do, how to present them, and how to interpret their consequences for decision-making and research planning.

3.1 Value of Information as a Decision-Making Tool

Value of Information is a concept from statistical decision theory, which is the science of making optimal decisions under uncertainty (Raiffa and Schlaifer 1961). As discussed in Chapter 1, health economic evaluations aim to support decision-making around which health-related intervention, from among the currently available alternatives, is optimal. This is identified by maximising the expected net benefit,

$$\max_d \mathrm{E}_{\boldsymbol{\theta}}\left[\mathrm{NB}_d(\boldsymbol{\theta})\right] = \max_d \mathcal{NB}_d = \mathcal{NB}_{d^*}.$$

so that the optimal intervention is $d^* = \arg\max_d \mathcal{NB}_d$. As explained in Chapter 1, in the Bayesian decision-theoretic perspective, we need to consider uncertainty about the parameters $\boldsymbol{\theta}$ to calculate the expectations $\mathrm{E}_{\boldsymbol{\theta}}$ of the net benefits $\mathrm{NB}_d(\boldsymbol{\theta})$ of different interventions d. Once these expectations are known, the optimal decision does not depend on how much uncertainty there is about the parameters – a different perspective from that taken in statistical hypothesis testing (Claxton 1999b).

So why is uncertainty still of interest for decision-making? This is because a different decision might be made if we had more data. Decision makers want to know the chance that the decision will be different, and the consequences of getting the decision wrong – hence whether more data should be obtained.

VOI measures are formal methods for estimating the value of obtaining additional information. When this value is high, decision makers may conclude that additional information is required *before* the decision is finalised. Conversely, when it is low, current information is deemed to be sufficient. To further focus future research, the value of information about specific model parameters can also be calculated and compared, to identify the parameters that have the largest impact on our ability to determine the optimal intervention, and where future research should be focused.

The expected value of obtaining additional information is calculated (Raiffa and Schlaifer 1961; Felli and Hazen 1998; Ades, Lu and Claxton 2004) as the difference between:

- The expected value of the optimal intervention based on current evidence.
- The expected value of the optimal intervention conditional on the additional information.

If the additional information does not change the optimal intervention (i.e., the one with the highest expected net benefit), then the value of the information is 0. However, if the intervention currently considered optimal is no longer optimal when the additional information is considered, then the information has value, because it can prevent the decision maker from incurring a loss by implementing an intervention that is non-optimal.

VOI is the expected value of information, not the actual value

VOI describes the expected value of more information. However, we *do not have this additional information at the time the VOI is calculated.* This is why the quantities in VOI calculations are framed as *expected* values, with respect to what is currently known about parameters $\boldsymbol{\theta}$. In practice (see later in this chapter, and in the next chapter) these expected values are calculated by using repeated *simulation* of artificial knowledge: either parameter values or future data. The *actual value* of real information is not known until that information is actually obtained.

As the value of information is measured as a difference in net benefit, VOI has the same units as the net benefit, i.e., either monetary units or health units. In our example from Chapter 2, the value of each intervention is measured in monetary units (£) so the VOI calculations will compute the value of information in £. Furthermore, this might be a value *per person* affected by the decision, or a *population* value, over all people who may be affected in the future – see Section 3.2.1.

3.2 Expected Value of Perfect Information

The first VOI measure we will introduce is the Expected Value of Perfect Information (EVPI), which measures the value of eliminating *all* uncertainty in *all* model parameters. This is an upper bound on the value of reducing decision uncertainty for the given decision model. If this value (expressed at a population level, see Section 3.2.1) is lower than the cost of conducting any feasible research study that aims to reduce uncertainty in the model parameters, then the decision can confidently be made without considering the collection of more information. Furthermore, EVPI could be compared across decision models (in the same policy-making setting) to inform prioritisation of further research.

To formally define EVPI, first imagine that we are able to learn the *exact* value of every model parameter, i.e., we would know that $\boldsymbol{\theta}$ is exactly equal to $\boldsymbol{\theta}'$. In this case, the value of the optimal decision would be

$$\max_d \mathrm{NB}_d(\boldsymbol{\theta}'),$$

and it would not be necessary to calculate the *expected* net benefit over $\boldsymbol{\theta}$, as there is no remaining parameter uncertainty in the model.[1]

Thus, the value of learning $\boldsymbol{\theta} = \boldsymbol{\theta}'$ is equal to the difference between $\max_d \mathrm{NB}_d(\boldsymbol{\theta}')$ and the value of the current optimal treatment (d^*) when $\boldsymbol{\theta} = \boldsymbol{\theta}'$;

$$\max_d \mathrm{NB}_d(\boldsymbol{\theta}') - \mathrm{NB}_{d^*}(\boldsymbol{\theta}')$$

However, as the model parameters are not known with certainty, we calculate the *expected* value of perfect information across all possible values of $\boldsymbol{\theta}'$;

$$\begin{aligned} \mathrm{EVPI} &= \mathrm{E}_{\boldsymbol{\theta}}\left[\max_d \mathrm{NB}_d(\boldsymbol{\theta}) - \mathrm{NB}_{d^*}(\boldsymbol{\theta})\right] \\ &= \mathrm{E}_{\boldsymbol{\theta}}\left[\max_d \mathrm{NB}_d(\boldsymbol{\theta})\right] - \max_d \mathrm{E}_{\boldsymbol{\theta}}\left[\mathrm{NB}_d(\boldsymbol{\theta})\right]. \end{aligned} \tag{3.1}$$

The expectation in this definition is taken with respect to the distribution that represents current beliefs about $\boldsymbol{\theta}$, the same distribution used for probabilistic analysis of the decision model.

Figure 3.1 provides a visual depiction of the EVPI definition in a very simple health economic model that has just one parameter θ. (Models are nearly always more complex, but this simple example helps to illustrate the principles of EVPI.) Two interventions are being compared. The net benefit for each intervention is a deterministic function of θ. If the true value of θ is less than 0.382, then intervention 2 (dashed black line) is optimal as it has the highest net benefit, whereas if θ is greater than 0.382 then intervention 1 (solid black line) is optimal. The expected net benefits of interventions 1 and 2 under current information are represented by horizontal lines, assuming that current knowledge about θ is represented by a uniform distribution between 0 and 1. Therefore, the current optimal intervention is intervention 1, and its expected net benefit (labelled E(NB1) = 0.66) gives the second term in Equation (3.1).

However, EVPI is calculated by finding the net benefit of the optimal intervention at each value of θ, represented by the grey line, and then taking the average weighted by the probability of each θ. This average is the horizontal line labelled E(Max), and is the first term in Equation (3.1). Therefore, EVPI

1. Recall from Chapter 1 that $\mathrm{NB}_d(\boldsymbol{\theta})$ itself is defined as an expected value: a summary of expected costs and effects over individuals.

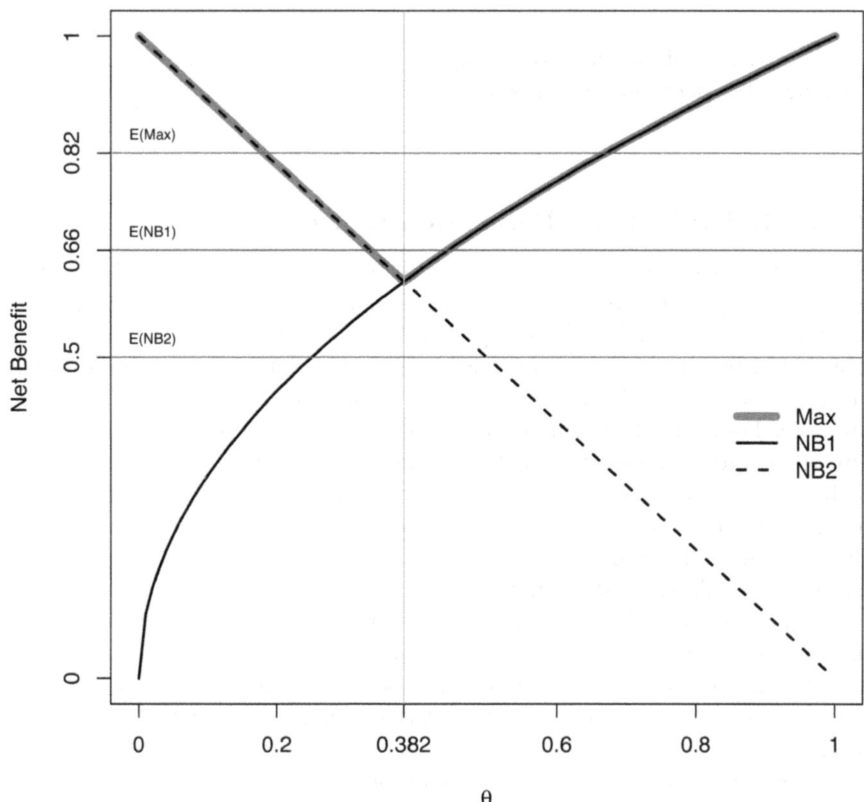

FIGURE 3.1
A visualisation of an EVPI calculation in a simple model that compares two interventions, with a single uncertain parameter θ. The net benefit of each intervention, and the "frontier" of the net benefit for the optimal intervention, are shown as functions of θ, labelled NB1, NB2 and Max. The mean of each of these functions is shown as a horizontal line, and the EVPI here is the difference between E(Max) and E(NB1).

is defined as the difference between the heights of E(Max) and E(NB1), in this case, 0.82 – 0.66. This quantity is always non-negative, as the decision only changes when the net benefit of another intervention is greater than that of the optimal intervention based on current information, as for $\theta < 0.382$ in Figure 3.1. This implies that further data cannot lead to worse decision-making, provided the new data are not systematically different from the old data (which we assume, as we calculate the expected VOI under current beliefs about $\boldsymbol{\theta}$). Conversely, the further information *only has value if it changes the decision*: for $\theta > 0.382$ nothing is gained, since we would still prefer intervention 1.

In Value of Information analysis, information only has value if it changes the decision being made.

3.2.1 Individual and Population EVPI

The EVPI, as defined above, is the expected value *per person*. However, a more useful quantity for research prioritisation is the *population* EVPI. This is defined by multiplying the per-person EVPI by the size of the population who would be impacted by the decision. This population size depends on the "time horizon" over which the decision is relevant. The population EVPI gives an upper bound for the value of any study aimed at reducing uncertainty in any of the model parameters. If this value is smaller than the expected costs of further research, then we can conclude that there would be little value in investigating any of the model parameters and therefore the decision on the optimal intervention can be taken using the current information without considering collecting additional data.

If, however, EVPI shows that there may be value in a future study, we might then like to determine which parameters are driving decision uncertainty and therefore where future research might be worthwhile. This can be achieved using the Expected Value of Partial Perfect Information (EVPPI) (Section 3.3).

More details about calculating population-level value of information are given in Section 3.4.4 through a worked example.

3.2.2 Calculating EVPI

It is rarely possible to compute EVPI analytically for several reasons. Firstly, the net benefit $\text{NB}_d(\boldsymbol{\theta})$ is often a complex function of the model parameters, implying that the uncertainty distribution for the net benefit is not available analytically. Secondly, the expected value of a maximum is generally challenging or impossible to compute. Thus, EVPI is usually estimated using Monte Carlo (MC) simulation, using the simulated values of the net benefit from the probabilistic analysis. As we now explain, this is easy to do.

From the definition of EVPI in Equation (3.1), the MC estimator of the EVPI is

$$\widehat{\text{EVPI}} = \frac{1}{S}\sum_{s=1}^{S} \max_d \text{NB}_d(\boldsymbol{\theta}_s) - \max_d \frac{1}{S}\sum_{s=1}^{S} \text{NB}_d(\boldsymbol{\theta}_s).$$

The second term of this MC estimator is the expected net benefit for the optimal intervention, which is calculated as the sample average of the simulated

TABLE 3.1
A "dataset" from probabilistic analysis of the chemotherapy model, containing the simulated values of three parameters, along with the Net Benefit values for two interventions, extended to demonstrate how to estimate the EVPI by MC simulation.

	Parameters				**Net benefits**			
Sample	π_1	π_2	c_{HC}	$\cdots$	**NB**$_1$	**NB**$_2$	$\mathbf{max}_d\,[\mathbf{NB}_d]$	d^*
1	0.439	0.391	829.7	$\cdots$	**468097**	466693	468097	1
2	0.457	0.254	822.4	$\cdots$	285619	**287594**	287594	2
3	0.561	0.297	824.0	$\cdots$	148989	**149935**	149935	2
$\vdots$	$\vdots$	$\vdots$	$\vdots$	$\ddots$	$\vdots$	$\vdots$	$\vdots$	$\vdots$
S	0.473	0.484	831.5	$\cdots$	**137315**	135387	137315	1
Mean	0.469	0.328	829.8	$\cdots$	267580	**267606**	267974	2

net benefit values. The first term requires that maximum net benefit is selected for each sampled value of $\boldsymbol{\theta}$. This is equivalent to taking the maximum net benefit for each *row* of the simulated "dataset" from probabilistic analysis.

This MC estimator is demonstrated in Table 3.1, for the chemotherapy model described in Chapter 2. For each row of the simulated dataset, we report the row-wise maximum net benefit in the $\max_d [\text{NB}_d]$ column before averaging over all the rows, giving a value of £267,974. The EVPI is then calculated as the difference between this value and the value of the optimal intervention (£267,606).

As the net benefit is calculated conditionally on a specific willingness-to-pay threshold (k), EVPI is also calculated conditionally on k. In Table 3.1, EVPI is calculated for a willingness-to-pay of £20,000 per QALY gained and is equal to £367 per person.

Calculating EVPI in this way can be done easily in any spreadsheet or statistical software. However, EVPPI and expected value of sample information (EVSI) are more computationally demanding, as we will see later in this chapter and in the next chapter, and in our view they are unsuited to spreadsheet software. In this book, we will implement all of these measures in `R`.

3.2.3 The `voi` `R` Package for VOI Calculation

The `voi` package has been developed to implement all practically useful methods for calculating measures of the value of information that we demonstrate in this book. It is available from the standard CRAN repository of `R` packages, and full documentation is also available at `https://chjackson.github.io/voi/`. In this section we illustrate its use for calculating and graphically representing the EVPI.

To run this code:

- The Chemotherapy example model should be installed, following the instructions in Section 2.3.5.
- The voi R package should be installed, using `install.packages("voi")`.

All code in the book is provided and linked from `https://chjackson.github.io/voi/articles/book.html`. If reading this book as a PDF file, don't paste code from the PDF, since some special characters may fail to copy.

```
> library(voi)
> source("04_analysis/01_model_run.R")
> # Specify willingness-to-pay
> wtp_fix <- 20000
> # Extract net benefit for this particular WTP
> nb <- m_net_benefit[ , , wtp_seq == wtp_fix]
> # Calculate EVPI from net benefit
> evpi(nb)
[1] 367.1811
```

In this code, the `evpi` function input is a matrix of samples from the PA distribution of the expected net benefit, with rows containing the simulations, and columns representing the different decision options.

The `voi` package can also compute EVPI for a range of different values of k, to explore EVPI as a function of willingness-to-pay. This requires the input to the `evpi` function to be in "cost-effectiveness analysis" form. This format is an R "list" object, that contains (at least) the following three named components:

- `c`: a matrix or data frame of samples from the PA distribution of the expected costs, with one column for each decision option.
- `e`: a matrix or data frame of samples from the PA distribution of the expected effects, with one column for each decision option.
- `k`: a vector of willingness-to-pay values.

While the cost-effectiveness analysis form can be created by the user, it is also the output of the `bcea` function in the BCEA package, a standardised package for post-processing the results of a probabilisitic health economic model (Baio, Berardi and Heath 2017).

Thus, EVPI can be computed across a range of willingness-to-pay values:

```
> chemotherapy_output <- list(e = m_costs_effects[, "Effects", ],
+                             c = m_costs_effects[, "Costs", ],
+                             k = seq(0, 50000, length.out = 501))
> EVPI <- evpi(chemotherapy_output)
> EVPI$evpi[EVPI$k == wtp_fix]
[1] 367.1811
```

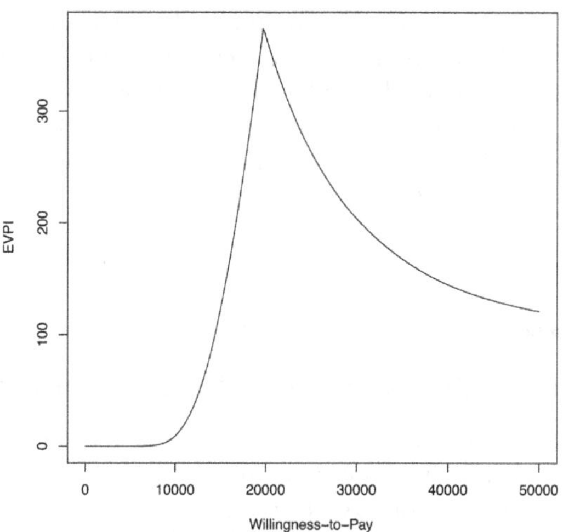

FIGURE 3.2
The estimated expected value of perfect information (EVPI) for willingness-to-pay values between £0 and £50,000 per QALY gained, for the Chemotherapy case study presented in Chapter 2.

3.2.4 Graphical Representations of EVPI

As in the above code, EVPI values for a specific willingness-to-pay can be extracted from the data frame `EVPI`, which contains two columns presenting the willingness-to-pay and the respective EVPI value. However, to present these results, EVPI is usually plotted across the different values for the willingness-to-pay, as in Figure 3.2:

```
> plot(EVPI,
+      xlab = "Willingness-to-Pay",
+      ylab = "EVPI",
+      main = "Expected Value of Perfect Information",
+      type = "l")
```

Figure 3.2 displays a commonly seen shape for the function of VOI measures against willingness-to-pay. The graph initially rises sharply, before reaching a peak at around £19,300 and dropping down. Finally, the right-hand side of the graph is relatively flat, and would likely start rising again as the willingness-to-pay threshold extends beyond £50,000 per QALY gained. The peak of the EVPI curve will typically be near to the ICER, where the optimal intervention switches between the strategies considered. This creates a discontinuity in the function gradient, and is associated with the point where the decision uncertainty is maximised. Thus, this common shape is seen when, as is the case in many health economic evaluations, a new intervention is more expensive and more effective than the standard-of-care. Finally, in this example, the value of information is maintained or increases after the peak, because the increase in willingness-to-pay leads to increases in the net benefit function and, through that, the VOI. In settings where the decision uncertainty is driven by costs alone, we would not see this increase, as the relative importance of uncertainty in the costs decreases as the willingness-to-pay increases.

3.3 Expected Value of Partial Perfect Information

The EVPI measures the expected value of eliminating uncertainty about all model parameters. While it is not generally possible in practice to eliminate all uncertainty, the population EVPI (Section 3.2.1) is still useful to decision makers as an upper bound on the value of further research, which allows research to be deemed unnecessary if its costs exceed this bound. However if we conclude that research is *potentially* valuable, we may want to go further and ask which model parameters might be worth knowing more about. This question can be answered with the Expected Value of Partial Perfect Information (EVPPI).

The EVPPI calculates the value of eliminating uncertainty about a specific parameter or set of parameters. This is the upper bound for the value of research that aims to reduce uncertainty about these parameters. Similar to EVPI, if the expected cost of any feasible research study focusing on these parameters is higher than the (population-level) EVPPI, then these parameters should not be the focus of future research. More directly, however, the EVPPI for a parameter or group of parameters can be interpreted as a measure of sensitivity of the decision to the parameter – the parameters with highest EVPPI are the biggest drivers of the decision uncertainty.

3.3.1 Definition of EVPPI

To formally define EVPPI, the parameters $\boldsymbol{\theta}$ underlying the decision model are split into two components $\boldsymbol{\theta} = (\boldsymbol{\phi}, \boldsymbol{\psi})$, where $\boldsymbol{\phi}$ comprise the *parameters of interest* for the EVPPI calculation (i.e., those that could be investigated further) and $\boldsymbol{\psi}$ are the remaining parameters. We sometimes call $\boldsymbol{\phi}$ the *focal* parameters and $\boldsymbol{\psi}$ the *non-focal* parameters.

We now define EVPPI as the value of obtaining perfect information about $\boldsymbol{\phi}$. This follows a similar argument to the definition of EVPI. If we were to obtain perfect information about $\boldsymbol{\phi}$, we would learn that $\boldsymbol{\phi} = \boldsymbol{\phi}'$. Taking into account this information, the value of the optimal intervention would be

$$\max_{d} \mathrm{E}_{\boldsymbol{\psi}|\boldsymbol{\phi}'} \left[\mathrm{NB}_d(\boldsymbol{\phi}', \boldsymbol{\psi})\right], \tag{3.2}$$

where the expectation $E_{\psi|\phi'}$ is over the uncertainty in the non-focal parameters ψ that would remain after learning that $\phi = \phi'$. Following from this, the value of learning that $\phi = \phi'$ is

$$\max_d \mathrm{E}_{\psi|\phi'} \left[\mathrm{NB}_d(\phi', \psi)\right] - \mathrm{E}_{\psi|\phi'} \left[\mathrm{NB}_{d^*}(\phi', \psi)\right], \tag{3.3}$$

which is only greater than 0 when the perfect information about the ϕ leads to a change in decision, i.e., the intervention that maximises the net benefit is no longer d^*.

EVPPI is then defined by taking the expectation of Equation (3.2) over all possible values of ϕ:

$$\begin{aligned}
\mathrm{EVPPI} =& \mathrm{E}_{\phi} \left[\max_d \mathrm{E}_{\psi|\phi'} \left[\mathrm{NB}_d(\phi', \psi)\right] - \mathrm{E}_{\psi|\phi'} \left[\mathrm{NB}_{d^*}(\phi', \psi)\right]\right] \\
=& \mathrm{E}_{\phi} \left[\max_d \mathrm{E}_{\psi|\phi} \left[\mathrm{NB}_d(\phi, \psi)\right]\right] - \max_d \mathrm{E}_{\phi,\psi} \left[\mathrm{NB}_d(\phi, \psi)\right] \\
=& \mathrm{E}_{\phi} \left[\max_d \mathrm{E}_{\psi|\phi} \left[\mathrm{NB}_d(\phi, \psi)\right]\right] - \max_d \mathrm{E}_{\theta} \left[\mathrm{NB}_d(\theta)\right]. \qquad (3.4)
\end{aligned}$$

In practice, calculating EVPPI is most useful when it demonstrates that there is little value in learning about a subset of parameters, as we can then rule out these parameters as a focus for future research. If EVPPI for a set of parameters is high, then we may want to estimate the value of a specific, feasible research study on these parameters, which can be calculated using the value of sample information (Chapter 4).

As the definition of EVPPI is more complex than the definition of EVPI, it is, again, rarely possible to compute EVPPI analytically. Thus, EVPPI must be computed by simulation in all practical applications. This is more challenging than computing EVPI, because the conditional expectation of the net benefit for a fixed value of ϕ in Equation (3.4) is not available directly from the PA samples. In Sections 3.3.2–3.3.4, we explain a range of methods for calculating EVPPI that each use simulation, but each use different ways of estimating this conditional expectation.

- Nested Monte Carlo simulation (Section 3.3.2). This is conceptually simple, but very computationally demanding.

- Algebraic methods (Section 3.3.3). These are computationally efficient and accurate, but only suitable for the simplest decision models.

- Nonparametric regression (Section 3.3.4). This is computationally efficient and applicable to any decision model, but relies on an approximation. Nonetheless, we have found this to be the most generally useful approach for EVPPI computation (Heath, Manolopoulou and Baio 2017).

Proof that EVPPI is bounded above by the EVPI

If there are two parameter subsets $\boldsymbol{\phi}$ and $\boldsymbol{\xi}$ such that $\boldsymbol{\phi}$ is larger but contains all of the parameters in $\boldsymbol{\xi}$, i.e., $\boldsymbol{\xi} \subset \boldsymbol{\phi}$, EVPPI of $\boldsymbol{\phi}$ is greater than or equal to the EVPPI of $\boldsymbol{\xi}$. To demonstrate this, assume that $\boldsymbol{\phi} = (\boldsymbol{\xi}, \boldsymbol{\xi}^c)$, meaning that $\boldsymbol{\xi}^c$ is the set of "additional" nuisance parameters arising within $\boldsymbol{\phi}$:

$$\begin{aligned}
\text{EVPPI}(\boldsymbol{\phi}) &= \text{E}_{\boldsymbol{\phi}}\left[\max_d \text{E}_{\boldsymbol{\psi}|\boldsymbol{\phi}}\left[\text{NB}_d(\boldsymbol{\theta})\right]\right] - \max_d \text{E}_{\boldsymbol{\theta}}\left[\text{NB}_d(\boldsymbol{\theta})\right] \\
&= \text{E}_{\boldsymbol{\xi}}\left[\text{E}_{\boldsymbol{\xi}^c|\boldsymbol{\xi}}\left[\max_d \text{E}_{\boldsymbol{\psi}|\boldsymbol{\phi}}\left[\text{NB}_d(\boldsymbol{\theta})\right]\right]\right] - \max_d \text{E}_{\boldsymbol{\theta}}\left[\text{NB}_d(\boldsymbol{\theta})\right] \\
&\geq \text{E}_{\boldsymbol{\xi}}\left[\max_d \text{E}_{\boldsymbol{\xi}^c|\boldsymbol{\xi}}\left[\text{E}_{\boldsymbol{\psi}|\boldsymbol{\phi}}\left[\text{NB}_d(\boldsymbol{\theta})\right]\right]\right] - \max_d \text{E}_{\boldsymbol{\theta}}\left[\text{NB}_d(\boldsymbol{\theta})\right] \\
&= \text{E}_{\boldsymbol{\xi}}\left[\max_d \text{E}_{\boldsymbol{\xi}^c|\boldsymbol{\xi}}\left[\text{E}_{\boldsymbol{\psi}|(\boldsymbol{\xi},\boldsymbol{\xi}^c)}\left[\text{NB}_d(\boldsymbol{\theta})\right]\right]\right] - \max_d \text{E}_{\boldsymbol{\theta}}\left[\text{NB}_d(\boldsymbol{\theta})\right] \\
&= \text{E}_{\boldsymbol{\xi}}\left[\max_d \text{E}_{(\boldsymbol{\psi},\boldsymbol{\xi}^c)|\boldsymbol{\xi}}\left[\text{NB}_d(\boldsymbol{\theta})\right]\right] - \max_d \text{E}_{\boldsymbol{\theta}}\left[\text{NB}_d(\boldsymbol{\theta})\right] = \text{EVPPI}(\boldsymbol{\xi})
\end{aligned}$$

where the third of these five steps follows from Jensen's inequality, as the function max($\cdot$) is convex. This property implies that it is always at least valuable to gain information about a larger number of parameters than a smaller number.

EVPPI values are not additive

Specifically, this means that

$$\text{EVPPI}(\boldsymbol{\xi}) + \text{EVPPI}(\boldsymbol{\xi}^c) \neq \text{EVPPI}(\boldsymbol{\phi}).$$

Furthermore, it is not possible to determine whether EVPPI($\boldsymbol{\xi}$) + EVPPI($\boldsymbol{\xi}^c$) will be smaller or larger than EVPPI($\boldsymbol{\phi}$). This is due to the definition of EVPPI, which includes a maximisation, and due to the complexity of health economic models, as they are typically a non-linear function of the underlying parameters. This non-additivity has important implications for EVPPI analysis, as parameters that appear to have small or no value when considered in isolation may be valuable when combined with other parameters. This means that it is important to consider EVPPI for sets of parameters, as well as individual parameters. As we will see in Section 3.4, it is sensible to group together those parameters which can be learned together in a future research study.

3.3.2 Calculating EVPPI by Nested Monte Carlo Simulation

The most conceptually simple, but most computationally expensive, method to compute EVPPI (Brennan et al. 2007) uses a "simulation within a simulation" (also known as a "two-level" or "nested" simulation) to estimate the conditional expectation $\text{E}_{\boldsymbol{\psi}|\boldsymbol{\phi}}[\text{NB}_d(\boldsymbol{\theta})]$.

- For each simulated value of the parameters of interest $\boldsymbol{\phi}_s$, $s = 1, \ldots, S$, R values must be sampled from the conditional distribution of the non-focal parameters $\boldsymbol{\psi} \mid \boldsymbol{\phi}_s$, $\boldsymbol{\psi}_{r,s}$, $r = 1, \ldots, R$.
- These R values $\boldsymbol{\psi}_{r,s}$ for each s are then used to estimate the "known-distribution" net benefit with $\boldsymbol{\phi} = \boldsymbol{\phi}_s$, for each decision d, as $\frac{1}{R}\sum_{r=1}^{R}[\text{NB}_d(\boldsymbol{\phi}_s, \boldsymbol{\psi}_{r,s})]$
- The intervention d with the maximum expected net benefit, conditional on $\boldsymbol{\phi} = \boldsymbol{\phi}_s$, is then determined.
- $\text{E}_{\boldsymbol{\psi}|\boldsymbol{\phi}}[\text{NB}_d(\boldsymbol{\theta})]$ is estimated by averaging the resulting maximum over the different ϕ_s.

Hence the EVPPI (Equation 3.4) is estimated as

$$\widehat{\text{EVPPI}} = \frac{1}{S}\sum_{s=1}^{S}\left[\max_d \frac{1}{R}\sum_{r=1}^{R}[\text{NB}_d(\boldsymbol{\phi}_s, \boldsymbol{\psi}_{r,s})]\right] - \max_d \frac{1}{S}\sum_{s=1}^{S}\text{NB}_d(\boldsymbol{\theta}_s). \quad (3.5)$$

This is straightforward to program, and a simple function to do it is provided in the `voi` R package. However, it requires $S \times R$ calculations of the net benefit for each decision option, where S and R can typically be in the order of 10 000 (Brennan et al. 2007). This represents an immense computational burden, even in settings where the net benefit is easy to compute. There are some economic models where the computational cost of calculating the net benefit will make this Monte Carlo estimator impractical and occasionally impossible to implement within realistic time frames. Nonetheless, as this is based on standard Monte Carlo expectations, it is asymptotically unbiased, and will be able to estimate EVPPI for any decision model provided that sufficient computational power is available.

Note that, for low dimensional $\boldsymbol{\phi}$, it is possible to reduce the computation time for this estimator slightly by using quadrature to estimate the outer expectation (Coyle, Buxton and O'Brien 2003). This allows us to cover the parameter space intelligently and therefore significantly reduce the number of simulations from $\boldsymbol{\psi} \mid \boldsymbol{\phi}_s$, requiring around $30 - 50$ rather than S simulations from the conditional distribution. However, for larger parameter subsets, this quadrature method becomes more expensive than full Monte Carlo sampling.

3.3.3 Calculating EVPPI Using Algebraic Methods

The computational complexity in calculating the EVPPI by MC simulation is directly related to the conditional expectation in the first term of

Equation (3.4); $\text{E}_{\boldsymbol{\psi}|\boldsymbol{\phi}}[\text{NB}_d(\boldsymbol{\theta})]$. Therefore, a range of methods have been developed to approximate

$$\text{E}_{\boldsymbol{\psi}|\boldsymbol{\phi}}\left[\text{NB}_d(\boldsymbol{\phi}, \boldsymbol{\psi})\right]$$

as a function of $\boldsymbol{\phi}$. If such a function is available, then the PA samples $\boldsymbol{\phi}_s$ for $s = 1, \ldots, S$ can be inputted into this function to estimate the conditional expectation without resorting to additional sampling. This drastically reduces the computational effort required to calculate the EVPPI and implies that it can be estimated directly from the PA samples.

One method for finding the analytical expression of this function, first introduced in Coyle and Oakley (2008), requires an approximately linear relationship between the *incremental* net benefit (INB) (the difference between the net benefit of two interventions, e.g., $\text{INB} = \text{NB}_1 - \text{NB}_0$) and the parameters of interest $\boldsymbol{\phi}$. It also requires that $\boldsymbol{\phi}$ can be modelled (at least approximately) using a Normal distribution. If both these assumptions hold, then analytical results based on the "unit loss integral" can be derived to calculate EVPPI in terms of the mean and variance of the conditional INB. Some additional sampling is often required to determine the mean and variance of $\text{E}_{\boldsymbol{\psi}|\boldsymbol{\phi}}\left[\text{INB}(\boldsymbol{\theta})\right]$. These values can be estimated based on only two (rather than S) simulations from the conditional distribution. However, the assumption of normality and linearity are restrictive, meaning that this method has rarely been used in practice.

Alternatively, it is possible to derive an analytic expression for the conditional expectation, provided that

1. The net benefit is a *linear* function of the "nuisance" or non-focal parameters $\boldsymbol{\psi}$.
2. $\boldsymbol{\psi}$ are *independent* of the parameters of interest $\boldsymbol{\phi}$.

(Felli and Hazen 1998). The independence between $\boldsymbol{\phi}$ and $\boldsymbol{\psi}$ implies that the conditional distribution $p(\boldsymbol{\psi} \mid \boldsymbol{\phi})$ is simply equal to the unconditional distribution $p(\boldsymbol{\psi})$ from which we have simulated in the PA. The assumption of linearity then allows us to write the net benefit as follows:

$$\text{NB}_d(\boldsymbol{\theta}) = \sum_{i=1}^{n_\psi} \psi^i f^i(\boldsymbol{\phi}),$$

where the superscript i indicates that ψ^i is the i-th element in the $\boldsymbol{\psi}$ vector, n_ψ is the number of these elements, and $f^i(\boldsymbol{\phi})$ is some known function of the parameters of interest. This means that the expectation can be written as

$$\text{E}_{\boldsymbol{\psi}|\boldsymbol{\phi}}\left[\sum_{i=1}^{n_\psi} \psi^i f^i(\boldsymbol{\phi})\right] = f^i(\boldsymbol{\phi}) \sum_{i=1}^{n_\psi} \text{E}_{\boldsymbol{\psi}}\left[\psi^i\right],$$

so only the mean values for $\boldsymbol{\psi}$, from the initial PA, are required to calculate the conditional expectation for each value of $\boldsymbol{\phi}$. Therefore, provided the linearity assumption holds, or other expressions that would allow for the calculation of each expectation separately, such as a multiplicative term between two independent elements of $\boldsymbol{\psi}$ (Brennan et al. 2007), it is possible to calculate EVPPI directly from the PA. However, this approximation deteriorates if the independence assumption is not valid (Coyle and Oakley 2008).

This idea has been extended to demonstrate how the model can be reparameterised such that this method can be used when its original formulation does not conform to these conditions (Madan et al. 2014). For example, if the net benefit is non-linear in $\boldsymbol{\psi}$, but $\boldsymbol{\psi}$ and $\boldsymbol{\phi}$ are independent, it would then simply be possible to re-parameterise the problem so that NB_d is linear in a new parameter set $\boldsymbol{\psi}' = f(\boldsymbol{\psi})$. In addition to this, Taylor series expansions can be used to approximate the expectation when the model is non-linear in the parameters of interest. The use of Taylor series expansions will provide an approximation to the true EVPPI, but it will be sufficiently accurate in many settings. The main disadvantage of these methods is that the analytic form of the health economic decision model must be available to determine (possibly complex) Taylor series expansions. This means that the computational cost of the MC simulation-based method has been replaced by the more mathematically challenging task of determining the functional form of the conditional expectation.

An example of these approaches is given by Welton et al. (2012) using a decision model from Ades, Lu and Claxton (2004) comparing universal testing and targeted testing for HIV screening. The net benefit under the two strategies is

$$\text{NB}_d(\boldsymbol{\theta}) = \begin{cases} 0 & d = 1, \text{ targeted} \\ N(1-a-b)\left[Me(1-h) - T(1-eh)\right] & d = 2, \text{ universal} \end{cases}$$

The epidemiological parameters are a, the proportion of the prenatal population born in sub-Saharan Africa (SSA), and b, the proportion who are previous or current injecting drug users (IDUs). The parameter e is the prevalence of HIV in the low risk group while h is the proportion of these patients who have already been diagnosed as HIV-infected. N is the overall size of the population at risk, T is the cost of testing and M is the net monetary benefit of detecting a case. The net benefit above is therefore the number of low-risk patients multiplied by the expected benefit, which accounts for some patients already being diagnosed, minus the expected costs of testing.

Parameters a, b, e and h were estimated by a multiparameter evidence synthesis and MCMC samples are available from their posterior distribution. They are also correlated. The economic parameters (where costs are in pounds) are $N = 105,000$ and $T = 3$, assumed to be known, while M is uncertain, but is assumed to have been previously estimated as

$$M = 600,012 - 54,296Y$$

where

$$Y \sim \text{Gamma}(0.56, 3)I(0, 2)$$

The first EVPPI calculation we consider is for a study investigating the focal parameters $\boldsymbol{\phi} = (a, b, e, h)$ and with non-focal parameters $\boldsymbol{\psi} = (M)$. Since the net benefit is linear in M, and M is not correlated with any of the $\boldsymbol{\phi}$, we can plug in the mean of M to get the conditional expectation of the net benefit for universal testing as

$$\mathrm{E}_{\boldsymbol{\psi}|\boldsymbol{\phi}}\left[\mathrm{NB}_d(\boldsymbol{\theta})\right] = N(1-a-b)\left[\mathrm{E}[M]e(1-h) - T(1-eh)\right]$$

and thus estimate the EVPPI(a, b, e, h).

Our second example is a trial estimating focal parameters $\boldsymbol{\psi} = M$ with non-focal parameters $\boldsymbol{\psi} = (a, b, e, h)$. Although the $\mathrm{NB}_d(\boldsymbol{\theta})$ is multi-linear in $\boldsymbol{\psi}$ and there are no correlations between $\boldsymbol{\phi}$ and $\boldsymbol{\psi}$, there are correlations between the elements of $\boldsymbol{\psi} = (a, b, e, h)$. In this case we can re-parameterise the problem so that the NB_d is linear in new parameters which are not correlated. For universal testing we can rewrite the net benefit as

$$\mathrm{NB}_2(\boldsymbol{\theta}) = NM\underbrace{(1-a-b)e(1-h)}_{\beta_1} - NT\underbrace{(1-a-b)(1-eh)}_{\beta_2}$$

These new parameters are

$$\beta_1 = (1-a-b)e(1-h)$$

$$\beta_2 = (1-a-b)(1-eh)$$

The net benefit is now linear in β_1 and β_2, so their correlation does not present a difficulty when calculating the conditional expectation of $\mathrm{NB}_2(\boldsymbol{\theta})$

In both of the examples above, the net benefit was a linear function, or could be re-parameterised into a linear function, of the non-focal parameters. An application where this is usually not the case is if the NB_d depends on probabilities, which may be defined using log-odds ratios δ and baseline log-odds μ, via the inverse of the logit transformation

$$p = \frac{\exp(\mu+\delta)}{1+\exp(\mu+\delta)}$$

Supposing that the log-odds ratio δ is the focal parameter ϕ, and the non-focal parameter $\psi = \mu$, then

$$p = \frac{\exp(\mu+\phi)}{1+\exp(\mu+\phi)}$$

and its expectation can be approximated by a linear function via a second-order Taylor approximation

$$\mathrm{E}_{\psi|\phi}\left[\frac{\exp(\psi+\phi)}{1+\exp(\psi+\phi)}\right] \approx H + \frac{H(1-H)(1-2H)\mathrm{Var}(\psi)}{2}$$

where

$$H = \frac{\mathrm{E}\left[\psi\right]+\phi}{1+\mathrm{E}\left[\psi\right]+\phi}$$

This means that we can use the mean of ψ to calculate H and, using also the variance of ψ, calculate the inner expectation of the EVPPI without nested simulation.

> **Algebraic methods for calculating EVPPI**
>
> **Benefits**
>
> - Can remove the need for nested simulations and give fast computation
> - Can be highly accurate, under specific conditions (linearity and independence)
>
> **Challenges**
>
> - Requires being able to write down the model algebraically
> - Only feasible for models that satisfy the linearity and independence conditions, or can be reparameterised or approximated in this form
> - Not amenable to general-purpose software – must be programmed differently for each model

3.3.4 Calculating EVPPI Using Nonparametric Regression

The most practical and widely used method for estimating EVPPI aims to estimate the functional form of $\mathrm{E}_{\boldsymbol{\psi}|\boldsymbol{\phi}}\left[\mathrm{NB}_d(\boldsymbol{\phi},\boldsymbol{\psi})\right]$ directly from the PA samples, without requiring access to the model structure. This is achieved by approximating the conditional expectation by fitting a regression with the net benefit of an intervention d as the outcome, and the parameters of interest $\boldsymbol{\phi}$ as predictors. We will start with a formal description of the general idea, before presenting several different ways of implementing it in practice. The regression principle has been used in several settings (e.g., Coyle and Oakley 2008; Oakley 2002) but the most generally applicable procedure was developed by Strong, Oakley and Brennan (2014), which our exposition here is based on. An overview is also given in Heath, Manolopoulou and Baio (2017).

The principle behind using regression to calculate EVPPI is to estimate the inner conditional expectation in Equation (3.4) by expressing the net benefit, for each simulation s from probabilistic analysis, as

$$\mathrm{NB}_d(\boldsymbol{\theta}_s) = \mathrm{E}_{\boldsymbol{\psi}|\boldsymbol{\phi}_s}\left[\mathrm{NB}_d(\boldsymbol{\phi}_s,\boldsymbol{\psi})\right] + \varepsilon_s,$$

where ε_s is an error term that has mean zero that captures the variability in the net benefit that is *not* explained by $\boldsymbol{\phi}$, i.e., the variability due to the non-focal parameters $\boldsymbol{\psi}$. The conditional expectation can be thought of as a function of $\boldsymbol{\phi}$ only, as the *conditional* expectation is only dependent on the value of $\boldsymbol{\phi}$. Therefore,

$$\mathrm{NB}_d(\boldsymbol{\theta}_s) = g_d(\boldsymbol{\phi}_s) + \varepsilon_s, \tag{3.6}$$

where $g_d(\cdot)$ is an unknown function of $\boldsymbol{\phi}$, which is then estimated by a regression model. Since this will not have a known functional form, flexible regression methods should be used, where we assume that $g_d(\cdot)$ is a smooth function of the parameters $\boldsymbol{\phi}$ whose EVPPI we are calculating, but make no other assumptions.

The EVPPI estimate is calculated using the following algorithm:

1. Obtain S PA samples for $\boldsymbol{\theta} = (\boldsymbol{\phi}, \boldsymbol{\psi})$ and $\mathrm{NB}_d(\boldsymbol{\theta})$, for each intervention.
2. For each intervention option d, fit a regression curve to the "dataset" comprising the sample of $\boldsymbol{\phi}_s$ as the observed "predictors," and the $\mathrm{NB}_d(\boldsymbol{\theta}_s)$ as the observed "response."
3. For each d and s, find the fitted value $g_d(\boldsymbol{\phi}_s)$ by evaluating the fitted regression curve at $\boldsymbol{\phi}_s$.
4. The EVPPI is estimated as

$$\widehat{\mathrm{EVPPI}} = \frac{1}{S}\sum_{s=1}^{S} \max_d g_d(\boldsymbol{\phi}_s) - \max_d \frac{1}{S}\sum_{s=1}^{S} g_d(\boldsymbol{\phi}_s).$$

In practice it is sufficient, and more efficient, to implement step (2) with the *incremental* net benefits INB_d (compared to some reference decision d, e.g., the standard of care) as the response, rather than the net benefits NB_d, requiring one fewer regression to be fitted. Note also that the fitted values $g_d(\boldsymbol{\phi}_s)$ are used in both terms of the formula for the EVPPI estimate, as this reduces simulation error.

Since all this method requires is the sample from the PA, and the choice of a regression model, it has been implemented in easily usable software. This includes web applications such as SAVI and BCEAWeb (as we discuss in Section 3.3.8), and the more general `voi` package. When we only require the EVPPI for one parameter $\boldsymbol{\phi}$, or the joint EVPPI for two or three parameters, the method produces estimates practically instantly, and estimates of EVPPI are robust to the choice of regression model. However, the challenge of specifying and computing an appropriate regression model increases as the number of parameters increases.

Regression-based methods for calculating EVPPI

Benefits

- Implemented in software (e.g., `SAVI` web app, `voi` `R` package).
- Automatic, fast and accurate computation for 1-2 parameter EVPPI.

Challenges

- Careful choice of regression model required to balance computational expense and accuracy for 3+ parameter EVPPI

Figure 3.3 illustrates the idea behind calculating EVPPI using nonparametric regression. Here we want to calculate the EVPPI of a single parameter ϕ, in a comparison of two treatments, a novel treatment $d = 2$ compared to the standard of care $d = 1$. Define the incremental net benefit of the novel treatment as $\text{INB}(\boldsymbol{\theta}) = \text{NB}_2(\boldsymbol{\theta}) - \text{NB}_1(\boldsymbol{\theta})$. With two treatments, the EVPPI, Equation (3.4), can be rewritten in the slightly simpler form $\text{EVPPI} = \text{INB}_{perfect} - \text{INB}_{current}$, the difference between the expected incremental net benefit under partial perfect and current information, where we define

$$\begin{aligned}\text{INB}_{perfect} &= \text{E}_{\boldsymbol{\phi}}\left[\max\left(0, \text{E}_{\boldsymbol{\psi}|\boldsymbol{\phi}}\left[\text{INB}(\boldsymbol{\phi}, \boldsymbol{\psi})\right]\right)\right] \\ \text{INB}_{current} &= \max\left(0, \text{E}_{\boldsymbol{\theta}}\left[\text{INB}(\boldsymbol{\theta})\right]\right)\end{aligned}$$

Figure 3.3 then shows all the ingredients of this formula and how they are estimated.

- The scatterplot points are simulations of the incremental net benefit $\text{INB}(\boldsymbol{\theta}_s)$ against the corresponding value of the parameter of interest $\boldsymbol{\phi}_s$, from the probabilistic analysis of the decision model.
- $\text{E}_{\boldsymbol{\theta}}\left[\text{INB}(\boldsymbol{\theta})\right]$ is estimated by the mean of all sampled values of $\text{INB}(\boldsymbol{\theta}_s)$ in the plot. Since this is positive, we prefer the novel treatment, and $\text{INB}_{current} = \text{E}_{\boldsymbol{\theta}}\left[\text{INB}(\boldsymbol{\theta})\right]$.
- The fitted nonparametric regression of the $\text{INB}(\boldsymbol{\theta}_s)$ on the $\boldsymbol{\phi}_s$ is shown by the thick black line in the plot. For each value of $\boldsymbol{\phi}$ (i.e., along the horizontal axis of the plot) this fitted value estimates $\text{E}_{\boldsymbol{\psi}|\boldsymbol{\phi}}\left[\text{INB}(\boldsymbol{\phi}, \boldsymbol{\psi})\right]$, the expected INB given "partial perfect information," i.e., knowledge of the value of $\boldsymbol{\phi}$. If the expected INB given this knowledge is negative, then we would no longer prefer the novel treatment.
- The broad grey line in the plot is the "frontier," our estimate of $\max\left(0, \text{E}_{\boldsymbol{\psi}|\boldsymbol{\phi}}\left[\text{INB}(\boldsymbol{\phi}, \boldsymbol{\psi})\right]\right)$ as a function of $\boldsymbol{\phi}$.

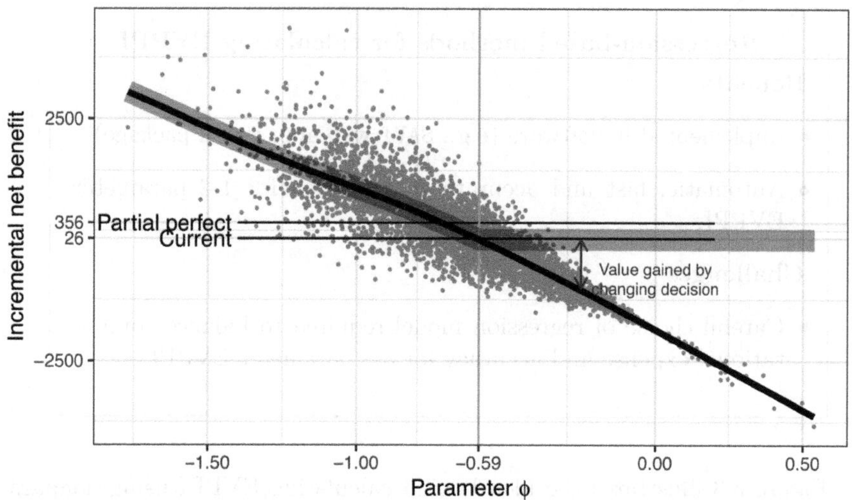

FIGURE 3.3
Illustration of computing EVPPI using nonparametric regression. The points are the S sampled values of the incremental net benefit (INB) and the parameter of interest ϕ. The fitted nonparametric regression function is shown by the medium-width black line, while the "frontier" is shown by the thick grey line. The expected INB under current information is the mean of all sampled INB values (26), while the expected INB under partial perfect information is the mean value of the "frontier" over all sampled values of ϕ (356). The EVPPI is the distance between these two INBs.

- Hence $\text{INB}_{perfect}$ is estimated by the average of the values of this frontier at the sampled values of ϕ_s, and the EVPPI follows as the difference between the estimates of $\text{INB}_{perfect}$ and $\text{INB}_{current}$.

This also illustrates the important principle of VOI methods: information *only has value if it changes the decision.* If we learn that $\phi > -0.59$, we switch our preference from the novel treatment $d = 2$ to the standard of care $d = 1$, and the expected value that we gain is the the same as the *loss we no longer incur* – the expected difference between the (negative) $\text{NB}_2 - \text{NB}_1$ and a loss of zero where $\text{INB} = 0$ (indicated in arrows in Figure 3.3).

3.3.5 Specific Methods for Nonparametric Regression

In this section, we shall explain the two different classes of regression models that have been used to calculate EVPPI: generalised additive models (GAM) and Gaussian process (GP) models. These methods differ in the way they define the regression function $g(\cdot)$ and in the computational methods used to fit the regression. While there are many other methods available for estimating a generic regression function, many from the field of *machine learning*, such as regression trees and neural networks, the use of these for VOI estimation has not yet been thoroughly investigated.

For clarity in these sections, the d subscript on the regression function $g(\cdot)$ is omitted.

Note: This section is technical, and requires some statistical expertise to understand! It is not necessary to understand these details in order to do most single-parameter EVPPI calculations, where the default methods implemented in software generally work well. However, in more complex situations, some appreciation of these details may be beneficial.

3.3.5.1 Generalised Additive Models

Generalised additive models (GAM) (Hastie and Tibshirani 1990) model the expectation of an outcome in terms of a sum of smooth functions of predictors. This is typically done using a linear function of *basis* terms $B_k(\cdot)$, where each basis term is a pre-specified function of the predictors $\boldsymbol{\phi}$,

$$g(\boldsymbol{\phi}) = \sum_{k=1}^{K} \alpha_k B_k(\boldsymbol{\phi})$$

and the coefficients α_k are estimated from the data. There are many ways of defining basis terms, and a popular approach is based on *splines* (Durrleman and Simon 1989; Wood 2003). Splines are defined using a set of parametric functions which are joined together at specific points known as *knots*. If the spline is made from cubic polynomials, for example, it is known as a cubic spline. In this book, we use the basis functions implemented in two widely used `R` packages.

1. The `mgcv` package (Wood 2017), which includes "thin plate" spline models (the package default, as described in Wood (2003)), cubic spline models, and many more.
2. The `earth` package (Milborrow 2011), that implements *multivariate adaptive regression splines* (MARS) (Friedman 1991). These are based on continuous, piecewise-linear functions, whose gradient changes at a series of knots.

Selecting and Fitting GAMs

GAMs can be made arbitrarily flexible by increasing the number of terms K, which in a spline model corresponds to increasing the number of knots. If the model is made too flexible, it "overfits," that is, it cannot generalise outside the observed data. Therefore we usually want to choose the model that optimises the ability of the model to predict outside the data. This choice is typically made using some kind of cross-validation. The `mgcv` package does this by deliberately choosing K to be large, while preventing overfitting by using penalised maximum likelihood estimation, with a *penalty* term that shrinks the estimated coefficients towards a smoother model. This ensures that deviations from smoothness are supported by the data, rather than encouraged by the model specification. The strength of shrinkage is controlled using a cross-validation method, described in Wood (2011). The `earth` package has a similar automated method for choosing the optimal piecewise linear regression form.

GAMs with Multiple Predictors

Note that ϕ can consist of one or more predictors. When calculating the joint EVPPI for two parameters, for example, ϕ will have two predictors. A flexible regression model for multiple predictors should consider the possibility that there may be an *interaction* between the predictors in their relation with the outcome. In the `mgcv` package, the user must specify which variables interact with each other, and smooth, flexible interaction models can be built as "tensor products." In `earth`, the user specifies a maximum degree of interaction to be considered in the automated model search (e.g., all two-way interactions).

However, as the number of predictors increases, it becomes more difficult to specify and fit an appropriate GAM, for several related reasons. Firstly, the number of potential models that can be specified increases, the most flexible including all multi-way interactions between all predictors. The number of coefficients to be estimated can then be very large. Then, to ensure that the coefficients are all identifiable from the available data, a very large dataset must be generated. Even if generating this dataset is computationally feasible, the process of estimating a large number of GAM coefficients from a large dataset may still be too computationally difficult. Strong, Oakley and Brennan (2014) and Heath, Manolopoulou and Baio (2017), for example, recommend that the method is limited to examples with up to four or five predictors (i.e., parameters for which we want the joint EVPPI).

Section 3.3.6 will give a demonstration of choosing between different GAMs for EVPPI estimation in practice.

3.3.5.2 Gaussian Process regression

An alternative class of flexible regression methods is based on Gaussian processes (GP) (Rasmussen and Williams 2006). These methods define the regression function $g(\phi)$ so that for any choice of $(\phi_1, \phi_2, \ldots, \phi_S)$, $(g(\phi_1), \ldots, g(\phi_S))$ has a multivariate normal distribution with some mean and covariance matrix. A Gaussian process regression model is defined by a "mean function" and a "covariance function," through which we can define the mean and covariance matrix of $(g(\phi_1), \ldots, g(\phi_S))$ for any set of $(\phi_1, \phi_2, \ldots, \phi_S)$.

A Gaussian process model can generally be made to fit data arbitrarily well. Informally, it does not require that the relation of the expected response

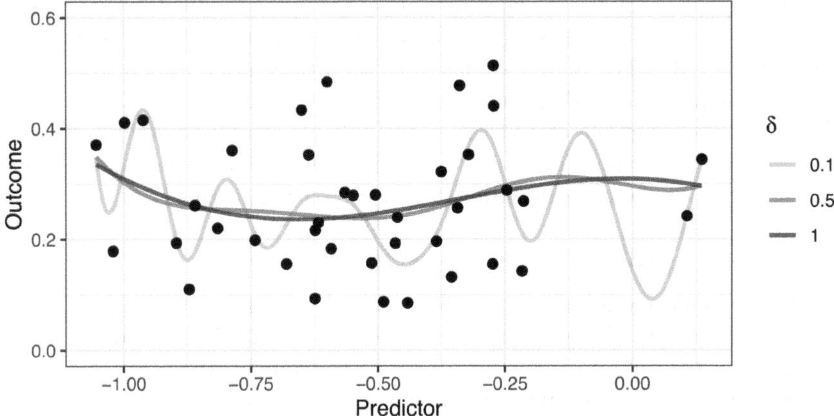

FIGURE 3.4
Illustration of Gaussian process regression for data with one outcome and one predictor. The data points are fitted by three alternative Gaussian process regression models with a linear mean function, no measurement error term ($\nu = 0$), and three different fixed values for the scale parameter δ.

with the predictors has a specific functional form, or that errors are normally distributed – the only requirement is that the output is a smooth function of the inputs. Figure 3.4 shows some examples of fitted Gaussian process regression models.

Example Mean and Covariance Functions

A commonly used mean function (e.g., Strong, Oakley and Brennan (2014)) defines the expectation of the outcome as a linear function of the predictors, so that the marginal expectation of $g(\boldsymbol{\phi}_s)$ is $1 + \sum_p \beta_p \phi_s^{(p)}$, where p indexes the different predictors, and β_p are their regression coefficients.

The covariance function $\mathcal{C}(r,s)$ describes the covariance between any two points $g(\boldsymbol{\phi}_r)$ and $g(\boldsymbol{\phi}_s)$. This is generally constructed so that the correlation between $g(\boldsymbol{\phi}_r)$ and $g(\boldsymbol{\phi}_s)$ is stronger when $\boldsymbol{\phi}_r$ is closer to $\boldsymbol{\phi}_s$, for any r, s. A typical covariance function (used in Strong, Oakley and Brennan (2014)) is the squared exponential function $\mathcal{C}_{\text{Exp}}$, defined by

$$\mathcal{C}_{\text{Exp}}(r,s) = \sigma^2 \exp\left[-\sum_{p=1}^{P}\left(\frac{\phi_r^{(p)} - \phi_s^{(p)}}{\delta_p}\right)^2\right] \tag{3.7}$$

where $\phi_r^{(p)}$ and $\phi_s^{(p)}$ are the r-th and the s-th data point (recall that in EVPPI estimation, these "data points" are the simulated values of the p-th of the decision model parameters of interest comprising $\boldsymbol{\phi}$). The "scale" parameter δ_p defines the smoothness of the relationship between the predictor and the response. For lower values of δ_p, there is a lower correlation between outcomes

$g(\boldsymbol{\phi}_r)$ and $g(\boldsymbol{\phi}_s)$ at $\boldsymbol{\phi}_r, \boldsymbol{\phi}_s$ that differ only through a small difference in a particular $\phi^{(p)}$ – hence $g()$ is driven more by the observed data points, as shown in Figure 3.4. For higher values of δ, the fitted model tends towards a linear regression, as defined through the mean function.

Then, by assuming the error term ϵ_s in (3.6) is normally distributed with a constant variance, the GP model describing the S simulations of the net benefit in terms of the corresponding $\boldsymbol{\phi}$ can be written in the form

$$(\text{NB}^{(1)}, \ldots, \text{NB}^{(S)}) \sim N(H\boldsymbol{\beta}, \sigma^2(\Sigma + \nu I))$$

where I is the identity matrix of size S, $\sigma^2\Sigma$ is the matrix with (r, s) element $C(r, s)$, and H is the "design matrix" whose first column is a vector of ones, and whose $p + 1$th column is the vector with sth element ϕ_s^p.

Two methods have been used for fitting GP regression models in the context of VOI estimation, both taking a Bayesian perspective.

Fitting Gaussian Process Regressions: Empirical Bayes Estimation

In Strong, Oakley and Brennan (2014), the "hyperparameters" δ, ν are estimated as the values which maximise the marginal posterior density (marginalised over β and σ). The posterior means of β and σ, hence the fitted values required to estimate the EVPPI, can then be derived as analytic functions of the estimated hyperparameters, as described fully in the paper.

The numerical optimisation to estimate the hyperparameters can be computationally challenging, particularly when the sample size S is large, as it involves the inversion of a $S \times S$ matrix. With a larger number of parameters (comprising $\boldsymbol{\phi}$), GP regression is not as problematic as GAM regression, but difficulties can still occur with estimating the hyperparameters, due to the higher dimension and complexity of the marginal posterior surface that is maximised over.

Fitting Gaussian Process Regressions by SPDE-INLA

An alternative estimation method, described in full in Heath, Manolopoulou and Baio (2016) and Baio, Berardi and Heath (2017), is computationally faster, though conceptually more complex. Instead of the squared exponential covariance function, this uses a specially constructed covariance function, obtained by approximating a "Matern" covariance function as a solution to a stochastic partial differential equation (SPDE). This allows an approximation to the full posterior distribution to be obtained by integrated nested Laplace approximation (INLA). To gain computational efficiency, this covariance function is only evaluated on an irregular grid of points. In general, the more points there are in this grid, the more accurate the approximation, but the greater the computational time. The `voi` package implements default values for the density of the grid that aims to balance these two concerns, but they can be adjusted for greater accuracy or faster run time.

The grid approximation used in this method was developed in spatial statistics, hence it has some constraints. Firstly, it is only applicable to calculating the joint EVPPI of $P = 2$ parameters or more, and not a single parameter. Secondly, if $P > 2$, it requires that the $P-$dimensional space of parameters can be projected to a two-dimensional space that contains all pertinent information about the relation between the parameters and the net benefit. This can be done using regression-based dimension reduction; "principal fitted components" methods are used in Heath, Manolopoulou and Baio (2016). While these two dimensions are often sufficient in practice to estimate EVPPI, there are some settings where two dimensions are not sufficient to capture the relationship. In this case, the mean of the GP model, $(H\boldsymbol{\beta})$, should be augmented to include non-linear terms, i.e., interaction terms between the different parameters for which we are computing EVPPI, and potentially higher-order functions of these parameters. (Note the `voi` package will warn the user if this extension may be necessary.)

Despite its computational efficiency, the SPDE-INLA method will be slower than GAM regression, so is not advised when GAM regression is possible. The MARS method of GAM construction may also be computationally more efficient than the SPDE-INLA methods for smaller subsets, but the SPDE-INLA approximation may be more efficient for large parameter subsets, i.e., those that compute the value of all parameters updated by a complex study.

3.3.6 Case Study and Guidelines for Estimating EVPPI Using Regression in R

Here we demonstrate how to use the regression methods to calculate EVPPI, using the chemotherapy model from Chapter 2 as an example. This section focuses on how to compute per-person EVPPI accurately. See Section 3.4 for a more fully developed EVPPI analysis in this case study, which focuses on interpretation and use of per-person and population EVPPI in health economic evaluation.

Throughout this section, we will calculate the per-person EVPPI at a specific willingness-to-pay of £20,000. Working with the `R` code repository explained in Chapter 2, a sample of size 5000 is taken from probabilistic analysis of the decision model.

> To run this code:
>
> - The Chemotherapy example model should be installed, following the instructions in Section 2.3.5.
> - The `voi` `R` package should be installed, using `install.packages("voi")`.
>
> All code in the book is provided and linked from `https://chjackson.github.io/voi/articles/book.html`. If reading this book as a PDF file, don't paste code from the PDF, since some special characters may fail to copy.

```
source("04_analysis/01_model_run.R")
nb <- m_net_benefit[ , , wtp_seq == 20000]
```

3.3.6.1 Using the `voi` Package to Calculate EVPPI

The function `evppi` from the `voi` `R` package is used throughout this section to compute the EVPPI. It is generally invoked in the form:

```
evppi(outputs, inputs, pars, ...)
```

There are three required arguments:

- `outputs`: this can take either of two forms:
  - (a) A matrix or data frame of samples from the uncertainty distribution of the expected net benefit, or
  - (b) A list of three components: a matrix or data frame of sampled costs, a matrix or data frame of sampled effects and a vector of willingness-to-pay values. Matrices or data frames have one column per decision option d, and one row per sample s from probabilistic analysis.
- `inputs`: a matrix or data frame of samples from the uncertainty distribution of the decision model parameters, with one column per parameter, and one row per sample s.
- `pars`: the parameters of interest ϕ for the EVPPI calculation. This can take either of two forms:
  - (a) A character vector. A single EVPPI value is returned, which is a single-parameter EVPPI if `pars` has one element, or a multi-parameter EVPPI if `pars` is a vector of multiple elements.
  - (b) A list of character vectors. A separate EVPPI value is calculated for each component of the list.

Other arguments to control the calculation methods can be supplied – some examples are given below.

By default, `evppi` uses a GAM to compute single-parameter EVPPI, or multi-parameter EVPPI for groups of four or fewer parameters. Unless otherwise specified, the GAM is the default spline model from `gam` in the `mgcv` package. Gaussian process regression (using the empirical Bayes method) is done by default for multi-parameter EVPPI calculations with groups of five or more parameters.

3.3.6.2 Calculating Single-Parameter EVPPI

We invoke `evppi` to calculate the single-parameter EVPPI for all the parameters in the model as follows. It returns a data frame containing the EVPPI estimates.

```
library(voi)
pars_all <- as.list(names(m_params))
ev_single <- evppi(outputs=nb, inputs=m_params, pars=pars_all)
ev_single
```

```
##                    pars    evppi
## 1      p_side_effects_t1  24.7915
## 2      p_side_effects_t2 260.9055
## 3            c_home_care   0.0000
## 4             c_hospital   0.0032
## 5                c_death   0.0000
## 6             u_recovery   7.7910
## 7            u_home_care 103.7762
## 8             u_hospital   2.6355
## 9     logor_side_effects 329.0098
## 10 p_hospitalised_total  21.8343
## 11                p_died  41.5116
## 12           lambda_home  10.0146
## 13           lambda_hosp  21.7919
## 14         rate_longterm  38.9335
```

The package also provides a `plot` method for these returned objects, which simply produces a dot-plot of the estimates, optionally sorted with the highest values at the top. This is produced using the `ggplot2` R package.

```
plot(ev_single, order = TRUE)
```

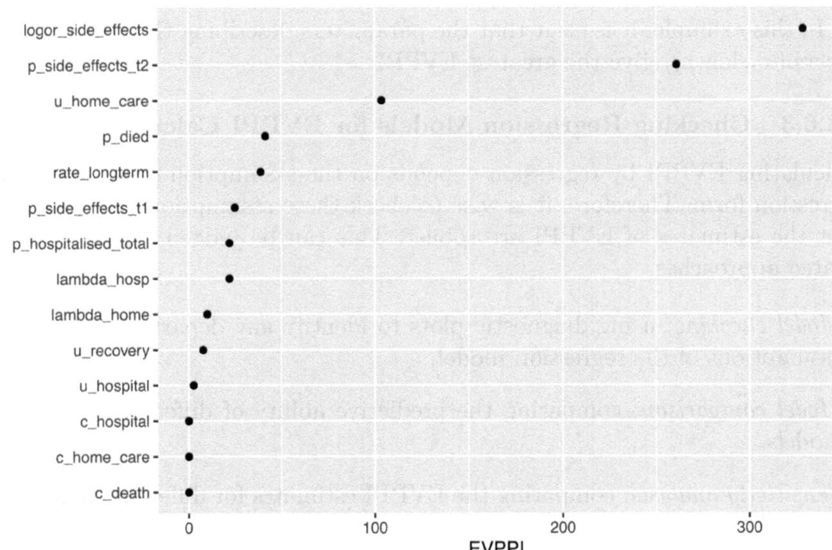

A similar calculation of single-parameter EVPPIs for a specified subset of parameters can be done with, for example,

```
evppi(outputs=nb, inputs=m_params,
      pars=list("logor_side_effects", "p_side_effects_t1", "u_hospital"))
```

We now calculate multi-parameter EVPPI for four groups of parameters: those associated with side effects ($\pi_0, \log(\rho)$), transition probabilites ($\Gamma_1, \Gamma_2, \lambda_1, \lambda_2$), costs ($c_{HC}, c_H, c_{death}$) and utilities ($q, q_{HC}, q_H$) respectively. To use `evppi` for this, a list of character vectors is constructed, where each list component is given a meaningful name. Since $p \leq 4$ for each of these components, the GAM method is used. This is computationally straightforward, taking no more than a few seconds.

```
par_groups <- list(
  "side_effects" = c("p_side_effects_t1","logor_side_effects"),
  "trans_probs" = c("p_hospitalised_total","p_died",
                    "lambda_home","lambda_hosp"),
  "costs" = c("c_home_care","c_hospital","c_death"),
  "utilities" = c("u_recovery","u_home_care","u_hospital")
)
ev_grouped <- evppi(outputs=nb, inputs=m_params, pars=par_groups)
ev_grouped
```

```
##           pars  evppi
## 1 side_effects 330.62
## 2  trans_probs  63.14
## 3        costs   7.01
## 4    utilities 105.27
```

In this example, it is clear that the parameters describing the risks of side effects ($\pi_0, \log(\rho)$) give the greatest EVPPI.

3.3.6.3 Checking Regression Models for EVPPI Calculation

Calculating EVPPI by regression depends on the assumption of a particular regression form. Therefore, it is wise to check these assumptions, and ensure that the estimates of EVPPI are robust. This can be judged through three related approaches:

- *Model checking*: using diagnostic plots to identify any departures from the assumptions of the regression model.
- *Model comparison*: comparing the predictive ability of different regression models.
- *Sensitivity analysis*: comparing the EVPPI estimates for different choices of method.

Here we illustrate some ways to do this in the example. We check the EVPPI for the relative risk of side effects, which had the largest EVPPI according to the default calculation above.

In the voi package, if the check=TRUE option is supplied, then evppi returns an object which contains the regression model objects themselves as attributes (this is not done by default, since the model objects may be large). The function check_regression then produces some standard diagnostic plots.

```
ev_single <- evppi(outputs=nb, inputs=m_params, pars=pars_all, check=TRUE)
check_regression(ev_single, pars = "logor_side_effects")
```

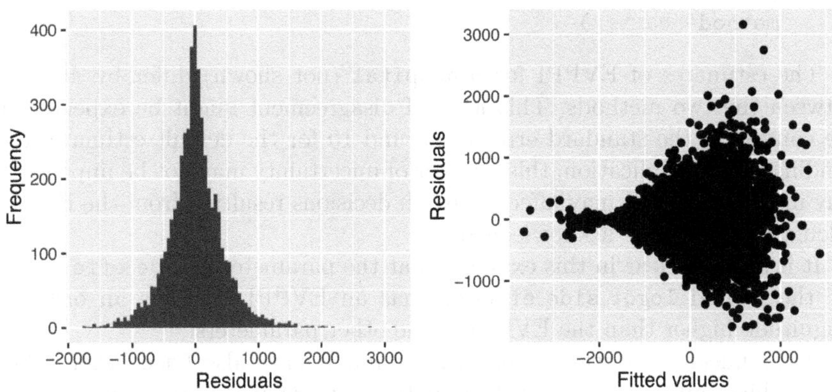

These plots illustrate the *residuals* from the model, defined as estimates of the error $\hat{\epsilon}_s = NB_s - \hat{g}(x_s)$ from the fitted regression model $\hat{g}$. The fundamental assumption of regression-based EVPPI calculation is that the expectation of this error $\mathrm{E}(\epsilon_s)$ is zero, for every s. The left plot shows that the residuals in this case have a mean of zero, and the right plot shows that the residuals are unrelated to the fitted values $\hat{g}(x_s)$.

Note that the variance of the errors appears to be increasing as $g(x_s)$ increases from 0 to 4000, however this is not necessarily a problem for estimating $\mathrm{E}(NB|x_s)$. Least squares estimates of regression models are unbiased as long as the errors have mean zero, even if the error variance is not constant. The consequence of a non-constant variance is to increase the standard error of the estimate.

Standard errors of the EVPPI estimates using the default GAM method can be produced by calling evppi with the se=TRUE option. These are produced by a simulating a sample of B alternative values for the regression coefficients, from a multivariate normal distribution defined by the estimate and covariance matrix of these coefficients, then calculating the standard deviation of the resulting sample of EVPPI estimates. This is moderately expensive (taking a few seconds in this case) so is not done by default. Note that this standard error only represents uncertainty in estimating the regression coefficients from the PA "dataset," and does not represent uncertainty due to the choice of regression function.

```
evppi(outputs=nb, inputs=m_params, pars=list("p_side_effects_t2","u_hospital"),
      se=TRUE)
```

We now consider whether other regression models would fit better, or give different estimates of the EVPPI. For example, we compare with a GAM based on multivariate adaptive regression splines (MARS, `method="earth"`).

```
evppi(outputs=nb, inputs=m_params,
      pars = list("p_side_effects_t2","u_hospital"))
evppi(outputs=nb, inputs=m_params,
      pars = list("p_side_effects_t2","u_hospital"),
      method="earth")
```

The estimates of EVPPI for `u_hospital` (not shown) differ by about 3 between the two methods. This level of disagreement might be expected in the context of the standard error of around 10 for the default estimate. Depending on the application, this amount of uncertainty may not be important. Any uncertainty that may affect research decisions resulting from the EVPPI calculation should be acknowledged.

It is at least clear in this example that the parameter `p_side_effects_t2` (or the related `logor_side_effects`) has an EVPPI which is an order of magnitude higher than the EVPPI of the other parameters.

If the uncertainty was of concern, we might have also tried other spline bases, which can also be specified through the `gam_formula` argument – see the `voi` package documentation. We might have also compared with a Gaussian process model, though given that these are substantially more computationally expensive, they are less useful in practice for producing single-parameter EVPPI.

3.3.6.4 Comparing Different Regression Specifications: Single-Parameter EVPPI

Each of the two classes of models (`mgcv` and `earth`) allows a wide variety of regression models to be built. Usually we want as flexible a model as possible, e.g., by accounting for all possible interactions. However as discussed in Section 3.3.5.1, this may not be computationally feasible, and we may want to simplify the model. This might be done in different ways.

As a simplified example of comparing different regression specifications, suppose we wanted to calculate the joint EVPPI for the two parameters related to side effects. We could specify a GAM with or without the interaction term. In the default method used by `evppi`, the regression model form can be controlled through the argument `gam_formula` (which is then passed to the `gam` function in `mgcv`). The EVPPI computation `e1` includes the interaction, and `e2` excludes the interaction.

```
(e1 <- evppi(outputs=nb, inputs=m_params, pars=par_groups[1], check=TRUE))
(e2 <- evppi(outputs=nb, inputs=m_params, pars=par_groups[1],
             gam_formula="s(p_side_effects_t1) + s(logor_side_effects)",
             check=TRUE))
check_regression(e1, plot=FALSE)
check_regression(e2, plot=FALSE)
```

```
##           pars evppi
## 1 side_effects   331

##           pars evppi
## 1 side_effects   332

## $AIC
## [1] 74754

## $AIC
## [1] 74838
```

Different models from the same regression family can be compared statistically. mgcv models can be compared using Akaike's information criterion (AIC), where lower values of AIC indicate that the model has better overall predictive ability. The check_regression method returns this statistic. In this example, the predictive ability is reduced by omitting the interaction terms, therefore we would prefer the EVPPI estimate from e1 (though the two estimates are not substantively different).

Likewise, the predictive ability of earth models can be compared using a generalised cross-validation statistic, returned from check_regression as gcv. These models by default include at most two-way interactions between predictors, but we can set degree=3 to also allow three-way interactions. This improves the gcv statistic here.

```
(e1 <- evppi(outputs=nb, inputs=m_params, pars=par_groups[1],
             method = "earth", check=TRUE))
(e2 <- evppi(outputs=nb, inputs=m_params, pars=par_groups[1],
             method = "earth", degree=3, check=TRUE))
check_regression(e1,plot=FALSE)
check_regression(e2,plot=FALSE)
```

```
##           pars evppi
## 1 side_effects   334

##           pars evppi
## 1 side_effects   333

## $gcv
## [1] 185678

## $gcv
## [1] 183384
```

There is currently no interface in voi to compare predictive ability between mgcv and earth models, but in principle this might be done using cross-validation (see, e.g., the caret R package (Kuhn 2022)).

3.3.6.5 Comparing Different Regression Specifications: Multi-Parameter EVPPI

As a final illustration, we demonstrate a more challenging multi-parameter EVPPI problem, for the six parameters describing the costs (c_{HC}, c_H, c_{death}) and utilities (q, q_{HC}, q_H) of each health state. The general approach we take is to do as much as is computationally feasible with each available method, compare predictive ability where possible, and acknowledge the extent of disagreement between the resulting EVPPI estimates.

GAMs using `mgcv` are instant to fit with interactions excluded (`ev1`), though take around half a minute with interactions of degree no more than 2 (`ev2`), and struggle with bigger models. MARS / `earth` regression models are instant to fit with interactions of degree 2. Judging from predictive model checks in these cases, interactions do not seem necessary, so we rule out `ev1` and `ev3`. Gaussian process modelling handles interactions more elegantly, by implicitly searching for the optimal extent of interaction. Here we fit GP models using the standard empirical Bayes method (`ev5`) and the SPDE-INLA approximation (`ev6`).

```
costs_utilities <- c(par_groups$costs, par_groups$utilities)
ev1 <- evppi(outputs=nb, inputs=m_params, pars=costs_utilities,
             method="gam",
             gam_formula = "s(c_home_care) + s(c_hospital) + s(c_death) +
                            s(u_recovery) + s(u_home_care) + s(u_hospital)",
             check=TRUE)
ev2 <- evppi(outputs=nb, inputs=m_params, pars=costs_utilities,
             method="gam", check=TRUE,
             gam_formula = all_interactions(costs_utilities, 2))
ev3 <- evppi(outputs=nb, inputs=m_params, pars=costs_utilities,
             method="earth", check=TRUE)
ev4 <- evppi(outputs=nb, inputs=m_params, pars=costs_utilities,
             method="earth", degree=2, check=TRUE)
ev5 <- evppi(outputs=nb, inputs=m_params, pars=costs_utilities,
             method="gp")
ev6 <- evppi(outputs=nb, inputs=m_params, pars=costs_utilities,
             method="inla")
```

The EVPPI estimates from these methods roughly agree, ranging from around 180 to 200. Note that beyond this we do not know which of these estimates is expected to be more accurate. Some theoretical work has been done evaluating the relative accuracy of different methods (e.g., the SPDE-INLA method is similar in accuracy to the empirical Bayes GP method, while being feasible at larger sample sizes), but in a real application we will not know the truth.

The sample size S was 5000 in this example. Note that very flexible regressions will become more difficult at larger sample sizes. If independent samples can be generated from the decision model, then large sample sizes are only necessary to represent highly skewed uncertainty distributions, that may occasionally arise, for example, for expected costs. Large sample sizes may also be necessary to represent posterior distributions in Bayesian statistical models, fitted with Markov Chain Monte Carlo methods that are slow to explore the posterior and produce highly correlated samples. In recent years, however, more efficient MCMC samplers have been developed that are less prone to this problem, and accessible software for these is available (Carpenter et al. 2016).

3.3.7 Other Methods for Calculating EVPPI

In practice, we have found that estimating EVPPI using regression methods is generally robust enough to indicate where further research may be beneficial in health economic evaluations, despite the uncertainties that can be encountered in larger problems. When a specific research study is being considered, the expected value of sample information (EVSI) is a more useful measure, and this will be covered in the next chapter.

Our review of computational methods for EVPPI is not exhaustive, however. Here we briefly describe some other methods that have been proposed.

Fang et al. (2022) presented two modifications that improve the efficiency of the standard Monte Carlo procedure (Section 3.3.2) to estimate EVPPI. The *multilevel Monte Carlo* method is based on efficiently rearranging the nested sums involved in the standard procedure, so they can be truncated in a way that optimises computational speed for a desired accuracy. The *quasi-Monte Carlo* method is based on replacing random samples of parameter values with deterministically constructed grids of values that are designed to optimally cover the parameter space. These methods are advocated for situations where the joint EVPPI for a large number of parameters is desired, hence the regression-based EVPPI computation methods are awkward. Fang et al. (2022) provide code to implement them, and show situations where they are more efficient than the standard Monte Carlo method.

Some specialised methods for single-parameter EVPPI computation were previously developed (Sadatsafavi et al. 2013; Strong and Oakley 2013). These can be awkwardly sensitive to tuning quantities, however, and we have found that in practice, GAM regression is robust and efficient enough for single-parameter EVPPI to render these older methods unnecessary. These methods are included in the `voi` package for academic interest.

3.3.8 EVPPI Analysis Through Web Interfaces

We end this section with a brief discussion of web-based interfaces for the calculation of EVPPI. In general, we would recommend the use of `R` (or other full programming languages) throughout a decision modelling analysis, for reasons of scientific reproducibility and flexibility (Baio and Heath 2017; Incerti et al. 2019). However, in practice, many health economic evaluations are based on spreadsheet or specialised decision modelling software. Web interfaces are helpful to make VOI methods accessible to modellers using any analysis software. We describe two of these web applications here:

- SAVI: `https://savi.shef.ac.uk/SAVI/`, Strong, Brennan and Oakley (2015).
- BCEAWeb: `https://egon.stats.ucl.ac.uk/projects/BCEAweb/`, Baio, Berardi and Heath (2017).

To use either of these applications, the following general procedure is followed.

- The user uploads text files (in CSV format) containing the sample of parameters $(\boldsymbol{\theta}_1, \ldots, \boldsymbol{\theta}_S)$ from probabilistic analysis of the decision model, and the corresponding sample of values from the expected costs and effects.
- The user specifies analysis settings (e.g., willingness-to-pay, treatment comparisons of interest).
- Standard cost-effectiveness summaries, such as cost-effectiveness planes, acceptability curves and EVPI, are displayed in the web interface.
- EVPPI is computed and displayed in the web interface, for parameters, or groups of parameters, selected by the user.
- A full report of the analysis can be produced for download (e.g., in PDF, Word or HTML format).

Both of these applications can calculate EVPPI using GAM (via `mgcv`) or the empirical Bayes Gaussian process method, while BCEAWeb additionally implements the SPDE-INLA method for Gaussian process regression. The apps differ in the exact nature and style of the cost-effectiveness summaries that can be produced, while SAVI has additional features for analysis of a payer's financial risks, using methods described in Grimm et al. (2017). Tuffaha et al. (2021) review these and some other web-based tools, including one based on a *minimal modelling* approach to prioritisation of further research (RANE, `https://shiny.york.ac.uk/rane/`, also see our Chapter 9), and one which is focused on estimating the potential value of studies of interventions in cardiovascular disease (VICTOR, `https://uwchoice.shinyapps.io/victor/`).

By contrast, the `voi` `R` package focuses only on VOI calculation rather than economic evaluation in general, and aims to provide a common computational engine for any useful VOI calculation method, around which different user interfaces could be built if desired. A unique feature of this package is its ability to compute the expected value of *sample* information using several methods, under a common interface – this is the subject of the next chapter.

3.4 An EVPPI Analysis: The Chemotherapy Case Study

In this section, we undertake an EVPPI analysis using the Chemotherapy case study from Chapter 2 within the `voi` package. We demonstrate how EVPPI analyses can be used for sensitivity analysis and for research prioritisation. The goal of sensitivity analysis is to determine which parameters and groups of parameters have the highest impact on decision uncertainty in our health economic model, while research prioritisation focuses on determining whether these drivers of uncertainty should potentially be the focus of future research to improve decision-making.

To run this code:

- The Chemotherapy example model should be installed, following the instructions in Section 2.3.5.
- The `voi` `R` package should be installed, using `install.packages("voi")`.

All code in the book is provided and linked from `https://chjackson.github.io/voi/articles/book.html`. If reading this book as a PDF file, don't paste code from the PDF, since some special characters may fail to copy.

3.4.1 Single-Parameter EVPPI

To undertake a sensitivity analysis using EVPPI, we typically begin by evaluating the single parameter EVPPI to identify parameters that provide a high contribution to decision uncertainty when considered alone. To achieve this, we use the `voi` package to compute EVPPI across a range of willingness-to-pay thresholds.

```
> ## Baseline Cost-Effectiveness Formatting
> chemotherapy_output <- list(e = m_costs_effects[, , "Effects"],
+                             c = m_costs_effects[, , "Costs"],
+                             k = seq(0, 50000, length.out = 501))
> ## Single parameter EVPPI
> pars_all <- as.list(names(m_params))
> ev_single <- evppi(outputs=chemotherapy_output, inputs=m_params,
+                    pars=pars_all)
```

In this analysis, we use the cost-effectiveness form for the `chemotherapy_output`, which estimates EVPPI on a grid of values for the willingness-to-pay between £0 and £50,000 per QALY gained. The EVPPI values change substantially across willingness-to-pay thresholds. For sensitivity analysis, it is often useful to initially evaluate EVPPI at a threshold equal to the ICER, since, informally, this is where decision uncertainty is maximised, leading to high EVPPI values, which makes it easier to detect differences in EVPPI values across parameters. Furthermore, the relative ranking of EVPPI values tends to be relatively stable across willingness-to-pay thresholds, so exploring these values at the ICER indicates how EVPPI might behave across different willingness-to-pay thresholds (Figure 3.5).

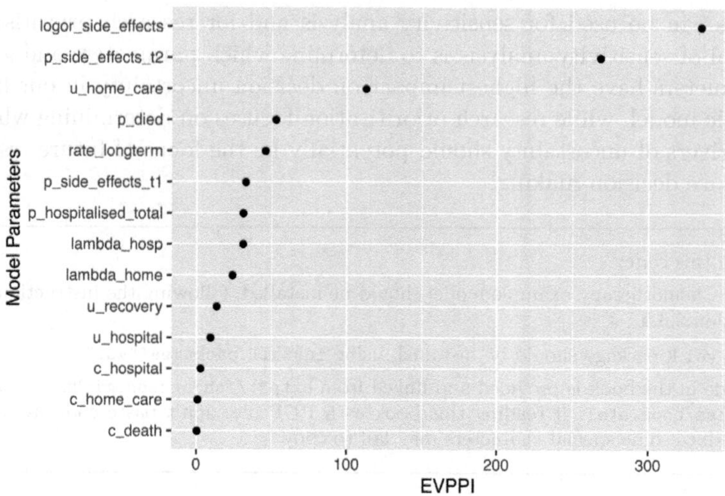

FIGURE 3.5
Single parameter EVPPI for all parameters in the Chemotherapy example at a willingness-to-pay of £19,700 per QALY gained.

```
> # Select the WTP on the grid that is closest to the ICER
> wtp.max <- chemotherapy_output$k[
+   which.min(abs(chemotherapy_output$k - ICER))
+   ]
> # Explore EVPPI at maximum uncertainty
> library(dplyr)
> ev_single %>%
+   filter(k == wtp.max) %>%
+   arrange(desc(evppi)) %>%
+   mutate(evppi = round(evppi, 2))
                   pars     k  evppi
1    logor_side_effects 19700 335.04
2     p_side_effects_t2 19700 267.92
3           u_home_care 19700 112.55
4                p_died 19700  52.69
5         rate_longterm 19700  45.77
6     p_side_effects_t1 19700  32.85
7  p_hospitalised_total 19700  31.17
8           lambda_hosp 19700  31.05
9           lambda_home 19700  23.95
10           u_recovery 19700  13.29
11           u_hospital 19700   9.28
12           c_hospital 19700   2.78
13          c_home_care 19700   0.78
14              c_death 19700   0.33
```

For the Chemotherapy example, we can clearly see that the odds ratio of side effects is associated with the highest EVPPI value, indicating that this is one of the key drivers of decision uncertainty as it is close to the EVPI of 374. Note that the probability of side effects for the novel intervention `p_side_effects_t2` is also associated with a high EVPPI value. This is likely to be because the probability of side effects from the novel intervention is defined as a function of the odds ratio of side effects and the probability of side effects for the standard treatment, indicating that learning either one of `p_side_effects_t2` or the odds ratio would provide a lot of information about the other. It is not surprising that the odds ratio is deemed to be the biggest driver of uncertainty about the intervention decision, as the key difference between the two interventions is that the novel intervention aims to reduce the side effects experienced due to chemotherapy at an elevated cost.

From this analysis, it seems that when considered separately the costs have minimal impact on the decision uncertainty, as all three cost parameters are associated with a low EVPPI. However, it is important to remember that low values for the single parameter EVPPI do not necessarily mean that this group of parameters will not be deemed valuable when their joint EVPPI is estimated. Therefore, we will consider which groups of related parameters might be valuable to learn better as a group (Section 3.4.3). The utilities and transition probabilities are associated with moderate EVPPI. The utility experienced during home care has a substantially higher EVPPI than that of the other utility parameters. This is likely to be because all individuals experience home care, and therefore the associated parameters are influential to the treatment decision.

When comparing EVPPI estimates between parameters, it is important to remember that the estimates are subject to computational uncertainty. The uncertainty that arises from estimating the nonparametric regression coefficients can be represented as a standard error for the EVPPI estimates (Section 3.3.6.3). In particular, we note that EVPPI is between 32.8 and 31.0 for the parameters ranked six to eight (`p_side_effects_t1`, `p_hospitalised_total` and `lambda_hosp`), a difference of 1.8, while the standard errors of these estimates are 6.5, 6.5 and 6.3, respectively.

```
> ## Calculate standard errors
> pars_small <- list("p_died", "rate_longterm", "p_side_effects_t1",
+                    "p_hospitalised_total", "lambda_hosp", "lambda_home")
> nb <- wtp.max * chemotherapy_output$e - chemotherapy_output$c
> ev_small_se <- evppi(outputs=nb, inputs=m_params, pars=pars_small,
+                      se = TRUE)
> ev_small_se
                  pars    evppi       se
1    p_side_effects_t1 32.84881 6.526196
2 p_hospitalised_total 31.18765 6.450910
3          lambda_hosp 31.04995 6.317019
```

Therefore, the ranking of these parameters is not certain, and we can only state that they each contribute roughly the same amount to the decision

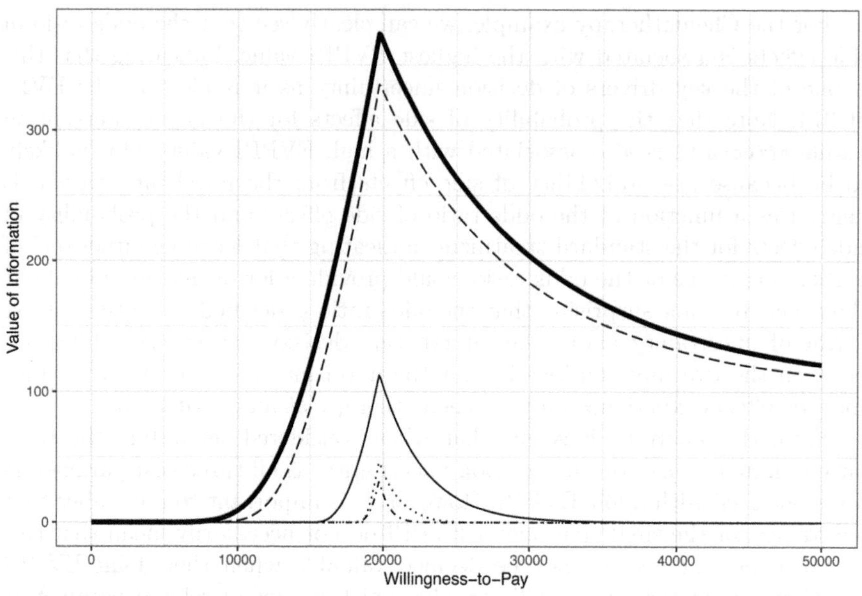

FIGURE 3.6
EVPPI plotted against willingness to pay for the most valuable single parameters in the chemotherapy model.

uncertainty, when considered separately. Note that it could be possible to reduce these standard errors by running more simulations from the PA, but greater precision may not be necessary in practice.

3.4.2 EVPPI across Willingness-to-Pay

Having explored the sensitivity of the decision to parameter uncertainty at the ICER, representing the maximal decision uncertainty, we now explore EVPPI across different values for the willingness-to-pay. As with EVPI, EVPPI values for each of the parameters will be sensitive to the chosen willingness-to-pay threshold. In the previous section, we compute the EVPPI for a range of willingness-to-pay thresholds between £0 and £50,000 per QALY gained.

In Figure 3.6, we display single parameter EVPPI across willingness-to-pay values for four model parameters. We can clearly see that the odds ratio of side effects has the highest value of information across all willingness-to-pay thresholds. We can also compare the EVPPI for the odds ratio of side effects to the EVPI (solid line) and can see that gaining perfect information about the relative risk has a similar value to eliminating all model uncertainty.

While the utility of home care is clearly the second most valuable single parameter, we can see that other two parameters have similar values for the EVPPI, indicating that the rankings are likely to be unstable. However, it is clear that above willingness-to-pay values of around £25,000 per QALY gained, each of these parameters have no value when learnt alone. Having undertaken this analysis, it is now important to consider the value of parameter subsets, as these can have larger value than any of the parameters individually.

3.4.3 Multi-Parameter EVPPI

A multi-parameter EVPPI analysis may be complex, as it may involve considering many parameter subsets. Instead of exhaustively considering all potential subsets, we typically focus on parameters that could be learnt at the same time, from a single research study.

We know that any parameter subset that includes the relative risk of side effects will have a high EVPPI, as it must be at least as large as the single-parameter EVPPI for this parameter. Thus, we consider a randomised controlled trial that would inform the relative risk of side effects, but where we also collect follow-up information for all individuals who experience side effects, which provides more information on the transition probabilities. Note that typically health economic decision models do not use randomised trial data to estimate the baseline risk of an event, because trial outcomes are often superior to outcomes in routine clinical practice. Therefore, we suppose that our hypothetical trial investigating the *relative risk* of side effects does not also inform the *baseline* risk of side effects for the standard chemotherapy.

```
> ## EVPPI Groups
> # Randomised Trial
> par_RCT <- list(
+   "side_effects" = c("logor_side_effects"),
+   "side_effects_and_follow_up" = c("logor_side_effects",
+                                 "p_hospitalised_total","p_died",
+                                 "lambda_home","lambda_hosp")
+ )
> ev_RCT <- evppi(outputs=chemotherapy_output, inputs=m_params,
+                 pars=par_RCT, method = "earth")
> ev_RCT %>%
+   filter(k == wtp.max)
                        pars     k    evppi
1               side_effects 19700 338.3852
2 side_effects_and_follow_up 19700 347.3137
```

There is some additional value in collecting the transition probability data alongside data on the relative risk. But this additional value is limited, perhaps unsurprisingly, as the EVPPI for the relative risk alone is close to the EVPI. In addition to considering a randomised trial, we could also consider a range of non-randomised study designs. For example, we could consider:

- A retrospective study of hospital administrative data that would provide information on the length of stay for patients in the hospital and whether they recovered or died. This would inform two parameters, **p_died** and **lambda_hosp**, as it would only track patients' movements once they were admitted into hospital.
- A registry study for all patients who receive the standard chemotherapy treatment, tracking outcomes and times that each of these outcomes would occur. This would provide information on the standard chemotherapy (**p_side_effects_t1**) and all transition probabilities.
- A study using population-level health administrative data to inform the long-term risk of death (**rate_longterm**) for patients who die after having recovered from side effects or did not experience them in the first place.
- A cost analysis investigating the costs for all health states. This could be achieved using administrative claims data (to estimate healthcare costs) and a patient questionnaire (to estimate out of pocket costs).
- A utility analysis to estimate the utility without experiencing side effects and for all health states. This could be achieved using surveys of the general public undertaking utility estimation tasks (e.g., time trade off questions) or through the administration of the standard utility measurement (e.g., EQ-5D questionnaires) to patients undergoing chemotherapy.

```
> # All Studies
> par_groups <- list(
+   "side_effects" = c("logor_side_effects"),
+   "hosp_trans_probs" = c("p_died","lambda_hosp"),
+   "trans_probs" = c("p_side_effects_t1", "p_hospitalised_total",
+                     "lambda_home", "p_died","lambda_hosp"),
+   "side_effects_and_follow_up" = c("logor_side_effects",
+                                    "p_hospitalised_total",
+                                    "p_died","lambda_home","lambda_hosp"),
+   "longterm_surv" = c("rate_longterm"),
+   "costs" = c("c_home_care","c_hospital","c_death"),
+   "utilities" = c("u_recovery","u_home_care","u_hospital")
+ )
> ev_groups <- evppi(outputs=chemotherapy_output, inputs=m_params,
>                    pars=par_groups)
>
> # Explore EVPPI at maximum uncertainty
> ev_groups %>%
+   filter(k == wtp.max) %>%
+   arrange(desc(evppi)) %>%
+   mutate(evppi = round(evppi, 2))
                        pars     k  evppi
1 side_effects_and_follow_up 19700 347.31
2               side_effects 19700 338.39
3                  utilities 19700 116.42
4                trans_probs 19700  79.45
```

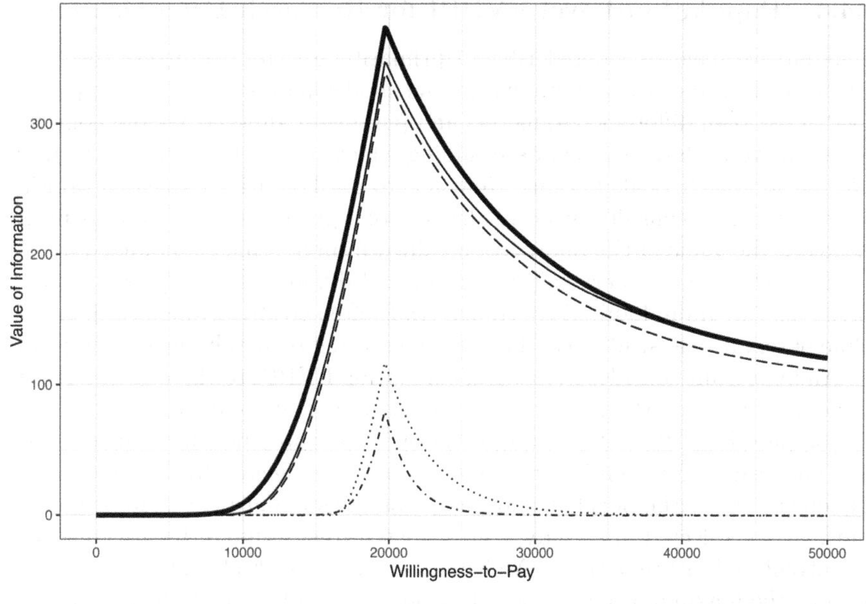

FIGURE 3.7
EVPPI across willingness-to-pay values for the considered groups of parameters.

```
5        hosp_trans_probs 19700  62.46
6           longterm_surv 19700  44.50
7                   costs 19700   0.00
```

This analysis highlights that we should potentially prioritise the non-randomised studies that focus on utilities and the transition probabilities over studies focusing on costs. We can also see that there is a relatively large difference between the value of a randomised study that would inform the odds ratio of side effects compared to the other studies. However, before we can make concrete recommendations about research prioritisation, we must also consider the size of the population impacted by the decision, and the cost of undertaking the proposed research (Section 3.4.4).

Figure 3.7 displays EVPPI across willingness-to-pay for the top four considered parameter subsets. The analysis across willingness-to-pay yields similar conclusions. The randomised controlled trials informing the odds ratio clearly offer higher value compared to the other proposed studies. The value of learning all the utilities, and the value of learning all the transition probabilities, are substantively identical in the context of the estimation error of EVPPI. Similarly to the single parameter analysis, the value of learning the costs is very low, demonstrating clearly that the cost parameters do not influence the decision uncertainty in this model.

3.4.4 Population-Level EVPPI for Research Prioritisation

Thus far, we have considered EVPPI primarily as a tool to assess the sensitivity of decisions to uncertainty in the model parameters, by comparing EVPPI between different parameters and parameter subsets. We did consider potential study designs when determining which parameter subsets are most relevant for EVPPI calculations. However, to use EVPPI for research prioritisation, we must consider whether the cost of a given research study is likely to exceed the value of the information gained from it, which is bounded above by the EVPPI. This is crucial, as the costs of different study types can vary drastically, with randomised controlled trials often costing millions of pounds, while observational studies can be performed using relatively meager budgets.

An important consideration when comparing EVPPI to the cost of a study is the unit in which the EVPPI is represented. The analyses we have done so far compute EVPPI at the individual level. This represents the value of reducing uncertainty for each individual who would be affected by the decision to implement either the novel chemotherapy or the standard care. However, the information collected in a study would be used to determine the interventions for current and future patients (McKenna and Claxton 2011). Therefore, to determine the maximum societal benefit from research targeting the parameters of interest, the individual level EVPPI should be multiplied by the number of individuals who will be affected by the decision, which we denote as P. This population size is unknown, and thus must be estimated to determine the value of future research.

To estimate the total population who would be affected by the decision, three quantities must be specified.

- P_y: the number of individuals who would be affected by the decision in a given year,
- Y: the number of years before the decision will be re-assessed,
- r: a discount rate

3.4.4.1 Number of Individuals P_y Affected by the Decision

For the chemotherapy example, P_y is the number of patients who would receive chemotherapy in a year, which we could estimate from prescription or administrative data. If the chemotherapy is only used as a first-line treatment, this would also coincide with the yearly incidence rate of new diagnoses for this cancer. Thus, in many examples, P_y will be the yearly incidence rate in the jurisdiction of interest. However, some interventions (e.g., therapy) may be limited by the number of practitioners available to provide the intervention and so P_y could be limited, or estimated accordingly. P_y could also change across different years if, for example, interventions require specialist training, which would limit the number of patients who could initially receive the interventions.

3.4.4.2 Number of Years Y before the Decision Will Be Re-assessed

This fixed "decision horizon" is chosen as a proxy for new technologies entering the market and future price changes that would alter the cost-effectiveness of the interventions (Philips, Claxton and Palmer 2008). It can loosely be interpreted as the length of time until the current decision will be reassessed due to external changes such as a new technology becoming available or the drug patent expiring and allowing for the development of generics. In practice, this replacement rate for interventions and the implementation of the optimal intervention will be incomplete (Grimm, Dixon and Stevens 2017b) making this decision horizon an even rougher approximation. However, this is likely sufficient for EVPPI analyses, as EVPPI is the theoretical maximum value for a study targeting the parameters of interest, and so this will provide an approximate maximum value for the study. Values between 5 and 15 years have commonly been used in practical VOI analyses. It is wise to do sensitivity analysis to the choice of Y if it may affect research decisions.

3.4.4.3 Discount Rate r

r is the discount rate used in the health economic model to discount the value of information to patients who will receive their intervention in the future (McKenna and Claxton 2011).

3.4.4.4 Total Population Affected P

Hence the *total population affected* P is calculated, based on these three quantities, as

$$P = \sum_{y=0}^{Y} \frac{P_y}{(1+r)^y}$$

A more in-depth discussion of estimating P is given in the context of EVSI estimation in Section 4.6.1.

Our chemotherapy example is a realistic but stylised case study. We suppose that the chemotherapy is used as a first-line treatment for breast cancer, which had a yearly incidence of around 46,000 in 2008 (Dewis and Gribbin 2009), which we will assume is stable over time, although time trends could be considered for higher accuracy. The total population P would then be around 253,000, 428,000 and 575,000 for decision horizons of 5, 10 and 15 years, respectively. We can then calculate the total population level EVPPI by fixing the willingness-to-pay threshold for the jurisdiction. In the UK, thresholds of between £20,000 and £30,000 per QALY gained are used (McCabe, Claxton and Culyer 2008) and we select the bottom end of this range.

```
> ## Population Level EVPPI
> pop_size <- 46000 * c(sum(1 / (1 + 0.035)^(0:5)),
+                       sum(1 / (1 + 0.035)^(0:10)),
```

```
+              sum(1 / (1 + 0.035)^(0:15)))
>
> ev_groups  %>%
+   filter(k == 20000) %>%
+   arrange(desc(evppi)) %>%
+   mutate(evppi_5Y = evppi * pop_size[1],
+          evppi_10Y = evppi * pop_size[2],
+          evppi_15Y = evppi * pop_size[3])
                     pars     k     evppi evppi_5Y evppi_10Y evppi_15Y
1 side_effects_and_follow_up 20000 341.31029 86587829 146273248 196526770
2               side_effects 20000 332.42383 84333402 142464834 191409940
3                  utilities 20000 108.00550 27400175  46287251  62189662
4                trans_probs 20000  69.24529 17567003  29676026  39871498
5           hosp_trans_probs 20000  51.56769 13082330  22100045  29692720
6              longterm_surv 20000  38.51966  9772145  16508133  22179655
7                      costs 20000   0.00000        0         0         0
```

The final step for research prioritisation is to consider the costs associated with collecting data based on each of the study designs. As EVPPI represents the maximum value that could be achieved through eliminating uncertainty in the parameters of interest, the population level EVPPI should substantially exceed the cost of research to consider prioritising the proposed study. In this example, we note that the EVPPI for the odds ratio for side effects is over £84 million, even with a 5-year decision horizon. While randomised controlled trials are expensive and can require budgets of over £100 million (Speich et al. 2018), these are relatively rare and often enrol a very large number of patients. Thus, there is likely to be net positive value in undertaking a randomised controlled trial to reduce uncertainty in the relative risk.

However, in addition to a randomised controlled trial, the costs related to the other proposed studies can also be considered:

- A retrospective study of hospital administrative data is likely to be relatively inexpensive, requiring salaries for the individual(s) undertaking the study and data access costs. However, despite this, the cost of this study is unlikely to approach the £13 million population-level EVPPI for the in-hospital transition probabilities with a 5-year decision horizon, so this study could offer significant value for money.

- A registry study to inform all transition probabilities and the baseline probability of side effects would be cheaper than a randomised controlled trial, but, if the registry needed to be created for the project, this would have substantially higher costs than the retrospective study. However, with a 5-year decision horizon, the expected value of this study exceeds £18 million, so this study is likely to have net positive value.

- A study using population-level health administrative data to determine the long-term rate of survival is also likely to be relatively inexpensive, requiring the same costs as above. Data access for population-level data may be higher, as the privacy requirements may be more stringent, and in many settings, highly qualified analysts are needed to manipulate data of that scale, i.e., many records for each member of the population. Despite this, the costs are highly unlikely to exceed £9.7 million, which indicates that this study is likely to provide value for money.

- Both the cost and utility analyses require questionnaires, which are likely to be more expensive to administer than the retrospective study. This means that the value of a study targeting costs may be limited, as the value of a study focusing on costs has no value, while the utilities would provide a maximum value of £27 million. Thus, a utility study should be prioritised and a cost study not considered.

Thus, due to the relatively large population size, most of the studies we have discussed would likely offer value to the decision maker, but some could potentially be prioritised on the basis of offering larger potential for a net positive value of research to the decision maker. However, to correctly guide these prioritisation decisions, we must use the expected value of sample information (introduced in the next chapter) to more closely investigate the relative values of these proposed studies, and determine the optimal use of research resources in this area.

4

The Expected Value of Sample Information

Anna Heath, Mark Strong, Christopher Jackson, Natalia Kunst, Nicky J. Welton, Hawre Jalal, and Edward C. F. Wilson

The previous chapter introduced two VOI measures, EVPI and EVPPI, that describe the expected value of *eliminating* uncertainty in the parameters, or a subset of the parameters, of a decision-analytic model. If the costs of further research exceed the EVPI, this research can be judged to be not worthwhile. Similarly if the costs of further research on specific parameters exceed the EVPPI for those parameters, this research can be deemed not worthwhile. EVPPI can also be used as a form of *sensitivity analysis* – to compare how much decision uncertainty is driven by uncertainty about different parameters.

In this chapter, we go further. If we conclude that research for particular parameters may be beneficial, we may want to ask whether a study *of a specific design and sample size* would be beneficial. Bigger studies, e.g., clinical trials with more participants, generally lead to better information, but on the other hand, the benefits of that extra information may not outweigh the costs of obtaining it.

This chapter explains a formal method for calculating the expected benefits of a specific research study. This is the *Expected Value of Sample Information* (EVSI). EVSI can be used either:

- By itself, for research prioritisation and design, by highlighting which studies have the potential to generate significant value for the decision maker.

- In combination with the costs of research for a given research study, to give the *Expected Net Benefit of Sampling* (ENBS). ENBS can be used to determine which studies have a net positive economic value, and determine the *optimal* study design to reduce uncertainty in a decision-analytic model.

Importantly, ENBS analyses can be used to determine whether the currently available evidence is sufficient for decision-making. If all feasible studies have negative ENBS, then the decision maker can conclude that decision uncertainty is not sufficient to proceed with future research. This may also be the

DOI: 10.1201/9781003156109-4

case if the ENBS is small but positive, as there may be other studies associated with higher potential value that should be prioritised (c.f., Chapter 10).

The range of studies that could be considered in an EVSI or ENBS analysis is almost limitless, which could make the task of determining the optimal study design challenging. However, the EVPPI analysis will usually have identified parameters for which completely eliminating uncertainty would be valuable. As EVSI cannot exceed EVPPI, these high value parameters and parameter subsets will usually be the focus of the proposed studies, as they are associated with the highest potential value for a study. Thus, an EVPPI analysis should be completed before moving forward with EVSI and ENBS calculations to design the most appropriate study. It is important to perform a comprehensive EVPPI analysis, which includes calculating the value of subgroups of parameters that could be investigated together, as these may have substantially higher potential value than each of the parameters considered individually.

In healthcare decision-making, we might consider a range of potential study designs, including:

- A randomised controlled trial investigating the relative efficacy of the interventions in the decision-analytic model. Note that randomised controlled trials often have a range of different outcomes, which could also inform other model parameters such as utilities or costs.

- An observational study (either retrospective or prospective) using administrative data to understand outcomes for the current standard care, i.e., baseline admission risk, baseline risk of complications, survival for the standard care, natural history of the disease.

- A registry trial for patients with a given disease, which would provide more detailed information on the current standard care or any other interventions used in current practice.

- A cost analysis using administrative claims data to determine costs associated with the standard care or other required care.

- A survey to understand quality of life or patient expenditure on healthcare. The breadth of surveys that could be considered are immense, but might include utility studies such as time trade off questions and utility measures using standard measurements such as the EQ-5D questionnaire.

- A combination of these strategies, e.g., a randomised controlled trial that collects utility data using a questionnaire.

- A systematic literature review to provide further evidence about the model parameters from published studies.

In Section 4.1 we will introduce the definition of EVSI. In Section 4.2, we will discuss how to define the sampling distributions for future study data, a necessary part of any EVSI calculation. We will then discuss different methods for calculating EVSI and how they can be implemented in the `voi` package (Sections 4.3–4.5). In general, EVSI calculations are more complex than EVPPI calculations – the most appropriate calculation method may depend on the proposed study, decision-analytic model and available expertise. Thus, we will provide guidance on which of these methods would be preferred in which context (Section 4.4). We will then (Section 4.6) discuss the definition of ENBS, and how it can be used to optimise the study design, and give some guidance for presenting the results of EVSI and ENBS analyses (Section 4.7). Finally, we will conclude with some examples of EVSI and ENBS analysis in the Chemotherapy case study (Section 4.8).

4.1 Definition of Expected Value of Sample Information

EVSI calculates the value of a specific study, aimed at *reducing* uncertainty in a subset of the decision-analytic model parameters. Similarly to EVPI and EVPPI, EVSI is defined as the difference between

- The value of the optimal intervention based on current evidence, and
- The expected value of the optimal intervention based on the updated evidence. For EVSI, this is the evidence obtained the study.

To formally define EVSI, denote by $\boldsymbol{X}$ the data that would arise from the study. We use the simplified notation $\boldsymbol{X}$ to represent any study data, of any complexity, that could be collected. For example, $\boldsymbol{X}$ may be a complex dataset that collects $\mathcal{O}$ outcomes from $\mathcal{M}$ individuals across $\mathcal{T}$ time points, e.g., a registry study that collects multiple outcomes from patients at 6-month intervals. Or $\boldsymbol{X}$ may simply be a single number representing an estimated treatment effect from a trial, e.g., a relative risk.

Nonetheless, irrespective of the complexity of the dataset $\boldsymbol{X}$, EVSI calculations assume that these data are used to update the information about the model parameters in a Bayesian manner (Bernardo and Smith 2009). From this point of view, the PA distributions for the parameters $p(\boldsymbol{\theta})$ represent the prior for the model parameters, and the data $\boldsymbol{X}$ are used to update this distribution to a posterior distribution for $\boldsymbol{\theta}$. This posterior distribution would then be used to represent parameter uncertainty in any subsequent probabilistic analysis. Thus, the distribution of the net benefit of each intervention is updated conditionally on the posterior distribution for $\boldsymbol{\theta}$. The value of the the optimal decision, conditional on the additional information contained in the data, is then

$$\max_d \mathrm{E}_{\boldsymbol{\theta}|\boldsymbol{X}} \left[\mathrm{NB}_d(\boldsymbol{\theta})\right], \tag{4.1}$$

where $\text{E}_{\boldsymbol{\theta}|\boldsymbol{X}}\left[\text{NB}_d(\boldsymbol{\theta})\right]$ is the expected posterior mean of the net benefit for intervention d.

The previous expression describes the value of the optimal decision based on a given dataset. However, when computing EVSI, it is not known what data will arise from the future study. Thus, in order to define EVSI, we first need to take the expectation of Equation (4.1) over all the possible datasets from the future trial,

$$\text{E}_{\boldsymbol{X}}\left[\max_d \text{E}_{\boldsymbol{\theta}|\boldsymbol{X}}\left[\text{NB}_d(\boldsymbol{\theta})\right]\right]$$

This expectation is taken with respect to the distribution of the future data $\boldsymbol{X} \sim p(\boldsymbol{X})$. From the perspective of an EVSI analysis, this is the *prior predictive* distribution of the data (See Section 1.6.3.1). In Bayesian analysis, the prior predictive distribution is the distribution of $\boldsymbol{X}$ that accounts for individual-level variation in the outcomes and uncertainty in the model parameters. Formally, the prior predictive distribution for the data is defined by specifying the sampling distribution of the proposed study data conditionally on the model parameters (or "likelihood"), $p(\boldsymbol{X} \mid \boldsymbol{\theta})$, and then integrating out the parameter uncertainty

$$p(\boldsymbol{X}) = \int p(\boldsymbol{X} \mid \boldsymbol{\theta}) p(\boldsymbol{\theta}) d\boldsymbol{\theta},$$

where $p(\boldsymbol{\theta})$ is the prior, which in this context is the distribution for the model parameters $\boldsymbol{\theta}$ from the probabilistic analysis (PA distribution).

Finally, EVSI can then be defined, as the difference between the expected value of the optimal decision conditionally on the data and the value of the current optimal intervention

$$\text{EVSI} = \text{E}_{\boldsymbol{X}}\left[\max_d \text{E}_{\boldsymbol{\theta}|\boldsymbol{X}}\left[\text{NB}_d(\boldsymbol{\theta})\right]\right] - \max_d \text{E}_{\boldsymbol{\theta}}\left[\text{NB}_d(\boldsymbol{\theta})\right], \tag{4.2}$$

where $\max_d \text{E}_{\boldsymbol{\theta}}\left[\text{NB}_d(\boldsymbol{\theta})\right]$ is the same as the second term in the definition for EVPI and EVPPI, i.e., the value of the optimal intervention based on current information.

Figure 4.1 gives a pictorial representation of the definition of EVSI. This figure aims to clarify the relationship between the model parameters, the decision-analytic model and the proposed data collection exercise.

(a) The top half of the picture represents the standard health economic decision-making process, where the (joint) PA distribution of the model parameters $\boldsymbol{\theta}$ is obtained from different information sources. The model parameters are then fed through the decision-analytic model to determine the current distribution of the net benefit for each intervention. The optimal intervention is found by taking the expectation of each net benefit and determining the maximum.

(b) The bottom half illustrates how EVSI is obtained. A potential study is designed by specifying a relationship between the model parameters and the future data. Potential future data is simulated based

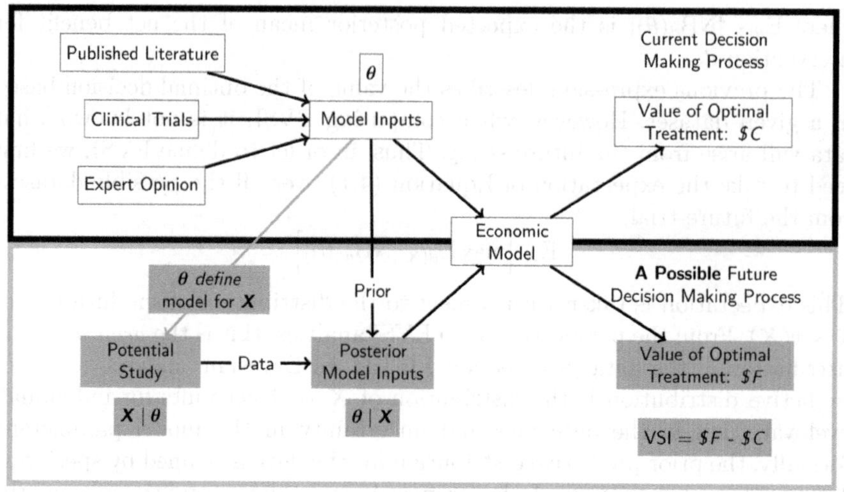

FIGURE 4.1
A pictorial description of the relationships between the parameters and data when defining EVSI, adapted from Figure 1, Heath and Baio (2018).

on current estimates of model parameters (the prior predictive distribution). This data updates the distribution of the model parameters. An updated distribution for the net benefit for each intervention is then found using the decision-analytic model. Finally, the optimal treatment conditionally on the new data is found, and the value of the sample information is the difference between these two values. The EVSI is then calculated as the average value over all the potential datasets.

4.2 Defining Distributions for Future Data in EVSI Analyses

The crucial distinction of an EVSI analysis from an EVPPI (or EVPI) analysis is that a distribution to describe the generation of future data from the proposed study must be specified. In practice, this involves an additional layer of simulation, which can greatly increase computational time and complexity. This distribution links the individual-level outcomes from the proposed study to the model parameters, which is required to define the prior predictive distribution of the data and determine the posterior distribution of the model parameters. For example, binary data are often described by a Bernoulli (or binomial) distribution, and numbers of events occurring in a given time period are often described by a Poisson distribution.

The most common method for defining the distribution of the future data is using a *parametric* data-generating process $p(\boldsymbol{X} \mid \boldsymbol{\theta})$ (Ades, Lu and Claxton 2004; Heath et al. 2022), which formalises a relationship between the proposed study data and the parameters. Thus, this section will focus on how to generate data for EVSI calculations using parametric data-generating processes. There are some settings where the parametric distribution of the data conditional on the model parameters does not need to be specified, e.g., using bootstrap simulation methods (Sadatsafavi, Marra and Bryan 2013; Heath et al. 2022), and for survival data (Vervaart et al. 2023) but these are not discussed.

The exact parametric assumptions for the data-generating process will change depending on the proposed study design, as they must reflect

(i) Which model parameters the study will inform.

(ii) What data should be collected to update these parameters.

As a general rule, the data-generating process should represent *individual-level* variation in study outcomes. As decision-analytic models are often built at the population-level, the parameters required to define the individual-level variation in outcomes are sometimes not required to define the decision-analytic model. Thus, these parameters may not be available in the current model.

If the parameters describing individual-level variation are not available, they may need to be extracted from the literature or estimated from available individual-level data in order to compute EVSI. For example, if the data are assumed to follow a normal distribution with the parameter θ_1 in the decision-analytic model being the mean, an assumption about the variance of the individual-level distribution of $\boldsymbol{X} \sim N(\theta_1, \sigma^2)$ would be required before estimating EVSI. Note that a similar situation occurs in power calculations (an alternative perspective to study design, based on hypothesis testing) where assumptions about individual level variation are also required. Crucially, σ^2 is not the variance of the PA distribution for θ_1 (which describes the uncertainty in knowledge of the parameter θ_1, rather than variations between individuals). The individual level variance σ^2 could be informed from previous studies of the same population reporting the same outcome. Since individual variability can differ between different populations, care is needed to match the inclusion criteria of the previous data to that of the proposed study. Thought is also required when the future study may collect information on multiple model parameters, where correlations between outcomes need to be reflected in the assumed parametric data-generating process.

Consider the chemotherapy example model from Chapter 2. For the following proposed future studies, we could define the following data-generating processes to link between the model parameters and the future data.

Study 1: A retrospective study of hospital administrative data.

This study would provide information on the length of stay for patients in the hospital and whether they recovered or died. This would inform the in-hospital transition probabilities, which are dependent on the model parameters Γ_2, the 1-year probability of in-hospital death and λ_2, the weekly conditional probability of recovering given that you survive. We assume these data are generated as follows:

- We first simulate the number of patients who died in hospital, from a total of N patients who were admitted into hospital. This can be simulated from a binomial distribution $Y_{dead} \sim Binomial(N, \Gamma_2)$.
- Based on this, we generate the time that each of the Y_{dead} patients die, from an exponential distribution, $T_{dead} \sim Exponential(r_{dead})$, where r_{dead} is a function of the transition probability to death γ_2 (and γ_2 is a function of the 1-year cumulative probability of in-hospital death Γ_2). More specifically, based on the Markov property and the use of a constant transition probability over the time horizon of the Markov model, we can transform from a probability to a rate $r_{dead} = -\log(1 - \gamma_2)$.
- Finally, we can use a similar distribution for the recovery times; $T_{recover} \sim Exponential(r_{recover})$, where $r_{recover}$ can be computed from the transition probability from hospital to recovery, i.e., $r_{recover} = -\log(1 - (\lambda_2(1 - \gamma_2)))$.

Study 2: A registry study

This study would involve all patients who receive the standard chemotherapy treatment, tracking outcomes and times that each of these outcomes would occur. This provides information on the proportion of patients experiencing side effects with the standard chemotherapy π_0 and all transition probabilities.

- The number of individuals experiencing side effects, from a total of M patients in the registry, is assumed to follow a binomial distribution $Y_{SE} \sim Binomial(M, \pi_0)$. Another suitable distribution, especially if individuals can experience multiple events, would be a Poisson distribution counting the number of events in the 1-year time period.
- The data for the in-hospital transition probabilities is assumed to be generated in the same way as in the retrospective study of hospital administrative data. For the transition probability from at-home care to hospital, we can generate the number of patients who are hospitalised due to side effects based on the 1-year probability of hospitalisation (Γ_1) as $Y_{Hosp} \sim Binomial(Y_{SE}, \Gamma_1)$.

Study 3: A cost analysis

This study would investigate the costs for all health states, providing information on c_{death}, the one-off cost of death, c_{HC}, the monthly cost of home care, and c_H, the monthly cost of hospital care.

Individual-level costs might then be simulated from a log-normal distribution with mean given by the parameter of interest, e.g., c_{death}. The standard deviation for each of these log-normal distributions should represent the individual-level variability in the costs experienced. In our study, this is not an input to the decision-analytic model, so must be estimated from an external source. Since this represents variability rather than parameter uncertainty, it is likely to be substantially larger than the standard error reported in Table 2.1.

Different data-generating mechanisms might be plausible depending on the model and study context. For example, we might assume the same cost distribution applies to all individuals, and generate a dataset of the corresponding sample size from this log-normal distribution. Or the number of individuals who experience a specific kind of of cost may itself be assumed to be randomly generated, e.g., out of a fixed number of individuals Y_{SE} who experience side effects, we might assume the number of people hospitalised Y_{Hosp} or who died Y_{dead} are binomially distributed, then generate two datasets with these respective sample sizes to inform c_H and c_{death}, respectively.

4.2.1 A General Algorithm to Simulate Study Data

These examples demonstrate how we can consider the relationships between the model parameters and the data. However, as we will see in Section 4.3, the majority of EVSI calculation methods require simulated datasets. These datasets should be simulated from the prior predictive distribution $p(\boldsymbol{X})$ to compute the outer expectation in the first term of Equation (4.2). Irrespective of the complexity of the assumed relationship between the parameters and the data $p(\boldsymbol{X}|\boldsymbol{\theta})$, plausible datasets can be generated from the prior predictive distribution by first simulating from the marginal distribution of the parameters $\boldsymbol{\theta}^* \sim p(\boldsymbol{\theta})$ and then simulating from the sampling distribution of the data based on the sampled parameter values $\boldsymbol{X}^* \sim p(\boldsymbol{X}|\boldsymbol{\theta}^*)$ (Heath et al. 2022). This process generates samples from the *joint* distribution of the data $\boldsymbol{X}$ and the parameters $\boldsymbol{\theta}$ as $p(\boldsymbol{X},\boldsymbol{\theta}) = p(\boldsymbol{X}|\boldsymbol{\theta})p(\boldsymbol{\theta})$. However, by definition, if we generate samples from the joint distribution of $p(\boldsymbol{X},\boldsymbol{\theta})$, and "ignore" the samples of $\boldsymbol{\theta}$, we generate datasets from the ("marginal") distribution of the data, $\boldsymbol{X} \sim p(\boldsymbol{X})$.

In practice, the probabilistic analysis for a standard decision-analytic model requires S samples of $\boldsymbol{\theta}$ and, thus, these samples are already available. Thus, a dataset generated from the distribution $\boldsymbol{X} \sim p(\boldsymbol{X}|\boldsymbol{\theta}^{(s)})$ is required for each set of simulated values from the probabilistic analysis $s = 1, \ldots, S$. To clarify the data-generating algorithm, the first two columns of Table 4.1 represent the standard probabilistic analysis with the parameter samples in the first section and the net benefits in the second. Throughout, this output from the decision-analytic model is known as the PA dataset.

In general, our study aims to record $\mathcal{O}$ study outcomes on $\mathcal{M}$ participants, resulting in $\mathcal{O} \times \mathcal{M}$ measurements in the study. Thus, a single study dataset is

TABLE 4.1
A representation of a probabilistic analysis sample with S samples for a set of P parameters and D decision options (columns 1 and 2). Simulated datasets are shown in column 3. The bracketed superscript indexes the parameter samples, corresponding net benefits, and simulated datasets.

Probabilistic analysis sample						**Simulated datasets**		
Parameters			**Net benefits**					
$\theta_1^{(1)}$	$\cdots$	$\theta_1^{(P)}$	$\text{NB}_1(\boldsymbol{\theta}_1)$	$\cdots$	$\text{NB}_D(\boldsymbol{\theta}_1)$	$X_1^{(1)}$	$\cdots$	$X_1^{(\mathcal{O}\times\mathcal{M})}$
$\theta_2^{(1)}$	$\cdots$	$\theta_2^{(P)}$	$\text{NB}_1(\boldsymbol{\theta}_2)$	$\cdots$	$\text{NB}_D(\boldsymbol{\theta}_2)$	$X_2^{(1)}$	$\cdots$	$X_2^{(\mathcal{O}\times\mathcal{M})}$
$\vdots$	$\ddots$	$\vdots$	$\vdots$	$\ddots$	$\vdots$	$\vdots$	$\ddots$	$\vdots$
$\theta_S^{(1)}$	$\cdots$	$\theta_S^{(P)}$	$\text{NB}_1(\boldsymbol{\theta}_S)$	$\cdots$	$\text{NB}_D(\boldsymbol{\theta}_S)$	$X_S^{(1)}$	$\cdots$	$X_S^{(\mathcal{O}\times\mathcal{M})}$

denoted as the vector $\boldsymbol{X} = \left(X^{(1)}, \ldots, X^{(\mathcal{O}\times\mathcal{M})}\right)$. The third column of Table 4.1 demonstrates that each row of the PA dataset $\boldsymbol{\theta}_s$ is used to sample from the conditional distribution of the data, $\boldsymbol{X}_s \sim p(\boldsymbol{X}|\boldsymbol{\theta}_s)$, to generate the samples $\boldsymbol{X}_1, \ldots, \boldsymbol{X}_S$ that follow the marginal distribution of the data $p(\boldsymbol{X})$. For cohort or registry studies that propose collecting the $\mathcal{O}$ individual level quantities at $\mathcal{T}$ different time points, the data can also be simulated using this algorithm but each simulated dataset will contain $\mathcal{O} \times \mathcal{M} \times \mathcal{T}$ measurements.

The relationships between the outcomes and the individuals in the study should be carefully considered when developing this parametric data generating process. For example, outcomes collected over time for an individual are likely to be correlated so should be generated appropriately. Alternatively, if data are to be collected from a cluster randomised trial, the individual-level outcomes for study participants within the same cluster should be correlated (Welton et al. 2014). Correlation within a study will have important implications for how informative the data and, through this, the value of undertaking the proposed study.

4.3 Calculating EVSI

Similarly to EVPI and EVPPI, EVSI can rarely be computed analytically, due to the complexity of the net benefit functions and the limited number of distributions that result in an analytic posterior distribution and prior predictive distribution. Thus, EVSI is typically approximated using simulation.

EVSI computation is more complex than the computation of EVPI and EVPPI, due to the necessary extra steps of simulating future data, and updating the prior distribution of parameters to the posterior distribution given these data. Therefore, there have been many different algorithms developed to calculate EVSI in many different situations. This section introduces a range of methods for calculating EVSI, which will range in complexity and applicability to different study designs, parameters of interest and decision-analytic model structures.

The theory behind each of the methods will be presented for interested readers. However, we also provide a summary of each of the algorithms to compute EVSI and a comprehensive discussion of the strengths, limitations and required inputs and expertise for each method presented in this section. Thus, this section presents both the theoretical underpinnings of each EVSI calculation method and the practical implementation of the methods. The following section discusses how to select the optimal method for a specific problem before Section 4.5 presents how these methods can be implemented in the `voi` package.

4.3.1 Calculating EVSI by Monte Carlo Simulation

The first method developed to calculate EVSI was based on a full nested Monte Carlo (MC) simulation scheme (Ades, Lu and Claxton 2004). This full simulation-based strategy follows a similar process to the calculation for EVPPI described in Section 3.3.2, although an extra step is needed to simulate the potential datasets, and simulation may also be required to update the prior to posterior for each potential dataset. This means that the computational complexity of EVSI calculation exceeds the complexity of nested EVPPI analyses, which is already prohibitively expensive in realistic decision-analytic models.

The nested MC simulation method begins by simulating datasets $\boldsymbol{X}$ using the method presented in Section 4.2, which generates a single dataset $\boldsymbol{X}_s$, for each set of simulated parameter values $s = 1, \dots, S$. For each of these simulated datasets, R simulated parameter values must then be sampled from the posterior distribution $p(\boldsymbol{\theta} \mid \boldsymbol{X}_s)$. In many settings, these R parameter values must be simulated using algorithms for Bayesian updating such as MCMC, as the posterior will not be available analytically. This may be very computationally expensive, with the consequence that the EVSI calculation is infeasible. If the R parameter values are available, the net benefit for each intervention can be computed. We can then average the net benefit over the R simulations to obtain the average posterior net benefit, hence to determine the expected value of the optimal intervention incorporating information from the new sample. Finally, EVSI can be estimated as:

$$\widehat{\mathrm{EVSI}} = \frac{1}{S}\sum_{s=1}^{S}\left[\max_d \frac{1}{R}\sum_{r=1}^{R}[\mathrm{NB}_d(\boldsymbol{\theta}_r \mid \boldsymbol{X}_s)]\right] - \max_d \frac{1}{S}\sum_{s=1}^{S}\mathrm{NB}_d(\boldsymbol{\theta}_s). \qquad (4.3)$$

It is possible to reduce the Monte Carlo error associated with this estimator, by using the nested simulations to estimate the value of the optimal intervention under current evidence:

$$\widehat{\text{EVSI}} = \frac{1}{S}\sum_{s=1}^{S}\left[\max_{d}\frac{1}{R}\sum_{r=1}^{R}[\text{NB}_d(\boldsymbol{\theta}_r \mid \boldsymbol{X}_s)]\right] - \max_{d}\frac{1}{S}\sum_{s=1}^{S}\frac{1}{R}\sum_{r=1}^{R}[\text{NB}_d(\boldsymbol{\theta}_r \mid \boldsymbol{X}_s)]. \tag{4.4}$$

In both settings, this estimator has a similar computational cost to the EVPPI estimator, as $R \times S$ model runs are required. However, if MCMC methods are used to estimate the posterior distribution of the model parameters, the inner simulations will normally have a higher computational cost than those for the EVPPI.

In addition to this, EVSI is often used to inform decisions about the design and sample size of a proposed study. This means that while a single EVPPI estimate is required to estimate the value of learning a model parameter exactly, multiple EVSI calculations will be required to compare the value of a range of different study designs, even when the studies all inform the same parameter. This introduces additional computational complexity to an already intensive procedure.

4.3.1.1 Improved Efficiency for the MC Estimator

If the PA distribution for the parameters $\boldsymbol{\theta}$ and the sampling distribution of the data are from a conjugate family, the complexity of the algorithm from Ades, Lu and Claxton (2004) can be reduced by determining the analytic posterior. This eliminates the additional complexity of the analysis compared to EVPPI calculations, as MCMC methods are not required, but still results in a nested simulation procedure. Similar to the EVPPI calculations outlined in Section 3.3.3, if an analytic form for the net benefit can also be determined, then EVSI can be calculated using a one-step procedure, where each row of the PA dataset $\boldsymbol{\theta}_s$ for $s = 1, \ldots, S$ is used to simulate a single potential sample $\boldsymbol{X}_s$ from $p(\boldsymbol{X} \mid \boldsymbol{\theta}_s)$ and the analytic posterior mean can then be used to determine the posterior net benefit for each intervention.

In addition to these relatively general efficient adjustments to the nested EVSI estimation, some researchers have used the central limit theorem to justify an approach that uses normal-normal conjugacy to provide analytic EVSI calculations (Eckermann, Karnon and Willan 2010; Willan, Goeree and Boutis 2012). This setting is only relevant when data on the net benefit itself are collected in the study. Thus, it is rarely used in practice. Additionally, methods are available for estimating EVSI for specific data collection exercises, such as cluster randomised clinical trials (Ades, Lu and Claxton 2004; Brennan and Kharroubi 2007b; Chen and Willan 2014; Welton et al. 2014). In general, these methods are likely to be accurate and computationally efficient in the settings for which they were designed. However, they have limited applicability and are typically based on finding analytical expressions/approximations for the posterior mean of the net benefit under specific assumptions about the parameters and the sampling distribution.

Finally, Sweeting and Kharroubi (2003) developed a method to approximate the posterior expectation using a weighted sum of the posterior density evaluated at a small number of points. As EVSI is based on the posterior expectation, this method can be used to approximate EVSI using a small number of posterior evaluations (Brennan and Kharroubi 2007a; Kharroubi, Brennan and Strong 2011). However, the formulae are relatively complicated and assume that the posterior distribution can be maximised and differentiated, which limits applicability. Quadrature can also be used to reduce the number of nested simulations required and efficiently approximate EVSI (Coyle, Buxton and O'Brien 2003).

4.3.2 General Purpose Efficient Methods for EVSI Computation

The key challenge of these methods that improve the efficiency of EVSI calculation is that they are limited in their applicability. Thus, we now discuss a range of more recently developed methods that are generally applicable to any model and proposed study. The flexibility of these methods makes them attractive for general-purpose EVSI estimation.

These methods differ substantially; however, they all focus on determining the expected posterior net benefit for each intervention, conditional on the data. This is because the computationally challenging part of EVSI estimation using MC simulation is precisely this estimation, as it requires $S \times R$ runs of the decision-analytic model. Overall, the goal of these methods is to reduce the computational burden of estimating the posterior mean net benefit. An initial simplification that each method makes to increase the numerical stability is to work in terms of the incremental net benefit or loss, defined, without loss of generality, as $\text{INB}_d(\boldsymbol{\theta}) = \text{NB}_d(\boldsymbol{\theta}) - \text{NB}_1(\boldsymbol{\theta})$ for $d = 2, \ldots, D$. This is because this means that we only need to calculate the expected posterior mean incremental net benefit for each of the $d = 2, \ldots, D$ decision options, resulting in $D-1$ rather than the D estimated posterior net benefit distributions.

To simplify the explanation of these methods in following sections, we denote the expectation of the incremental net benefit for intervention d, conditional on data $\boldsymbol{X}$, i.e., the target of the approximation methods, as

$$\mu_d^{\boldsymbol{X}} = \text{E}_{\boldsymbol{\theta}|\boldsymbol{X}}\left[\text{NB}_d(\boldsymbol{\theta})\right].$$

In a similar manner, we also denote the expectation of the incremental net benefit for intervention d, conditional on the parameters of interest ϕ, as

$$\mu_d^{\boldsymbol{\phi}} = \text{E}_{\boldsymbol{\theta}|\boldsymbol{\phi}}\left[\text{NB}_d(\boldsymbol{\theta})\right].$$

This conditional expectation is required to compute EVPPI for the parameters $\boldsymbol{\phi}$ and plays a crucial role in two of the general purpose EVSI approximations.

Based on this new notation, EVSI is defined as

$$\widehat{\text{EVSI}} = \frac{1}{S}\sum_{s=1}^{S}\max_d \mu_d^{\boldsymbol{X}_s} - \max_d \frac{1}{S}\sum_{s=1}^{S}\mu_d^{\boldsymbol{X}_s}, \tag{4.5}$$

where $\mu_d^{\boldsymbol{X}_s}$ is the estimated posterior expectation of the incremental net benefit for the dataset $\boldsymbol{X}_s$, $s = 1, \ldots, S$. Note finally that $\mu_1^{\boldsymbol{X}_s} = 0$ for all $\boldsymbol{X}_s$.

The following sections will introduce four general-purpose methods for estimating EVSI:

- The moment-matching method (Section 4.3.3).
- The regression-based method (Section 4.3.4).
- The importance sampling method (Section 4.3.5).
- The Gaussian approximation method (Section 4.3.6).

Each section starts by explaining the principles and mathematical details behind each method, and finishes with a concise step-by-step summary of how to carry out an EVSI calculation using the method. Section 4.4 then compares the relative advantages of each of these four methods and gives guidance to support the selection of an appropriate method in practice.

4.3.3 Moment-Matching Method

The moment-matching method relies on nested Monte Carlo simulation, but reduces the number of required simulations from S to Q, between 30 and 50, and can be augmented to calculate EVSI across different sample sizes with negligible extra computational cost. Overall, the moment-matching method is based on rescaling simulations of μ_d^{ϕ}, which are available from EVPPI calculation, to estimate the distribution of $\mu_d^{\boldsymbol{X}}$ (Heath and Baio 2018; Heath, Manolopoulou and Baio 2018). This is computationally efficient, as the simulations of μ_d^{ϕ} are usually available before undertaking EVSI calculations, and therefore, the moment-matching method can focus solely on estimating the rescaling factors. The rescaling factors are chosen such that the distribution of $\mu_d^{\boldsymbol{X}}$ has the correct mean and variance, where the mean and variance of the distribution of the posterior mean incremental net benefit are estimated through simulation. In this sense, this method estimates $\mu_d^{\boldsymbol{X}}$ from μ_d^{ϕ} by matching the first two statistical moments.

4.3.3.1 Expectation and Variance for Posterior Expected Incremental Net Benefit

To estimate the mean and variance of $\mu_d^{\boldsymbol{X}}$, we use the formulae for conditional iterated expectation (Weiss 2006). These formulae specify that the mean of $\mu_d^{\boldsymbol{X}}$ is given by

$$\mathrm{E}_{\boldsymbol{X}}\left[\mu_d^{\boldsymbol{X}}\right] = \mathrm{E}_{\boldsymbol{X}}\left[\mathrm{E}_{\boldsymbol{\theta}|\boldsymbol{X}}\left[\mathrm{INB}_d(\boldsymbol{\theta})\right]\right] = \mathrm{E}_{\boldsymbol{\theta}}\left[\mathrm{INB}_d(\boldsymbol{\theta})\right],$$

i.e., the average of the posterior expected incremental net benefit is equal to the mean of the $\mathrm{INB}_d(\boldsymbol{\theta})$, which can be obtained using the PA samples. This is because the future data are generated by the same parameters that inform our decision under current information. Therefore, we do not expect the net benefits to change *on average* after "observing" new data. The consequence of collecting new data, however, is to reduce the *uncertainty* around the net benefit.

Specifically, the variance of the posterior expected mean for the incremental net benefit is given by

$$\text{Var}_{\boldsymbol{X}}\left[\mu_d^{\boldsymbol{X}}\right] = \text{Var}_{\boldsymbol{X}}\left[\text{E}_{\boldsymbol{\theta}\mid\boldsymbol{X}}\left[\text{INB}_d(\boldsymbol{\theta})\right]\right] = \text{Var}_{\boldsymbol{\theta}}\left[\text{INB}_d(\boldsymbol{\theta})\right] - \text{E}_{\boldsymbol{X}}\left[\text{Var}_{\boldsymbol{\theta}\mid\boldsymbol{X}}\left[\text{INB}_d(\boldsymbol{\theta})\right]\right].$$

In other words, the variance of the posterior mean incremental net benefit is equal to the prior variance of the incremental net benefit minus the average posterior variance of the incremental net benefit, over all possible future samples.

Thus, to calculate the mean and variance of $\mu_d^{\boldsymbol{X}}$ by simulation, we need to estimate the average posterior variance over all possible samples $\boldsymbol{X}$, provided samples of the net benefit given the current level of uncertainty and $\mu_d^{\boldsymbol{\phi}}$ are available. The moment-matching method saves computation compared to pure Monte Carlo, because the expected posterior variance can be obtained using only a small number (Q) of nested simulations (Heath and Baio 2018) Furthermore, the expected posterior variance can be estimated across different sample sizes with only a single nested simulation per sample size (Heath, Manolopoulou and Baio 2019). This significantly reduces the computational complexity of EVSI estimation and allows for simple optimisation over sample size.

4.3.3.2 Nested Simulation to Estimate the Variance of the Posterior Net Benefit

The expected variance of the posterior net benefit can be estimated by nested simulation with a small number of simulations, usually between 30 and 50 (Heath and Baio 2018). This is because the posterior variance is relatively *stable* across different future samples $\boldsymbol{X}$ and, thus, the complete range of possible datasets does not need to be considered. Furthermore, the moment matching method intelligently picks the samples that should be used to ensure that the expected posterior variance can be well approximated (Heath, Manolopoulou and Baio 2018).

Specifically, we need to calculate the expected posterior variance:

$$\text{E}_{\boldsymbol{X}}\left[\text{Var}_{\boldsymbol{\theta}\mid\boldsymbol{X}}\left[\text{INB}_d(\boldsymbol{\theta})\right]\right] = \text{E}_{\boldsymbol{\theta}}\left[\text{E}_{\boldsymbol{X}\mid\boldsymbol{\theta}}\left[\text{Var}_{\boldsymbol{\theta}\mid\boldsymbol{X}}\left[\text{INB}_d(\boldsymbol{\theta})\right]\right]\right],$$

where the two outer expectations on the right-hand side estimate the expectation over the prior predictive distribution of $\boldsymbol{X}$. In general, we choose a small number of $\boldsymbol{\theta}$ values while ensuring that the full range of $\boldsymbol{\theta}$ values is explored before sampling *one* future sample conditional on each value of $\boldsymbol{\theta}$. This estimates the expected posterior variance accurately using quadrature (Davis and Rabinowitz 2007). As $\boldsymbol{X}$ is related to the parameters of interest $\boldsymbol{\phi}$ only, we select Q evenly spaced values for $\boldsymbol{\phi}$ from the PA samples; $\boldsymbol{\phi}^q$, $q = 1, \ldots, Q$. These

Q values are then used to simulate one future sample from $\boldsymbol{X} \mid \phi$ for each ϕ value, resulting in Q simulated datasets. A posterior distribution is then found for each simulated future sample and the variance of the incremental net benefit calculated, σ_q^2, $q = 1, \ldots, Q$. This method can be used irrespectively of the underlying model structure, provided it is computationally feasible to find the posterior distribution (which might require MCMC simulation) Q times.

To estimate EVSI across different sample sizes using the moment-matching method, we add a further sample size component to the data simulation process and variance calculation. Suppose that we are trying to estimate EVSI between sample sizes $N_{\min}$ and $N_{\max}$. Thus, we create a vector $\boldsymbol{N} = (N_{\min} = N_1, N_2, \ldots, N_Q = N_{\max})$ of Q sample sizes between these two limits. Each potential dataset $\boldsymbol{X}_q$ is simulated conditional on the vector (ϕ_q, N_q), rather than solely on (ϕ_q), represented pictorially in Figure 4.2. Once these datasets $\boldsymbol{X}_1, \ldots, \boldsymbol{X}_Q$, with different sample sizes, have been simulated, the extended moment-matching method proceeds in the same manner as the standard moment-matching method; each dataset updates the distribution of the model parameters, the decision-analytic model is then run to find the posterior for the incremental net benefit conditional on this updated distribution for the parameters and the posterior variance of the incremental net benefit is then calculated, giving σ_q^2 for $q = 1, \ldots, Q$.

4.3.3.3 Practical Considerations for the Moment-Matching Method

There are two practical considerations for estimating EVSI across different sample sizes. Firstly, the variance of the posterior distribution is estimated less precisely in general for smaller sample sizes. Thus, it is recommended to space the N_q values evenly on the square root scale between $N_{\min}$ and $N_{\max}$ to simulate more studies with smaller sample sizes, hence getting better information about how the variance relates to the sample size. Secondly, the variance is only estimated accurately if ϕ_q and N_q are uncorrelated. Thus, we recommend randomly reordering the values of ϕ_q before estimating the posterior variance, to break any unintended correlations. Figure 4.2 demonstrates the process for selecting ϕ^q and N^q pictorially, the equally spaced values for ϕ are displayed in the left-hand side of Figure 4.2. Darker colours indicate larger values for ϕ and are randomly permuted independently for each column of the ϕ_q, $q = 1, \ldots, Q$ matrix. This creates a new matrix where each row is then combined with a sample size N_q, spaced on the square root scale, to generate a potential future sample $\boldsymbol{X}_q$.

Within the `voi` package, the values for ϕ_q and N_q are selected automatically based on the maximum and minimum values chosen for the sample size and the simulations for the model parameters. The user is still advised to select the number of nested simulations Q, with the default equal to 50. In general, the larger the value for Q, the more accurate the EVSI estimation, but the longer the required computation time.

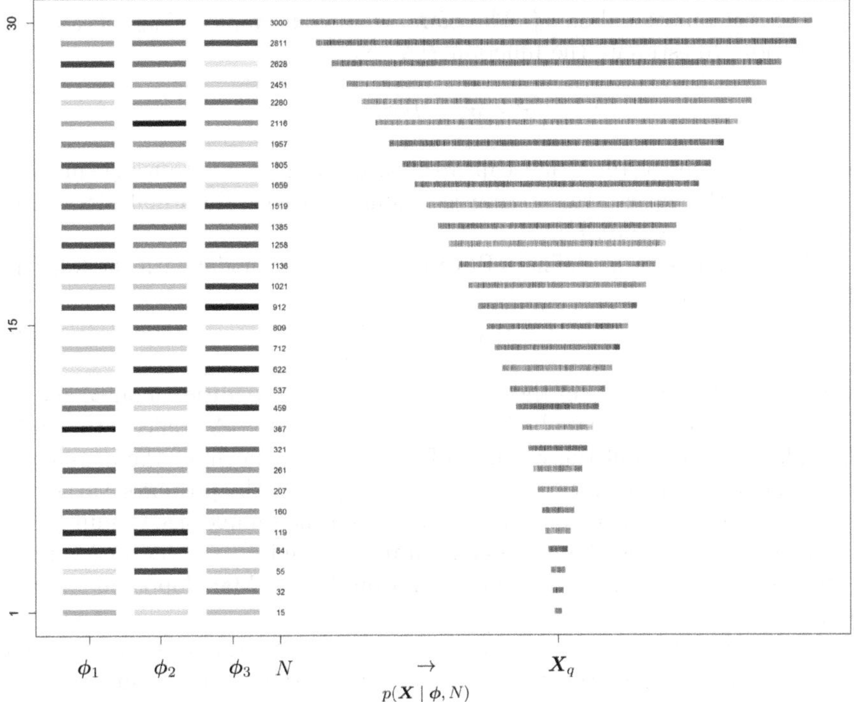

FIGURE 4.2
A pictorial representation of the sampling procedure for the future samples in the extended moment matching method. Darker lines represent higher values for the parameters. Here, the parameters of interest ϕ_1 to ϕ_3 have been re-ordered independently and are then used to generate random samples $\boldsymbol{X}_q$ for increasing sample sizes as q increases. These sample sizes are equally spaced on the square root scale. Reproduced from Heath, Manolopoulou and Baio (2019).

4.3.3.4 Estimating the Variance of the Posterior Mean

If EVSI is estimated for a single sample size, the variance of the posterior mean for the incremental net benefit $\sigma^2_{\boldsymbol{X}}$ can be calculated as the difference between the prior variance of the incremental net benefit σ^2 and the average of the estimated posterior variances

$$\sigma^2_{\boldsymbol{X}} = \sigma^2 - \frac{1}{Q}\sum_{q=1}^{Q}\sigma^2_q.$$

However, an alternative strategy is required when estimating EVSI across sample sizes, and we wish to estimate the expected posterior variance conditional on any sample size N. Specifically, we aim to estimate a function $f(N)$ that calculates the variance of the posterior mean for the incremental net benefit conditional on σ_q^2. Therefore, we use the values of σ_q^2, obtained by simulation, to estimate the function $f(N)$ as

$$\sigma^2 - \sigma_q^2 = f(N_q) + \varepsilon_q,$$

where ε_q is an error term that captures variation in σ_q^2 due to using simulated datasets with sample size N_q and is assumed to have a normal distribution centred on 0, $\varepsilon_q \sim N(0, \sigma_\varepsilon^2)$.

To estimate $\sigma_{\mathbf{X}}^2$, we define a Bayesian non-linear model with the following functional form,

$$f(N) = \sigma_{\boldsymbol{\phi}}^2 \frac{N}{N+h}, \tag{4.6}$$

where $\sigma_{\boldsymbol{\phi}}^2$ is the variance of the expected net benefit conditional on $\boldsymbol{\phi}$, and h is a parameter to be estimated. This functional form for $f(N)$ reflects the underlying knowledge about the variance of the posterior mean for the incremental net benefit as it increases as N increases is equal to the posterior variance of the expected net benefit conditional on $\boldsymbol{\phi}$ when the sample size is infinite. Priors must be set for h, the regression parameter, and σ_ε^2, the residual variance – the priors in the `voi` package are taken from Heath, Manolopoulou and Baio (2019).

Note that if there are more than two decision options in a specific decision-analytic model, then the variances in this section will become variance-covariance matrices. The mean across different matrices can be calculated by the mean of each element of the variance-covariance matrix separately. Similarly, the relationship between sample size and each element of the variance-covariance matrix can be estimated separately and combined to estimate EVSI. These transformations are implemented by default in the `voi` package.

4.3.3.5 Moment-Matching Using Linear Transformations

Finally, once $\sigma_{\mathbf{X}}^2$ has been calculated, either using the standard formula if EVSI is estimated for a single sample size or as the fitted values from the regression equation for a specific sample size N, we use a linear transformation of $\mu_d^{\boldsymbol{\phi}}$ to estimate $\mu_d^{\mathbf{X}}$. This involves estimating the constants a and b such that $a\,\mu_d^{\boldsymbol{\phi}} + b$ has the correct mean and variance:

$$\begin{aligned}
\mathrm{E}_{\boldsymbol{\theta}}\left[a\ \mathrm{INB}_d(\boldsymbol{\theta}) + b\right] = \mathrm{E}_{\mathbf{X}}\left[\mu_d^{\mathbf{X}}\right] &\Rightarrow a\mathrm{E}_{\boldsymbol{\theta}}\left[\mathrm{INB}_d(\boldsymbol{\theta})\right] + b = \mathrm{E}_{\boldsymbol{\theta}}\left[\mathrm{INB}_d(\boldsymbol{\theta})\right] \\
\mathrm{Var}_{\boldsymbol{\theta}}\left[a\ \mathrm{INB}_d(\boldsymbol{\theta}) + b\right] = \mathrm{Var}_{\mathbf{X}}\left[\mu_d^{\mathbf{X}}\right] &\Rightarrow a^2\mathrm{Var}_{\boldsymbol{\theta}}\left[\mathrm{INB}_d(\boldsymbol{\theta})\right] = \sigma_{\mathbf{X}}^2.
\end{aligned}$$

Solving for a and b yields

$$a = \sqrt{\frac{\mathrm{Var}_{\mathbf{X}}\left[\mu_d^{\mathbf{X}}\right]}{\mathrm{Var}_{\boldsymbol{\theta}}\left[\mathrm{INB}_d(\boldsymbol{\theta})\right]}} = \frac{\sigma_{\mathbf{X}}}{\sqrt{\mathrm{Var}_{\boldsymbol{\theta}}\left[\mathrm{INB}_d(\boldsymbol{\theta})\right]}} \quad \text{and} \quad b = \mathrm{E}_{\boldsymbol{\theta}}\left[\mathrm{INB}_d(\boldsymbol{\theta})\right](1-a), \tag{4.7}$$

which depend on the expectation and variance of the incremental net benefit and its expected posterior variance.

These constants allow for a relatively simple interpretation of the approximation of $p\left(\mu_d^{\boldsymbol{X}}\right)$. The constant a is related to the reduction in the incremental net benefit variance that will be obtained by learning $\boldsymbol{X}$. This means that the more information $\boldsymbol{X}$ contains about the incremental net benefit, the higher the value of a. The constant b is then the mean of the incremental net benefit weighted by one minus this reduction in variance, where the weight is related to how much information is contained in $\boldsymbol{X}$. Thus, the density of the $\mu_d^{\boldsymbol{X}}$ is estimated as a convex combination of our initial beliefs about the incremental net benefit and the mean of the incremental net benefit before data are collected.

In general, the higher the sample size of $\boldsymbol{X}$, the more information it contains and therefore the higher the value of a. This implies that as the sample size in the data collection exercise increases, our approximation for $p(\mu_d^{\boldsymbol{X}})$ approaches the distribution (from the original PA samples) for the expectation of the incremental net benefit conditional on $\boldsymbol{\phi}$. In other words, an infinite future sample implies that the distribution of $\mu_d^{\boldsymbol{X}}$ is exactly equal to the distribution of $\mu_d^{\boldsymbol{\phi}}$, and EVSI is equal to EVPPI, as required.

4.3.3.6 Calculating EVSI Using the Moment-Matching Method: Summary

To summarise, the moment-matching method estimates EVSI through the following steps:

1. Perform PA simulation to obtain $\boldsymbol{\theta}_s$ and $\text{INB}_d^{\boldsymbol{\theta}_s}$, $s = 1, \ldots, S$.
2. Estimate EVPPI to obtain simulations of $\mu_d^{\boldsymbol{\phi}}$, the inner expectation in the first term of Equation (3.4).
3. Extract Q, with $30 < Q < 50$, sample quantiles from the simulations of $\boldsymbol{\phi}$, denoted by $\boldsymbol{\phi}_q$. To estimate EVSI across different sample sizes, Q proposed sample sizes are also required, denoted by N_q for $q = 1, \ldots, Q$.
4. For $q = 1, \ldots, Q$:
   (a) Simulate a future dataset, with sample size N_q if required, from the sampling distribution $p(\boldsymbol{X} \mid \boldsymbol{\phi}_q)$.
   (b) Use Bayesian methods to update the distribution of the model parameters.
   (c) Rerun the PA to update the distribution of the net monetary benefits.
   (d) Calculate the *variance* of the net monetary benefit, denoted by σ_q^2.
5. Calculate the rescaling factor σ^2

(a) If estimating EVSI for a single sample size, $\sigma^2 = \text{Var}\left[\text{INB}_d^{\theta_s}\right] - \frac{1}{Q}\sum_{q=1}^{Q}\sigma_q^2$.

(b) If estimating EVSI across sample sizes, σ^2 for a given N is estimated by estimating h in the non-linear regression function $\frac{N}{N+h}$ with $\text{Var}\left[\text{INB}_d^{\theta_s}\right] - \sigma_q^2$ as the dependent variable and N_q as the independent variable.

6. Rescale the simulations of μ_d^{ϕ} so their variance is equal to σ^2 using

$$\sqrt{\frac{\sigma^2}{\text{Var}(\mu_d^{\phi})}}\mu_d^{\phi} + \left(1 - \sqrt{\frac{\sigma^2}{\text{Var}(\mu_d^{\phi})}}\right)\text{Mean}\left(\mu_d^{\phi}\right),$$

where $\text{Mean}\left(\mu_d^{\phi}\right)$ is the mean of μ_d^{ϕ} simulations and $\text{Var}(\mu_d^{\phi})$ is the sample variance of μ_d^{ϕ} simulations.

7. These rescaled simulations estimate $\mu_d^{\boldsymbol{X}}$.

8. The EVSI is estimated as $\widehat{\text{EVSI}} = \frac{1}{S}\sum_{s=1}^{S}\max_d \mu_d^{\boldsymbol{X}_s} - \max_d \frac{1}{S}\sum_{s=1}^{S}\mu_d^{\boldsymbol{X}_s}$.

4.3.4 Regression-Based Method

The regression-based method for EVSI calculation aims to estimate the functional form for $\mu_d^{\boldsymbol{X}}$ using nonparametric regression (Strong et al. 2015). This method is similar to the regression-based method for EVPPI calculation (see Section 3.3.2) but instead of regressing on the parameter itself, we regress on *summaries of the study data* that encapsulate information about the parameter provided by the study. To estimate $\mu_d^{\boldsymbol{X}}$, the regression-based method recognises that we can express the observed net benefit $\text{NB}_d(\boldsymbol{\theta})$ as a sum of the conditional expectation that we require, and a mean-zero error term,

$$\text{NB}_d(\boldsymbol{\theta}) = \text{E}_{\boldsymbol{\theta}|\boldsymbol{X}}\left[\text{NB}_d(\boldsymbol{\theta})\right] + \varepsilon. \tag{4.8}$$

where the error ε is a function of both the study data $\boldsymbol{X}$ and the parameters $\boldsymbol{\theta}$. We can demonstrate that ε has zero mean, by rearranging Equation (4.8)

$$\varepsilon = \text{NB}_d(\boldsymbol{\theta}) - \text{E}_{\boldsymbol{\theta}|\boldsymbol{X}}\left[\text{NB}_d(\boldsymbol{\theta})\right], \tag{4.9}$$

and then taking expectations with respect to both the distribution of the study data $\boldsymbol{X}$ and the distribution of the model parameters $\boldsymbol{\theta}$,

$$\begin{aligned}\text{E}_{\boldsymbol{X},\boldsymbol{\theta}}(\varepsilon) &= \text{E}_{\boldsymbol{X},\boldsymbol{\theta}}\left[\text{NB}_d(\boldsymbol{\theta})\right] - \text{E}_{\boldsymbol{X},\boldsymbol{\theta}}\left[\text{E}_{\boldsymbol{\theta}|\boldsymbol{X}}\left[\text{NB}_d(\boldsymbol{\theta})\right]\right] \\ &= \text{E}_{\boldsymbol{\theta}}\left[\text{NB}_d(\boldsymbol{\theta})\right] - \text{E}_{\boldsymbol{X}}\left[\text{E}_{\boldsymbol{\theta}|\boldsymbol{X}}\left[\text{NB}_d(\boldsymbol{\theta})\right]\right],\end{aligned}$$

noticing that the first term of Equation (4.9) is a function only of $\boldsymbol{\theta}$, and the second term a function only of $\boldsymbol{X}$ ($\boldsymbol{\theta}$ having been integrated out). Then, by the law of total expectation, $\mathrm{E}_{\boldsymbol{X}}\left[\mathrm{E}_{\boldsymbol{\theta}|\boldsymbol{X}}\left[\mathrm{NB}_d(\boldsymbol{\theta})\right]\right] = \mathrm{E}_{\boldsymbol{\theta},\boldsymbol{X}}\left[\mathrm{NB}_d(\boldsymbol{\theta})\right]$, and hence $\mathrm{E}[\varepsilon] = 0$. Note that this holds for all PA distributions for $\boldsymbol{\theta}$ and irrespective of the relationship between $\boldsymbol{\theta}$ and $\boldsymbol{X}$.

Next, as for the calculation of EVPPI, we recognise that the expectation $\mathrm{E}_{\boldsymbol{\theta}|\boldsymbol{X}}\left[\mathrm{NB}_d(\boldsymbol{\theta})\right]$ is an unknown function of the study data $\boldsymbol{X}$. We denote this function $g_d(\boldsymbol{X})$, and therefore can specify a regression equation

$$\mathrm{NB}_d(\boldsymbol{\theta}_s) = g_d(\boldsymbol{X}_s) + \varepsilon_s, \tag{4.10}$$

where $\boldsymbol{X}_s$ is the dataset simulated conditional on the parameters $\boldsymbol{\theta}_s$. In contrast to EVPPI calculation where $\boldsymbol{\phi}$ is a set of model parameters, which will typically have a low dimension, the data $\boldsymbol{X}$ will usually be high dimensional as they represent the individual level outcomes in the study. However, the information provided by the data about the parameters $\boldsymbol{\theta}$ can often be expressed in a simpler form, as a low-dimensional summary statistic $T(\mathbf{X}) = \{T_1(\mathbf{X}), \dots, T_p(\mathbf{X})\}$. In the simplest models, there exist *sufficient* statistics, where the posterior distribution is independent of $\boldsymbol{X}$ given $T(\mathbf{X})$. In more general models, we might hope that given $T(\boldsymbol{X})$, the posterior mean of the net benefit is approximately independent of $\boldsymbol{X}$. This is the idea behind the regression-based method, where we fit a nonparametric regression model between the simulated net benefit values and the summary statistic,

$$\mathrm{NB}_d(\boldsymbol{\theta}_s) = g_d(T(\boldsymbol{X}_s)) + \varepsilon_s. \tag{4.11}$$

4.3.4.1 Selecting a Summary Statistic $T(\cdot)$

As discussed in Section 4.2, study data $\boldsymbol{X}$ aim to record $\mathcal{O} \times \mathcal{M}$ outcomes. Furthermore, the study data may be informative for one or more model parameters. In settings where the study data only update a single model parameter ϕ, a natural choice for $T(\boldsymbol{X})$ is the sample estimator for ϕ. For example, if we are calculating the expected value of a two-arm, binary outcome trial to update beliefs about an odds ratio, then we choose $T(\boldsymbol{X})$ to be the sample odds ratio.

If the study updates more than one model parameter, then we could consider q summary statistics $T(\boldsymbol{X}) = \{T_1(\boldsymbol{X}), \dots, T_q(\boldsymbol{X})\}$, where each $T_i(\boldsymbol{X})$ is a sample estimator for each of the q parameters of interest. For example, to calculate the expected value of a study to learn about the shape and scale parameters of a Weibull distribution from censored time-to-event data, then $\{T_1(\boldsymbol{X}), T_2(\boldsymbol{X})\}$ could be specified as the sample estimates of the shape and scale parameters for the distribution. Alternatively, the number of events and the total time at risk have also been identified as a suitable summary statistic for survival models (Vervaart et al. 2022). Note that when $T(\boldsymbol{X})$ is multivariate, g must be a multivariate smoothing function, with suitable regression methods discussed in Section 3.3.5.

To support the selection of appropriate summary statistics, we have developed a range of applied examples in Sections 4.5 and 4.8. These study designs consider a range of different study designs, parameters of interest and distributions for the future data.

4.3.4.2 Calculating EVSI Using the Regression-Based Method: Summary

Similar to EVPPI, EVSI is computed with the regression-based method using the following algorithm:

1. Perform a probabilistic analysis, obtaining S samples for $\boldsymbol{\theta} = (\boldsymbol{\phi}, \boldsymbol{\psi})$ and $\text{INB}_d(\boldsymbol{\theta})$ for each intervention.
2. For each simulated parameter set $\boldsymbol{\theta}_s$, simulate a dataset from the assumed distribution of the data $p(\boldsymbol{X} \mid \boldsymbol{\theta}_s)$ and summarise in a low-dimensional summary statistic $T(\boldsymbol{X}_s)$
3. For each intervention option $d = 2, \ldots, D$, fit a regression curve to the "dataset" comprising the sample of $T(\boldsymbol{X}_s)$ as the observed "predictors," and the $\text{INB}_d(\boldsymbol{\theta}_s)$ as the observed "response."
4. For each d and s, find the fitted value $g_d(T(\boldsymbol{X}_s))$ by evaluating the fitted regression curve at $T(\boldsymbol{X}_s)$.
5. The EVSI is estimated as

$$\widehat{\text{EVSI}} = \frac{1}{S}\sum_{s=1}^{S} \max\{0, \max_d g_d(T(\boldsymbol{X}_s))\} - \max\{0, \max_d \frac{1}{S}\sum_{s=1}^{S} g_d(T(\boldsymbol{X}_s))\}.$$

Just as with EVPPI, implementing the third step using the *incremental* net benefits only requires us to fit $D-1$ regression models, rather than D if we were using the net benefit directly. This reduces the error and requires us to take the maximum of a set defined by 0 (the incremental net benefit for $d = 1$) and the estimated incremental net benefit for interventions $d = 2, \ldots, D$.

4.3.5 Importance Sampling Method

The importance sampling method for EVSI estimation (Menzies 2016) uses an alternative method for estimating the posterior expectation conditional on a specific sample $\boldsymbol{X}_s$. The importance sampling method is based directly on the definition of posterior expectation for a specific sample

$$\text{E}_{\boldsymbol{\theta}|\boldsymbol{X}_s}[\text{NB}_d(\boldsymbol{\theta})] = \int_{\boldsymbol{\Theta}} \frac{\text{NB}_d(\boldsymbol{\theta})\, p(\boldsymbol{X}_s \mid \boldsymbol{\theta})}{p(\boldsymbol{X}_s)} d\boldsymbol{\theta}.$$

If we assume that simulations of $\boldsymbol{\theta}$ are available to estimate this integral by Monte Carlo simulation, the integral can be re-written as

$$\int_{\boldsymbol{\Theta}} \frac{\text{NB}_d(\boldsymbol{\theta})\, p(\boldsymbol{X}_s \mid \boldsymbol{\theta})}{p(\boldsymbol{X}_s)} d\boldsymbol{\theta} = \sum_{k=1}^{K} \text{NB}_d(\boldsymbol{\theta}_k)\, w_k(\boldsymbol{X}_s),$$

where $w_k(\boldsymbol{X}_s)$ are weights that are proportional to the sampling density (informally known as the "likelihood") of the data, $p(\boldsymbol{X}_s \mid \boldsymbol{\theta}_k)$ for $k = 1, \ldots, K$. More specifically, the weights can be estimated as

$$w_k(\boldsymbol{X}_s) = \frac{p(\boldsymbol{X}_s \mid \boldsymbol{\theta}_k)}{\sum_{l=1}^{K} p(\boldsymbol{X}_s \mid \boldsymbol{\theta}_l)},$$

where the sum on the bottom of this fraction estimates the normalising constant of the Bayesian distribution by ensuring that the weights sum to 1. Thus, the posterior expectation for a specific simulated dataset $\boldsymbol{X}_s$, $\mu_d^{\boldsymbol{X}_s}$ can be estimated by reweighting simulated values of $\text{NB}_d(\boldsymbol{\theta})$ by the normalised likelihood. This is equivalent to the sampling importance resampling (SIR) algorithm developed by Rubin to estimate posterior distributions (Rubin 1988).

In practice, as we already have simulations of $\text{NB}_d(\boldsymbol{\theta})$ from the probabilistic analysis, we typically set $K = S$ and reuse these simulations. Thus, to implement the importance sampling method, the likelihood of the data, given the parameters, needs to be specified and calculated for each set of parameter values from the PA. This process then needs to be repeated for each simulated dataset, $\boldsymbol{X}_s$ to compute the conditional expectation for that dataset. This means that the importance sampling method typically requires $S \times S$ calculations of a likelihood function;

- For each row of the PA dataset, a single dataset is simulated; $\boldsymbol{X}_s$.
- For each simulated dataset, compute the likelihood of this dataset conditional on every simulated parameter set from the PA distribution.
- Calculate the normalised likelihoods for the single simulated dataset and calculate the weighted sum.

Note that it is often computationally easier to define and calculate the likelihood on the log scale. This is because calculating the log likelihood for independent datasets, i.e., as you would typically assume arise from a study, requires the log likelihood for each individual to be summed together. In contrast, working on the *natural* scale, rather than the log scale, involves multiplications rather than sums, which can result in numbers that are too small for a computer to represent ("underflow"). However, the weight definition requires the likelihood on the natural scale, so once the log likelihood has been computed, it should be exponentiated before calculating the weights through normalisation. In some practical settings, where both S and the proposed study sample size are large, this can still lead to numerical issues, as the sum of the exponentiated log likelihoods can be large.

In theory, once the weights have been computed, the importance sampling method only requires the summation of simulated net benefit values for each simulated dataset. However, as the proposed data collection exercise gets larger, the weights will become more extreme, i.e., the likelihood values for most simulated parameter sets will become approximately 0 and only a very small number of net benefit values will contribute to the estimation of the posterior expectation (Menzies 2016). This leads to an overestimation of EVSI when using the original simulated values of the net benefit. To adjust for this and maintain accuracy for the importance sampling method, the simulations of the net benefit conditional on the parameters of interest $\boldsymbol{\phi}$ only as opposed to the simulations of the net benefit itself should be used to estimate EVSI. This is similar to the moment matching method, which rescales $\mu_d^{\boldsymbol{\phi}}$ to compute EVSI.

4.3.5.1 Calculating EVSI Using the Importance Sampling Method: Summary

1. Perform PA simulation to obtain $\boldsymbol{\theta}_s$ and $\text{INB}_d^{\boldsymbol{\theta}_s}$, $s = 1, \ldots, S$.
2. Estimate EVPPI to obtain simulations of $\mu_d^{\boldsymbol{\phi}}$, the inner expectation in the first term of Equation (3.4).
3. For each $s = 1, \ldots, S$:
   (a) Simulate a dataset $\boldsymbol{X}_s$ from $p(\boldsymbol{X} \mid \boldsymbol{\phi}_s)$.
   (b) Compute the likelihood of $\boldsymbol{X}_s$ conditional on every PA simulation for $\boldsymbol{\phi}$, i.e., calculate the likelihood L_k of $\boldsymbol{X}_s$ for each parameter set $\boldsymbol{\theta}_k$ with $k = 1, \ldots, S$.
   (c) Compute $w_k(\boldsymbol{X}_s) = \frac{L_k}{\sum_{l=1}^S L_l}$ so $w_k(\boldsymbol{X}_s)$ sums to 1.
   (d) Calculate the weighted sum of $\mu_d^{\boldsymbol{\phi}_k}$, $\sum_{k=1}^S w_k(\boldsymbol{X}_s)\mu_d^{\boldsymbol{\phi}_k}$, where $\mu_d^{\boldsymbol{\phi}_k}$ is the expectation of the net benefit conditional on the k-th simulated value of $\boldsymbol{\phi}$.
4. Each weighted sum estimates $\mu_d^{\boldsymbol{X}_s}$.
5. The EVSI is estimated as $\widehat{\text{EVSI}} = \frac{1}{S}\sum_{s=1}^{S} \max_d \mu_d^{\boldsymbol{X}_s} - \max_d \frac{1}{S}\sum_{s=1}^{S} \mu_d^{\boldsymbol{X}_s}$.

4.3.6 Gaussian Approximation Method

The Gaussian approximation (GA) method relies on a simple approximation of the distribution of the posterior mean of the parameter(s) of interest, which is then used to compute EVSI (Jalal and Alarid-Escudero 2018). The GA method defines the distribution of the posterior mean for each (univariate) parameter $\phi^{\mathbf{X}}$ as a weighted average of the prior distribution and a "point mass" on the prior mean, such that

$$\phi^{\mathbf{X}} = \sqrt{v}\phi + (1 - \sqrt{v})\bar{\phi},$$

where $\bar{\phi}$ is the prior mean, and v is the variance reduction, defined as

$$v = \frac{N}{N + N_0},$$

where N is the sample size of the proposed new sample, and N_0 is the prior sample size. Notice that $\phi^{\mathbf{X}}$ denotes the posterior mean, which is a random variable, and has a distribution centred on the prior mean of ϕ and a reduced variance. Similar to the distribution of the posterior expectation for the net benefit, $\mu_d^{\mathbf{X}}$, the behaviour of the variance of the posterior mean for $\boldsymbol{\phi}$ in relation to N is different from what one may expect. This is because the posterior mean is determined across the range of potential datasets (i.e., the posterior mean prior to the actual data collection). For example, at the extreme where we propose to collect no new data, so $N = 0$, then $v = 0$ and the variance of the posterior mean will be 0. Thus, as we did not collect any new data, we are certain that our posterior mean will be exactly equal to the best estimate under our prior knowledge, which is the prior mean. Conversely, if we propose to collect a dataset with an infinitely large sample size $N = \infty$, then $v = 1$ and $\text{Var}(\phi^{\mathbf{X}}) = \text{Var}(\boldsymbol{\phi})$. This is because if we collect an infinitely large dataset, then we are sure that we will *cover* the entire prior space, and the distribution of the posterior mean will exactly replicate the prior distribution. As N increases between $N = 0$ and $N = \infty$, v will increase between 0 and the prior variance.

It can be shown (Jalal and Alarid-Escudero 2018) that this approximation for the distribution of the posterior mean works well for different distributions for the parameters and the data in both conjugate (e.g., beta-binomial) and non-conjugate settings. The performance of the GA method relies on the data and the prior being approximately Gaussian. Due to the central limit theorem, this is likely to be a fairly good approximation provided the sample size of the proposed study and the prior sample size are sufficiently large. Thus, as a general rule of thumb, the GA method improves once the value of N_0 and N exceed 30. For sample sizes and effective sample sizes less than 30, it is advisable to consider whether the Gaussian approximation is appropriate.

Once we obtain $\phi^{\mathbf{X}}$ separately for each of the parameters in the subset of interest $\boldsymbol{\phi}$, we now need to estimate the distribution of the posterior mean net benefit, $\mu_d^{\mathbf{X}}$, from the distribution of the posterior mean for the individual parameters. This can be achieved by creating a model that links the parameter values to the expected net benefit, i.e., the regression models that are used to compute EVPPI (see Section 3.3.4);

$$\text{NB}_d(\boldsymbol{\theta}) = g_d(\boldsymbol{\phi}) + \epsilon,$$

where $g_d(\cdot)$ is the statistical regression function that defined the model net benefits as a function of the parameters of interest $\boldsymbol{\phi}$. Once this statistical model is available, the posterior expectation of each of the parameters can be plugged into the regression function. Thus, we can obtain the expected net benefits given the distribution of the posterior mean for the parameters of interest, such that

$$\mu_d^{\mathbf{X}} = g_d(\boldsymbol{\phi}^{\mathbf{X}}),$$

where $\boldsymbol{\phi}^{\mathbf{X}}$ denotes the estimated distribution of the posterior mean for all the parameters of interest and is estimated by rescaling each simulated parameter separately.

Essential to the GA method is the correct estimation of N_0, the prior sample size, or the effective sample size, which represents the amount of information in the prior. For some distributions, obtaining N_0 is quite straightforward. For example, for a Beta distribution, N_0 can be estimated by the sum of the shape and scale parameters. In some other cases, N_0 can be obtained from expert opinions or provided as the sample size in a published study. However, if these cases are not relevant, the following formula can be used to compute N_0 numerically

$$N_0 = n\left(\frac{\text{Var}(\phi)}{\text{Var}(\phi^{\mathbf{X}})} - 1\right),$$

where $\text{Var}(\phi^{\mathbf{X}})$ can be estimated by running a full nested simulation for a data collection exercise of sample size n. In this case, S datasets of size n would be simulated before the model parameters are updated using full posterior analysis (typically using MCMC methods). This estimation procedure is likely to be more efficient than running a nested simulation procedure to estimate EVSI, as the decision-analytic model does not need to be rerun. The advantage is that it only needs to be run for a single sample size n before EVSI can be approximated across a range of sample sizes. More information about the effective sample size can be found in Morita, Thall and Müller (2008).

The GA method can be extended to complex designs and multivariate priors, but the general theory behind the method remains the same. The following algorithm clarifies the implementation of the GA method for studies that update multiple independent parameters.

4.3.6.1 Calculating EVSI Using the Gaussian Approximation Method: Summary

1. Perform simulation from probabilistic analysis, to obtain $\boldsymbol{\theta}_s$ and $\text{INB}_d^{\boldsymbol{\theta}_s}$, $s = 1, \ldots, S$.
2. Fit $D-1$ regression models with $\text{INB}_d^{\boldsymbol{\theta}_s}$ as outcomes and $\boldsymbol{\phi}_s$ as covariates. Note that the description in the previous section works on the net benefit scale, but using the incremental net benefit increases computational efficiency.
3. For each element of $\boldsymbol{\phi}$ denoted by ϕ^p for $p = 1, \ldots, P$:
   (a) Determine the prior effective sample size n_0^p.
   (b) For a proposed study with N participants, compute a weighted sum of the simulations of ϕ^p and $\bar{\phi^p}$, the mean of the p-th parameter;
   $$\phi^p\sqrt{\frac{N}{N+n_0^p}} + \bar{\phi^p}\left(1 - \sqrt{\frac{N}{N+n_0^p}}\right).$$
4. Using the regression models from Step 2, predict the model outcomes for the rescaled $\boldsymbol{\phi}$ simulations.

5. The fitted values from Step 4 estimate $\mu_d^{\mathbf{X}_s}$.
6. The EVSI is estimated as $\widehat{\text{EVSI}} = \frac{1}{S}\sum_{s=1}^{S} \max_d \mu_d^{\mathbf{X}_s} - \max_d \frac{1}{S}\sum_{s=1}^{S} \mu_d^{\mathbf{X}_s}$.

4.4 Selecting an EVSI Estimation Method

The four general-purpose methods presented above offer different advantages and disadvantages. Each method may be more suitable for different decision-analytic models, proposed data collection exercises and trial design goals. To support the selection of an appropriate EVSI estimation method, the following section outlines the input and expertise required to use each of these methods and their respective strengths and limitations (Kunst et al. 2020).

4.4.1 Inputs and Expertise Required for EVSI Calculation

The four EVSI approximation methods discussed in this book have diverse requirements and make different assumptions. All four methods require the simulations from a probabilistic analysis of a decision-analytic model. The moment matching and importance sampling methods also require calculation of the EVPPI for the parameter(s) to be evaluated and updated in the proposed study. If studies informing different parameters or groups of parameters are considered, then EVPPI would need to be computed for each. In terms of expertise, there are three skills, which we briefly outline below, that the analyst may require to compute EVSI. None of the methods require all three skills, but each method requires the mastery of at least one of the skills. Table 4.2 summarises the required skills and inputs for each method.

Firstly, the approximation methods may use *regression* to model the relationship between the incremental net benefit and key predictors. As explained in Sections 3.3.4 and 3.3.5, these are usually nonparametric regression methods that allow us to capture relatively complex relationships. The moment-matching method may also require regression methods if EVSI is estimated across different sample sizes.

Secondly, more general statistical modelling skills may be needed. All approximation methods require the simulation of potential datasets (except in the Gaussian approximation method when the effective sample size is already available). The regression-based method requires the user to summarise the datasets, which require expertise in standard methods for the statistical analysis of study results, for example, maximum likelihood procedures. The Gaussian approximation may also require this, as summary statistics may be used to estimate the effective sample size. The importance sampling method then

TABLE 4.2
The skills and inputs required to compute EVSI with the regression-based (RB), importance sampling (IS), Gaussian approximation (GA) and moment-matching (MM) methods. ✓ indicates that the skill/input is required. * indicates that the skill/input *may be* required.

#	Requirements	RB	IS	GA	MM
	Inputs				
1	Full decision-analytic model				✓
2	Probabilistic [sensitivity] analysis	✓	✓	✓	✓
3	Simulations of the expected net benefit conditional on ϕ (required to compute EVPPI)		✓		✓
	Skills				
1	Regression methods	✓		✓	*
2	Statistical modelling, e.g., with likelihoods	✓	✓	*	
3	Bayesian inference			*	✓

requires that the analytic likelihood for $\boldsymbol{X}$ must be specified and coded as a function.

Finally, the moment-matching and Gaussian approximation methods may require *Bayesian* updating, often requiring specialised software such as `BUGS` (Lunn et al. 2000), `JAGS` (Plummer 2015) or `Stan` (Carpenter et al. 2016). Bayesian updating is also required by the nested Monte Carlo method. Note that the Gaussian approximation method only requires full Bayesian updating when the prior effective sample size cannot be estimated by a more efficient method.

4.4.2 Strengths and Limitations of EVSI Estimation Methods

As each EVSI approximation method is different, the most suitable method for EVSI calculation will depend on the decision-analytic model, the sampling distribution for the data, the expertise of the analyst and the amount of computation time available. We now discuss the strengths and limitations of these methods to help analysts select the most appropriate method. A summary is presented in Table 4.3.

4.4.2.1 Moment Matching Method

Strengths

This method uses the same nested simulation structure as the nested Monte Carlo methods, so if the nested Monte Carlo method has already been developed, it can be easily adapted for this method. Furthermore, this method

TABLE 4.3
Selected strengths and limitations of the four EVSI approximation methods. GA denotes the Gaussian approximation method, IS denotes the importance sampling method, MM denotes the moment matching method, and RB denotes the regression-based method. ✓ indicates that the characteristic is required for the given method, * indicates that the characteristic *may be* required for the given method.

#	Characteristics	RB	IS	GA	MM
	Strengths				
1	Estimates EVSI for complex studies collecting a large number of outcomes		✓		✓
2	Only requires the samples from probabilistic analysis	✓	✓	✓	
3	Uses nonparametric (typically GAM) regression	✓		✓	
4	Estimates EVSI for different study sizes with same computational cost			✓	✓
5	Quantifies uncertainty in estimate	✓		✓	✓
	Limitations				
1	Requires simulated study data	✓	✓	*	✓
2	Requires accurate EVPPI estimation		✓		✓
3	Can be computationally challenging to estimate EVSI for proposed studies with a large sample size		✓		
4	Requires simulated study data to be summarised in a low dimensional statistic	✓		*	
5	Struggles if proposed study has more than five/six outcomes	✓		✓	
6	May estimate inaccurate EVSI if the proposed study has small prior effective sample size			✓	
7	May estimate inaccurate EVSI if the proposed study has small sample size				✓

can estimate EVSI for multiple alternative sample sizes with a fixed additional computational cost, and estimate uncertainty around the final EVSI estimate. Finally, the moment matching method can estimate EVSI irrespective of the number of outcomes considered, $N_{\mathcal{O}}$.

Limitations

The moment-matching method requires simulated study data and relies on performing a PA Q times, with $30 < Q < 50$. Thus, the decision-analytic model must be rerun a substantial number of times, which is challenging in computationally expensive decision-analytic models, such as microsimulation models. It also assumes that the model itself is available to the analyst, rather than simply the results of the PA. If the original PA simulation size is small, this method is inaccurate, and will be inaccurate if the sample size of the proposed study is less than 10. Finally, this method requires an accurate estimate of EVPPI and is more accurate for studies that will have large impact on the underlying uncertainty in the decision-analytic model, i.e., EVPPI needs to be high compared to EVPI, ideally greater than 40%.

4.4.2.2 Regression-Based Method

Strengths

The decision-analytic model does not need to be rerun to produce EVSI estimates, so EVSI can be computed by an analyst with access to the PA dataset only. The flexible regression methods required can also be easily implemented in the `voi` package, making this method relatively simple to implement once the *summary statistics* are available. Furthermore, EVSI estimates across different study sample sizes can be obtained at a constant computational cost. Finally, if the model is judged to fit well, an estimate of the uncertainty in the EVSI estimate can be obtained.

Limitations

For complex studies where the number of collected outcomes $N_{\mathcal{O}}$ is greater than five or six, it can be challenging to fit a sufficiently accurate regression model. The simulated study data must also be correctly summarised to obtain accurate EVSI estimates. This can be challenging and time-consuming in more complex studies. The relationship between the incremental net benefit and $T(\boldsymbol{X})$ must also be well-approximated by the regression model to ensure accurate EVSI estimation.

4.4.2.3 Importance Sampling Method

Strengths

This method only requires the PA results to compute the EVSI and access to the original decision-analytic model is not necessary. Furthermore, the computational cost of the importance sampling method does not depend on the number of study outcomes $N_{\mathcal{O}}$, once the likelihood function for the potential data to be collected has been defined.

Limitations

The accuracy of the EVSI estimation relies heavily on the appropriate specification of the analytic likelihood and the accurate estimation of the EVPPI. The importance sampling method can also be computationally problematic when the sample size of the future data is large, as the likelihood tends to 0 with increasing sample size, leading to inaccurate EVSI estimation (Heath et al. 2020). Finally, uncertainty in the EVSI estimation procedure cannot be estimated with this method.

4.4.2.4 Gaussian Approximation Method

Strengths

Estimation of EVSI with this method does not require rerunning the decision-analytic model, and access to PA results is sufficient. The Gaussian approximation method can estimate EVSI across different proposed study sample sizes at minimal computational cost, once the prior effective sample size has been estimated. Thus, the optimal sample size for proposed data collection can be obtained cheaply. Finally, a measure of uncertainty in the EVSI estimation procedure is available from this method.

Limitations

This estimation method can lead to inaccurate EVSI estimation if the prior effective sample size is small. Additional simulation methods may also be required to estimate the prior effective sample size if it cannot be obtained directly. The GA method also relies on a regression model, therefore it may perform poorly if this regression model does not capture the relationship between the incremental net monetary benefit and the parameters of interest. Moreover, if the number of collected outcomes is over five or six, the proposed GAM regression methods become challenging (Strong, Oakley and Brennan 2014). Although there are many other nonparametric regression procedures available (as discussed in Section 3.3.5).

To support the selection of the most appropriate method, Table 4.4 highlights which methods would be appropriate, based on features of the decision-analytic model and proposed study. The key to EVSI estimation is that it may not be possible to compute EVSI for all given situations using a single method. Thus, it is crucial to understand which method is most appropriate, and which method is easiest for the given analyst to implement.

4.5 Calculating EVSI in the `voi` Package

To facilitate the use of these EVSI computation methods when they are most appropriate, they are implemented in the `voi` package through the `evsi` function. The aim of the `voi` package is to implement different VOI computation methods under the same interface, and allow VOI measures to be computed with the minimum of user input necessary for each computation method.

TABLE 4.4
The recommendations for EVSI approximation methods based on features of the decision-analytic problem and proposed study. GA, Gaussian approximation method, IS, importance sampling method, MM, moment-matching method; RB, regression-based method; ✓ indicates that the method is appropriate for the given feature.[1] EVSI estimation does not require rerunning the decision-analytic model. This may be advantageous if the analyst does not have access to the decision-analytic model or if the decision-analytic model is computationally expensive.[2] The IS method can be computationally challenging when estimating EVSI for proposed studies with a sample size over 1,000 (Heath et al. 2020) because this method re-estimates EVSI across different sample sizes.

#	Features	RB	IS	GA	MM
1	Designing complex studies collecting a large number of outcomes (more than six outcomes)		✓		✓
2	Using only probabilistic analysis results, without needing access to the decision-analytic model[1]	✓	✓	✓	
3	Identifying optimal sample size by examining different study sizes with same computational cost			✓	✓
4	Including uncertainty quantification in the estimate	✓		✓	✓
5	Proposing studies with a large sample size[2]	✓		✓	✓
6	Proposing studies with small prior effective sample size	✓	✓		✓
7	Proposing studies with small sample size	✓	✓	✓	

For each method, EVSI can be calculated with a single function call, specified in a similar way. Currently, the `voi` package can be used to compute EVSI using the moment matching, importance sampling and regression-based methods. Code to implement the Gaussian approximation method is available alongside the primary publication (Jalal and Alarid-Escudero 2018) and in the `dampack` package (Alarid-Escudero et al. 2021).

Each of these methods, as discussed in Section 4.4, requires different inputs and, thus, the information that must be supplied to the `evsi` function changes depending on which methods is used. However, all methods require:

- The matrix of PA simulations for the parameters (`inputs`).
- The corresponding health economic outputs (`outputs`).

As is required for the `voi` package, these outputs can either be in "net benefit" form, which provides the net benefit for each intervention at a given willingness-to-pay threshold or "cost-effectiveness" form, which provides the costs and effects alongside a range of willingness-to-pay thresholds.

4.5.1 Built-in Study Designs in the voi Package

The `voi` package can be used to calculate EVSI for a range of built-in study designs, which can be implemented using the `study` argument in the `evsi` function. For example, to compute EVSI for a two-arm randomised controlled trial investigating the effect of the novel intervention and the standard of care on the probability of experiencing side effects, the `trial_binary` study type can be used.

> To run the code in this section:
>
> - The Chemotherapy example model should be installed, following the instructions in Section 2.3.5.
> - The `voi` `R` package should be installed, using `install.packages("voi")`.
>
> All code in the book is provided and linked from `https://chjackson.github.io/voi/articles/book.html`. If reading this book as a PDF file, don't paste code from the PDF, since some special characters may fail to copy.

4.5.1.1 GAM Regression Method

This is illustrated here using the GAM regression-based method, with sample sizes between 50 and 500,

```
> evsi_builtin_rb <- evsi(outputs = chemotherapy_output,
+                         inputs = m_params,
+                         study = "trial_binary",
+                         pars = c("p_side_effects_t1",
+                                  "p_side_effects_t2"),
+                         n = seq(50, 500, by = 50),
+                         method = "gam")
```

In this code, we must also define the parameters that will be updated by the proposed study using the argument `pars`, the sample size (per arm, assumed equal for each arm) for which EVSI should be calculated using the argument `n`, and the computational method in `method`.

4.5.1.2 Importance Sampling Method

The same code can be used to compute EVSI using the importance sampling method by changing the `method` argument to `"is"`.

```
> evsi_builtin_is <- evsi(outputs = chemotherapy_output,
+                         inputs = m_params,
+                         study = "trial_binary",
+                         pars = c("p_side_effects_t1",
```

```
+                          "p_side_effects_t2"),
+                      n = seq(50, 500, by = 50),
+                      method = "is")
```

4.5.1.3 Moment Matching Method

To compute EVSI using the built-in study designs and the moment matching method, we require the following additional inputs, since the decision-analytic model must be rerun to estimate the posterior variance.

- A function to generate the PA distributions for the parameters (`par_fn`).
- A function to run the decision-analytic model (`model_fn`).
- The parameters of the prior distributions used in Bayesian analysis of the study data (`analysis_args`).

The built-in `"trial_binary"` study design requires the use of Beta priors for the probability of the event in each arm. This is slightly awkward for the Chemotherapy example, where the probability of side-effects under the novel treatment was specified via a log odds ratio parameter given a normal distribution. Thus, we have converted this probabilistic analysis distribution to a Beta prior for the probability, by translating the mean and SD (from the Monte Carlo sample of `p_side_effects_t2`) to the Beta shape and scale, using the function `betaPar` from our decision-analytic model.

```
> # Beta prior for standard care is set using the number of events
> beta_params_t1 <- c(1 + n_side_effects,
+                     1 + n_patients - n_side_effects)
> # Beta prior for the novel intervention is approximated from the
> # mean andstandard deviation of the PA distribution for the
> # probability of side effects.
> beta_params_t2 <- betaPar(mean(m_params$p_side_effects_t2),
+                          sd(m_params$p_side_effects_t2))
> # EVSI calculation with moment matching method
> evsi_builtin_mm <- evsi(outputs = chemotherapy_output,
+                     inputs = m_params,
+                     study = "trial_binary",
+                     pars = c("p_side_effects_t1", "p_side_effects_t2"),
+                     n = seq(50, 500, by = 50),
+                     method = "mm",
+                     model_fn = calculate_costs_effects,
+                     analysis_args = list(a1 = beta_params_t1[1],
+                                          b1 = beta_params_t1[2],
+                                          a2 = beta_params_t2$alpha,
+                                          b2 = beta_params_t2$beta),
+                     par_fn = generate_psa_parameters)
```

4.5.1.4 Available Built-in Study Designs

There are currently three built-in study designs in the `voi` package:

- `"binary"`: This calculates EVSI for a study that is investigating a binary outcome that informs a single-model parameter. In our example, this is the 5-year probability of death. In this case, the sample size of the study, i.e., the number of participants for which we will collect this single binary outcome, will be included in the `n` argument in the `evsi` function, and the parameter of interest will be specified using the `pars` argument.

- `"trial_binary"`: This study is used in the example above and calculates EVSI for a two-arm trial with binary outcomes. Note that this in-built study assumes that the information collected in the study will update information for both the probabilities. In some decision-analytic models, we assume that RCTs will only update information about the relative treatment effect and, therefore, the information does not update the uncertainty in the baseline probability of an event. If this assumption is required, then a bespoke study design must be used instead of a built-in design.

- `"normal_known"`: This study measures the value of collecting a normally distributed outcome, with a known standard deviation, to give information about the mean of the normal distribution. For this study design, the known standard deviation of the individual level distribution must be loaded into the `evsi` function using the `aux_pars` argument, e.g., `evsi(..., aux_pars=list(sd=2))`.

4.5.2 Bespoke Study Designs

EVSI can also be computed in the `voi` package for user-written study designs. This requires the user to write a *data generation* function to simulate data for the proposed trial design.

4.5.2.1 Constructing the Data Generation Function

The data generation function is an `R` function which:

- Takes as input a data frame of simulated parameter values, with names matching those in `inputs`.

- Outputs a data frame with the simulated data. This will have S rows, i.e., the number of rows equals the number of PA simulations.

The number of columns in the output corresponds to different "data points" in the study. How a "data point" is defined depends on the EVSI computation method and how the data are analysed. For the regression-based method, there should be one column per summary statistic, in which case the number of columns is usually small. For other computation methods, this depends on how the likelihood, or procedure used to do Bayesian inference from the study data, is constructed – e.g., there might be a large number of columns if each data point corresponds to a single individual in a trial. Finally, the data generation function can also take additional arguments, such as the sample size of the study.

To demonstrate how to define bespoke study designs in the `voi` package, we will develop a study investigating the utilities for individuals treated in hospital. The model parameter to be updated in this study is `u_hospital`, which represents the mean utility for patients in hospital. In addition to `u_hospital`, we must determine the individual-level variation in the utility measurements, which we specify as the individual-level standard deviation. In this example, we also account for uncertainty in the individual-level standard deviation in our data generation process using a uniform distribution with a minimum of 0.00001 and a maximum of 0.4. We generate the utility measurements from a normal distribution, truncated above by 1, for each individual. The truncated normal distribution represents the distribution of the utility as values above 1 cannot be simulated by allowing individuals to experience health states with negative utility, i.e., health states that are considered worse than death by the individual.

For the moment matching and importance sampling methods, the data for this study are simulated and saved at the individual level with the following simulation function:

```
> # Data generation function - individual data (for other EVSI methods)
> utility_datagen_fn_indiv <- function(inputs, n = 500){
+   # Load the data
+   X_hospital <- matrix(nrow = nrow(inputs), ncol = n[1])
+   for(i in 1:nrow(inputs)){
+     m_hospital <- inputs[i, "u_hospital"]
+     sd_hospital <- inputs[i, "sd_iid_hospital"]
+     X_hospital[i, ] <- truncnorm::rtruncnorm(n[1],
+                                               mean = m_hospital,
+                                               sd = sd_hospital,
+                                               a = -Inf, b = 1)
+   }
+   data_save_dat <- data.frame(cbind(X_hospital = X_hospital))
+   return(data_save_dat)
+ }
```

In this function, the data for `n` individuals are simulated from a truncated normal distribution, with one utility value per individual saved in the `X_hospital` object. As the individual level standard deviation for the utility is required to simulate the data, it must also be provided in the `inputs` matrix, using the column heading `sd_iid_hospital`. Note that we will discuss later how to add **sd_iid_hospital** to the **inputs** matrix to ensure that the data can be simulated.

For the regression-based method, each simulated dataset must be summarised into a low-dimensional summary statistic. In this example, we use the mean utility as a data summary:

```
> # Data generation function - aggregate data (for regression method)
> utility_datagen_fn_agg <- function(inputs, n = 500){
+   dat_indiv <- utility_datagen_fn_indiv(inputs, n = n)
+   X_hospital_mean <- rowMeans(dat_indiv)
+   data_save_dat <- data.frame(X_hospital_mean = X_hospital_mean)
+   return(data_save_dat)
+ }
```

In this code, we reuse the individual-level simulation function developed for the moment-matching and importance sampling methods before using **rowMeans** to calculated the row-specific mean, i.e., a single mean per dataset. These are then outputted to create a dataset with S rows and 1 column for use in the regression-based method.

4.5.2.2 GAM Method with Bespoke Study Designs

Once the data generation function has been specified, EVSI can be computed using the regression-based method with GAM regression using the **method = "gam"** argument. However, before computing EVSI, we have to add the **sd_iid_hospital** column to the PA dataset, which is saved in the **R** object **m_params** for the Chemotherapy example:

```
> ## Add the individual level standard deviation
> m_params$sd_iid_hospital <- runif(nrow(m_params), 0.00001, 0.4)
>
> ## Regression Based Method
> evsi_utility <- evsi(outputs = chemotherapy_output,
+                      inputs = m_params,
+                      pars = c("u_hospital"),
+                      n = seq(50, 1000, by = 200),
+                      method = "gam",
+                      datagen_fn = utility_datagen_fn_agg)
```

In this code, we first add the **sd_iid_hospital** parameter by simulating from the uniform distribution, as defined above. We then estimate EVSI for five dataset sizes between 50 and 1,000 $(50, 250, 450, 650, 850)$. Compared to calculating EVSI for a built-in study design, the **evsi** function requires the same inputs for a user-written design, except that instead of using the **study** argument, we supply the **utility_datagen_fn_rb** function to generate the data from our bespoke study, through the **datagen_fn** argument.

4.5.2.3 Moment Matching Method with Bespoke Study Designs

These calculations are more advanced, and some knowledge of Bayesian inference is assumed.

For the moment-matching and importance sampling methods, additional inputs are required alongside the data generation function. For the moment-matching method, the posterior distribution for the parameters of interest needs to be updated conditional on the generated data. The decision-analytic model then needs to be rerun, using samples from the updated posterior distribution as inputs. Thus, the user must specify:

- A function to analyse the data and provide updated distributions for the model parameters using the `analysis_fn` argument.
- A function to rerun the decision-analytic model, which is also required for the built-in studies using the `model_fn` argument.

The analysis function should have three arguments:

- The data generated from the data generation function (`data`).
- A list with any constants required by the Bayesian analysis, e.g., prior parameters or settings to control computation (`args`). Alternatively, constants can also be hard-coded inside the analysis function, but defining them through `args` allows different EVSI calculations to be performed with different settings – if this is done, the user should supply these settings through the `analysis_args` argument to `evsi`.
- The name of the parameters of interest (`pars`).

For the previous example, we show how the JAGS software is used to do the Bayesian analysis, using an analysis function defined as:

```
> # Analysis function based on JAGS
> utility_analysis_fn <- function(data, args, pars){
+   # Create the data list for JAGS
+   data_jags <- list(X_hospital = as.vector(data),
+                     n = args$n,
+                     alpha_hospital = betaPar(
+                       args$u_hospital_mu,
+                       args$u_hospital_sd
+                     )$alpha,
+                     beta_hospital = betaPar(
+                       args$u_hospital_mu,
+                       args$u_hospital_sd
+                     )$beta)
```

```
+
+   trial <- "
+   model {
+     for(i in 1:n){
+       X_hospital[i] ~ dnorm(u_hospital, tau_hospital)T(, 1)
+     }
+     u_hospital ~ dbeta(alpha_hospital, beta_hospital)
+     sd_hospital ~ dunif(0.00001, 0.4)
+     tau_hospital <- 1 / sd_hospital ^ 2
+   }
+   "
+
+   filein <- file.path(tempdir(),fileext="datmodel.txt")
+   cat(trial, file=filein)
+
+   # Perform the MCMC simulation with JAGS.
+   bugs.data <- jags(
+     data =  data_jags,
+     parameters.to.save = pars,
+     model.file = filein,
+     n.chains = 1,
+     n.iter = args$n.iter,
+     n.thin = 1,
+     n.burnin = 250,
+     quiet=TRUE, progress.bar = "none")
+
+   u_hospital <- bugs.data$BUGSoutput$sims.matrix[, "u_hospital"]
+   return(data.frame(u_hospital = u_hospital))
+ }
```

This analysis function uses JAGS to update the distribution of the mean utility for hospitalised individuals. JAGS (Just Another Gibbs Sampler) is software developed to sample from the generic Bayesian posterior distributions using Markov Chain Monte Carlo (MCMC) methods (Plummer 2003). JAGS uses similar syntax to WinBUGS, an alternative Bayesian software (Lunn et al. 2000). JAGS uses a declarative form to specify the structure of the Bayesian model using a series of relationships, i.e., between the model parameters and the data. In the general syntax, the symbol ~ indicates that a parameter/data point follows the distribution named after the symbol. For example,

```
sd_hospital ~ dunif(0.00001, 0.4)
```

indicates that the standard deviation of the utility `sd_hospital` is assigned a uniform prior distribution between 0.00001 and 0.4. This matches the distribution used to generate the data, i.e., the distribution of `sd_iid_hospital`. Further to this, the syntax `<-` indicates a deterministic relationship between the two elements. For example,

```
    tau_hospital <- 1 / sd_hospital ^ 2
```

defines the precision of individual level utilities `tau_hospital` as a function of the standard deviation, for which we have a prior distribution. This is because the normal distribution in the JAGS software is parameterised in terms of precision, rather than standard deviation as is used in `R`. Finally, note that distributions in JAGS are defined using the `d` prefix, i.e., `dbeta` and `dnorm`. For more information on the available distributions in JAGS and their parameterisations, see Plummer (2003).

Once the model has been declared, as we see in the `trial` text object in the `utility_analysis_fn` function above. The JAGS program must be run to update the prior distributions for the parameters with the observed data. Within the `R2jags` package in `R`, this is achieved through the `jags` function, which takes as inputs:

- `data`: A named list of the data that have been collected. For EVSI, this would be the simulated data loaded into the analysis function in the `data` argument of the function. We also load additional data elements including the sample size of the study and the parameters for the prior distributions. Within the analysis function to compute EVSI, we load these additional elements in the `args` argument so they can then be included in the data list for use within JAGS.

- `parameters.to.save`: A vector of the names of the parameters you want JAGS to monitor. In this example, these are provided in the `pars` argument of the function as these are the parameters of interest for your health economic model. JAGS does not save the parameter simulations for all the model parameters because in complex models, this would lead to memory challenges.

- `model.file`: This is the text of the model file, which we created in the proceeding lines using the declarative model structure discussed above.

- Elements that outline how the MCMC algorithm will proceed including the number of chains `n.chains`, the number total number of MCMC iterations, which we load in the `args` argument of the analysis function and the burn in for the MCMC chains `n.burnin`. Many resources exist to discuss these key choices, including Lunn et al. (2000), Kruschke (2015) and Plummer (2003).

From this description, it is evident that the `args` argument of the `utility_analysis_fn` function is crucial for loading several elements that are required for the Bayesian analysis of the data. In particular, to estimate EVSI within the `voi` package, the `analysis_args` argument needs to include the elements, `n`, the initial trial sample size, `u_hospital_mu` and `u_hospital_sd`, the mean and standard deviation of the PA distribution for the in-hospital utility and `n_iter`, the number of MCMC iterations used by JAGS to estimate the posterior distribution of the parameters. For example,

```
> analysis_args <- list(n = 30,
+                       u_hospital_mu = u_hospital_mu,
```

```
+                         u_hospital_sd = u_hospital_sd,
+                         n.iter = 2000)
```

Finally, EVSI can be computed using the `evsi` function. Here we supply a function to evaluate the decision-analytic model, through the `model_fn` argument. This is set to be the function `calculate_costs_effects`, which we defined in Chapter 2. Note also that `m_params` is augmented to include `sd_iid_hospital` as for the regression-based method:

```
> evsi_utility <- evsi(outputs = chemotherapy_output,
+                       inputs = m_params,
+                       pars = c("u_hospital"),
+                       pars_datagen = c("u_hospital","sd_iid_hospital"),
+                       n = seq(30, 1000, by = 200),
+                       method = "mm",
+                       datagen_fn = utility_datagen_fn_indiv,
+                       model_fn = calculate_costs_effects,
+                       analysis_args = analysis_args,
+                       analysis_fn = utility_analysis_fn,
+                       par_fn = generate_psa_parameters,
+                       Q = 50)
```

Another new feature is shown here: `pars_datagen` indicates the parameters in `inputs` needed to *generate* the study data, while `pars` indicates the parameters that are *learnt* from the study data – these are different in this case, as the individual-level utility SD `sd_iid_hospital` does not inform the decision-analytic model.

4.5.2.4 Importance Sampling with Bespoke Study Designs: Constructing the Likelihood Function

For the importance sampling method, the user must provide a function to compute the likelihood. For use within the `evsi` function, the likelihood function should compute a vector of likelihood values for a single dataset across all the parameter simulations in the PA dataset. For this example, the likelihood function should be written as follows:

```
> # Likelihood function
> utility_likelihood <- function(data, inputs){
+   # Load the data
+   ll <- numeric(nrow(inputs))
+   data_vec <- unlist(data)
+
+   for(i in 1:nrow(inputs)){
+     m_hospital <- inputs[i, "u_hospital"]
+     sd_hospital <- inputs[i, "sd_iid_hospital"]
+     ll[i] <- exp(
+       sum(
+         log(
```

```
+           truncnorm::dtruncnorm(data_vec,
+                                 mean = m_hospital,
+                                 sd = sd_hospital,
+                                 a = -Inf, b = 1)
+         )))
+   }
+   return(ll)
+ }
```

In this example, the likelihood can be computed using the `dtruncnorm` function, which provides the density of the truncated normal distribution. We compute the log likelihood for each of the data points in the simulated data, before summing and taking the exponential. (Note that this is the log likelihood rather than the likelihood – working on the log scale is more numerically stable). Note again, that the individual level variance is loaded in the `inputs` so that is corresponds directly to the individual level standard deviations that were used to simulate the data.

Based on this function for the likelihood, EVSI can be computed using the importance sampling method with the following code, where the likelihood is loaded in the `likelihood` argument:

```
> evsi_utility <- evsi(outputs = chemotherapy_output,
+                      inputs = m_params,
+                      pars = c("u_hospital"),
+                      pars_datagen = c("u_hospital", "sd_iid_hospital"),
+                      n = seq(50, 1000, by = 200),
+                      method = "is",
+                      nsim = 1000,
+                      datagen_fn = utility_datagen_fn_indiv,
+                      likelihood = utility_likelihood)
```

It is important to note that the importance sampling method can be relatively slow, especially for large PA dataset sizes as the likelihood must be computed $S \times S$ times, where S is the size of the PA dataset. EVSI must also be computed separately for each sample size, which can add further computational time. Thus, it is not advised to use the importance sampling method if you wish to compute EVSI across a large range of different sample sizes.

4.6 Expected Net Benefit of Sampling

Thus far, this chapter has discussed the definition of EVSI and how to compute EVSI in theory and in practice. We now move to discuss how EVSI can be used to inform research design decisions. The key metric used to support research design using EVSI is known as the Expected Net Benefit of Sampling (ENBS). ENBS measures the net total benefit of undertaking a specific research study, accounting for the number of individuals who would be impacted by any change in decision and the cost of undertaking research. Thus, ENBS is defined as the difference between (i) the population-level EVSI and (ii) the cost of undertaking the proposed study (Conti and Claxton 2009), both of which are clarified below. In general, studies with ENBS greater than 0 have some potential to offer value for money for the decision maker and studies with higher values for the ENBS should be prioritised, subject to the overall research budget constraint and competing research priorities (see Chapter 10). Furthermore, ENBS can be estimated for proposed studies with different design characteristics, e.g., sample size, length of follow-up, and these characteristics can then be optimised.

4.6.1 Population-Level EVSI

As with EVPPI and EVPI, the EVSI calculations discussed so far in this chapter have focused on estimating EVSI for an individual, i.e., the value of reducing uncertainty for an individual who would be affected by the decision. However, the information collected in the proposed study would determine the interventions for current and future patients (McKenna and Claxton 2011). Therefore, to determine the societal benefit from the proposed research study targeting the parameters of interest, the individual level EVSI should be multiplied by the number of individuals who will be affected by the decision, P. Again, as with EVPPI and EVPI analyses, the size of this population is unknown and must be estimated to compute ENBS. See also Section 3.4.4 for an introduction to how P is estimated – here we give a slightly more in-depth perspective.

A common method for estimating P begins by first determining the number of individuals who would be affected by the decision in a given year (P_y). Usually, we consider P_y as the yearly incidence rate of the disease in question but there are a range of settings in which this may be inappropriate. Firstly, P_y could be adjusted to account for the delayed uptake and slow decommissioning of a technology (Grimm, Dixon and Stevens 2017b) (also discussed in Chapter 12). Secondly, P_y could be adjusted to account for limitations in the number of available providers. It is also possible to consider "catch up" interventions that implement the optimal intervention for a wider "prevalent" population in the first year after approval (Philips, Claxton and Palmer 2008). Overall, the definition and size of the population should be carefully considered and reflect the best available evidence about the use of the technology.

Once we have determined P_y, the number of years Y before the decision will be re-assessed should be estimated. As discussed previously, this "decision horizon" is a proxy for new technologies entering the market and future price changes that would alter the cost-effectiveness of the interventions (Philips, Claxton and Palmer 2008). Values between 5 and 15 years have commonly been used in practice and sensitivity analyses to the choice of Y are usually performed. Note that in some disease areas, such as oncology, the decision horizon may be short, while other disease areas are less fast moving and treatments will be used for a longer period of time. Medical devices are particularly prone to short decision horizons as improvements in technology are common. From Y and P_y, the total population can finally be computed as

$$P = \sum_{y=0}^{Y} \frac{P_y}{(1+r)^y},$$

where r is a discount rate used to discount the Value of Information (VOI) to patients who will receive their intervention in the future (McKenna and Claxton 2011).

There are a couple of important elements to note here. Firstly, there is no intrinsic reason why annual population estimates, with annual discounting, are required. For some applications and decision horizons, monthly or weekly estimates could be used, with corresponding monthly or weekly discount rates applied. These may provide a more accurate estimate of the present-value equivalent of the overall population that can benefit from the reduced uncertainty, as they do not assume that all people in a given year receive the benefit at the start of that year. In the plotting discussed in Section 4.7, continuous discounting is used, as it easily allows for time horizons of any length. Secondly, in most VOI calculations, Y and P_y are assumed to be fixed and known, which is an unrealistic assumption. Section 4.7 explores how these values can be subject to scenario analyses, but alternative methods to exploring uncertainty in these values could also be considered (Coyle 2004).

In practice, multiplying EVSI by P to prioritise research implies that, all else being equal, studies in high prevalence diseases will have higher value. When optimising the expected net benefit of sampling, high prevalence disease requires research that gathers information from more individuals compared to lower prevalence diseases. In other words, decisions that affect more individuals will require higher sample sizes (resulting in lower statistical uncertainty) to achieve value, as the impact of an incorrect decision would be larger. From a VOI perspective, this is because VOI measures are concerned with the risk of implementing an inefficient treatment that appears efficient under current information. If an inefficient treatment is only going to be used by a small number of patients, then this non-optimal use of resources may not be severe, even if the treatment turns out to be economically inefficient. Therefore, the more widespread the disease, the more people will use the treatment and the greater the opportunity cost for incorrect decision-making.

4.6.2 Research Costs

Estimating the cost of a proposed research study is a task that is undertaken more commonly than estimating the total population that would be impacted by a research decision. As discussed in Section 3.4.4, study costs are typically required for funding applications, and can be approximated with relative certainty. In contrast to Section 3.4.4, the research costs required for ENBS should be estimated as accurately as possible, as they will be used to support decision-making around study design and which study(ies) to commission.

To design research, we typically consider direct costs that fall into two key categories: fixed and variable costs. Fixed costs include all costs that will be incurred irrespective of the size of the study/the number of patients enrolled. These costs are related to the conduct and dissemination of the study. For example, fixed costs may include:

- Analyst costs.
- Database management and extraction costs.
- Research assistants and managers salaries.
- Publication costs.
- Research oversight costs.

Variable costs are then all costs that are directly attributable to patient enrolment in the study. In some study designs, these may be minimal, e.g., retrospective observational studies may not cost extra to enrol more patients and the sample size limits will be due to data availability, or may be exceed the fixed costs substantially, e.g., the per patient enrolment costs in clinical trials will include individual data collection costs, sample processing, pharmacy costs, etc. These costs are what are traditionally considered research costs in funding applications and can be defined in collaboration with study experts.

In addition to these costs, ENBS analyses can consider opportunity costs of research, especially in prospective studies. Firstly, EVSI analyses generally assume that the current optimal treatment will be implemented while the research is ongoing. If this is not the case, the study costs should include an opportunity cost equal to the difference between the current optimal intervention and the intervention that will be used in the general population while the research is ongoing. Secondly, any patients enrolled in the study will not benefit from the research results. This means that the study costs should include EVSI for every individual enrolled in the prospective study. Finally, in prospective interventional studies, individuals who receive the non-optimal intervention will experience an opportunity loss, which should be considered as part of the study costs (Conti and Claxton 2009). Note that in large patient populations, these last two study costs may have negligible impact on ENBS, particularly opportunity costs related to EVSI. However, for rare diseases, removing the patients enrolled in the study from the potential population who would benefit from research may have substantial impact on ENBS.

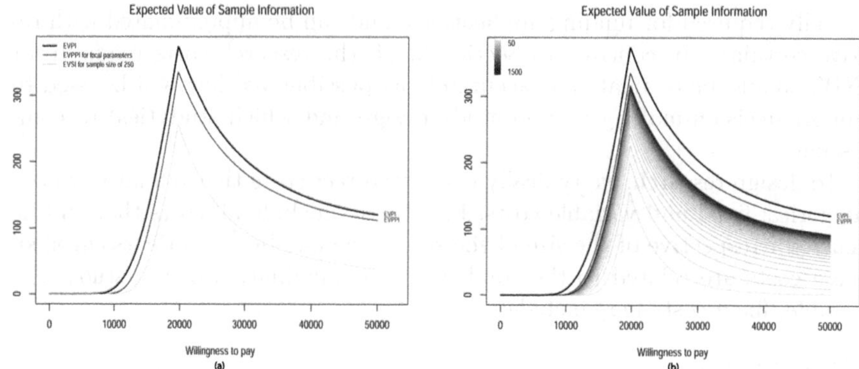

FIGURE 4.3
The EVSI plotted across different willingness-to-pay thresholds, (a) shows the EVSI for one sample size and (b) across a large number of sample sizes.

4.7 Presenting EVSI and ENBS Analyses

To support decision-making and research design with EVSI and ENBS, it is important to appropriately present the results of these analyses. In general, visualising the results of EVSI and ENBS analyses is more complex than for both the EVPI and EVPPI as there are more variables in an EVSI analysis, e.g., willingness-to-pay, sample size, study design, trial costs, decision horizon. Therefore, this section presents a suite of graphics that can be used to explore different aspects of the EVSI/ENBS analysis.

Firstly, EVSI can be plotted across different values of the willingness-to-pay, similar to the graphics seen for the EVPI and EVPPI, displayed in Figure 4.3 (a). In this figure, we can compare the relative sizes of EVSI, EVPPI and EVPI and determine whether the proposed data collection exercise is able to efficiently update information about the model parameters. We would hope that EVSI is close to EVPPI for all willingness-to-pay thresholds. Generally, the number of EVSI curves that can be plotted together will be relatively small, meaning that this graphical representation may be limited if the aim is to compare a range of different potential designs.

One potentially useful graphical technique is to use colour gradients to display increasing sample sizes, as in Figure 4.3 (b). This allows us to ascertain how quickly EVSI approaches EVPPI. However, even using colours may make this graphic challenging to read as a large number of alternative sample sizes are considered. While this graphic aligns most closely to the standard presentation for EVPI and EVPPI, it may be most commonly used when the optimal study design has already been determined and the aim is to present sensitivity analyses around the willingness-to-pay for this threshold.

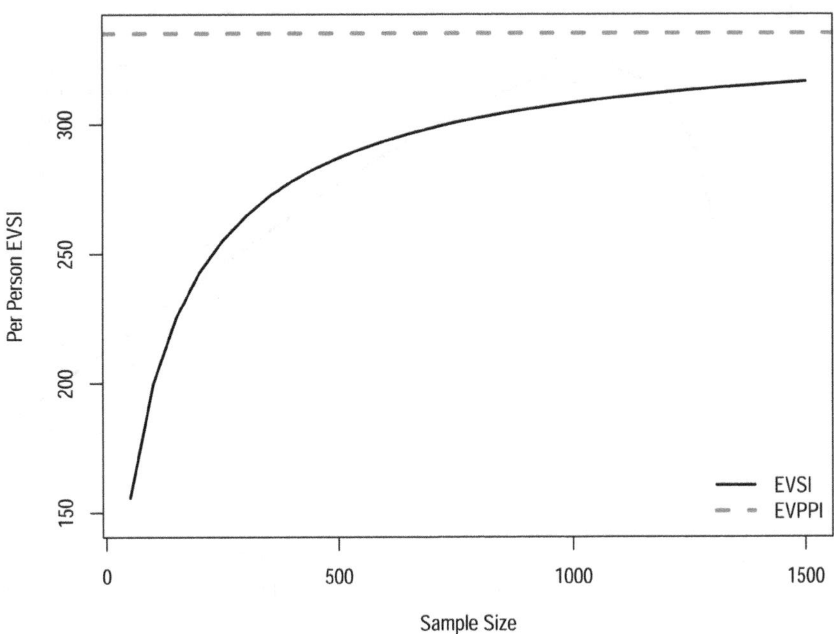

FIGURE 4.4
The EVSI plotted across different values of the study sample size for a specified willingness-to-pay value. The graphic also shows uncertainty bands for the EVSI.

> R code to produce all the plots in this section is linked from `https://chjackson.github.io/voi/articles/book.html`.

4.7.1 Study Sample Size Considerations

EVSI is often estimated for different sample sizes, which can be used to determine the optimal sample size for the proposed study. For a given willingness-to-pay threshold, EVSI across different sample sizes can be presented to demonstrate how quickly EVSI approaches the maximum value, i.e., EVPPI. Figure 4.4 displays EVSI across different sample sizes for a willingness-to-pay of £20000.

It is also possible to present ENBS across different sample sizes (Figure 4.5), conditionally on assumptions about the population size who would be affected by the decision, the cost of the study and a fixed willingness-to-pay

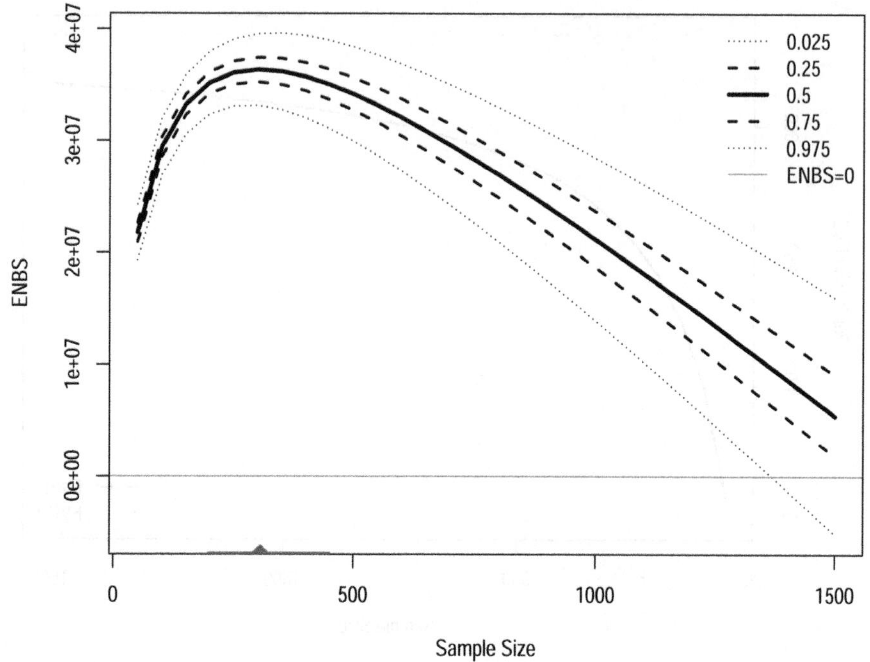

FIGURE 4.5
The ENBS across different sample sizes. The alternative curves represent the uncertainty in the value of the ENBS arising from uncertainty in knowledge about the study costs and the EVSI estimation procedure.

threshold. Figure 4.5 demonstrates the proposed plot for ENBS, which allows the user to identify the optimal sample size for the proposed study. ENBS initially increases, before reaching a peak at an optimal value, and descending when the cost of enrolling additional patients in the trial exceeds the additional value given by the patient. Figure 4.5 also includes interval estimates for ENBS, which allows the user to incorporate uncertainty about the cost of the proposed study, both fixed and per-patient costs. Known values for these costs can also be considered, in which case, the ENBS values are assumed to be known with certainty.

In Figure 4.5, for this example, ENBS is always positive, although it approaches 0 for larger sample sizes. Thus, even though there is an optimal sample size, a trial of any of the considered sizes would have an economic benefit, for this willingness-to-pay and patient population. In the figure, the optimal sample size is marked by a triangle on the x-axis, 305 patients in this example. As can be seen, ENBS is typically relatively flat at the optimal value, which means that there is a relatively large range of sample sizes that would allow for approximately the same value to the decision maker. To highlight this, we have identified the range of sample sizes that would provide a value within 5% of the optimal value; 200 to 450.

In practice, identifying this high value region from the ENBS analysis may be more useful, as it would allow considerations such as feasibility to inform the decision-making around study design. It could also help to reconcile the ENBS analysis with other approaches to study design such as statistical power. For example, if the sample size required to achieve adequate statistical power is within the high value region for from the ENBS analysis, then the proposed research study is feasible from a standard statistical and health economic perspective.

4.7.2 Sensitivity Analyses for Optimal Study Design

Determining the optimal study design using ENBS is challenging as the willingness-to-pay, decision horizon and population to benefit from the decision are not known with certainty. As such, it is recommended that sensitivity analyses are used to explore the impact of these assumptions on the ENBS analysis. In this section, we present two key plots that have been suggested to present the results of these sensitivity analyses.

Firstly, to undertake sensitivity analyses to the willingness-to-pay, conditional on the decision horizon and population to benefit, we can use the Curve of Optimal Sample Size (COSS) (Conti and Claxton 2009; Jutkowitz et al. 2019). This graphical representation plots the optimal sample size of the proposed study against willingness-to-pay. Figure 4.6 displays an example of the COSS, which provides the optimal sample size as a solid line and then displays the interval where the ENBS is within 5% of the optimal value as two dashed lines. In general, as the decision uncertainty increases, close to the ICER for our model, the optimal sample size of the future study increases. The length of the interval where ENBS remains within 5% of the optimal changes substantially over the willingness-to-pay thresholds, indicating that the ENBS curve becomes flatter for specific willingness-to-pay thresholds.

One potential drawback with this representation of the COSS is that there is no indication of whether the optimal sample size results in a study with positive or negative ENBS. Thus, further exploration at the optimal sample size is required to determine whether a study of this size should be performed. Secondly, the COSS is limited by the sample sizes considered in the original EVSI estimation; between 50 and 1,500 in this example, which means that the true optimal sample size may not be identified. Thus, to obtain an accurate plot for the COSS, it is important to consider a wide range of sample sizes in the original EVSI estimation. It is also important to include a relatively fine grid of sample sizes in the original EVSI estimation, as otherwise the COSS does not exhibit the smooth shape seen in Figure 4.6. Finally, the COSS does not currently account for the fact that studies enrolling more participants would take longer to complete and therefore research the number of patients who would benefit from the decision.

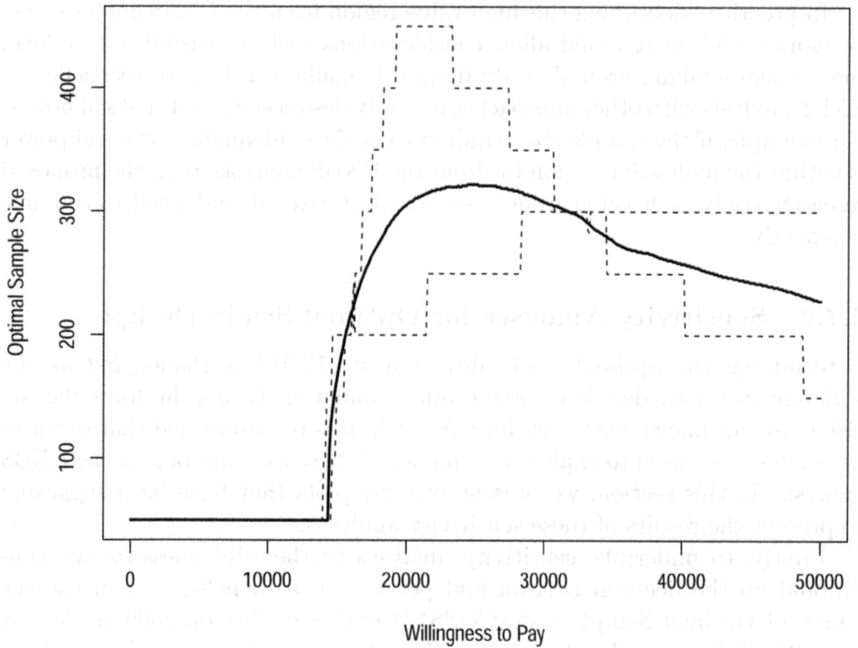

FIGURE 4.6
The optimal sample size of the study of interested plotted against willingness-to-pay threshold. The dashed lines present the interval within which the expected net benefit of sampling remains within 5% of its optimal value.

Figure 4.7 displays another representation of sensitivity analyses for the optimal study design. Specifically, Figure 4.7 displays the probability of a cost-effective trial, where uncertainty arises from imperfect knowledge of study costs. In this plot, we perform a sensitivity analysis to incidence population and time horizon, which are used on the two axes, and then display a heat-map for the probability of a cost-effective trial. Lighter colours represent higher probabilities of a cost-effective trial, with white representing a 100% probability of a cost-effective trial. If study costs are known with certainty, Figure 4.7 would be black and white, with black representing non-cost-effective studies.

Practically, the investigator must specify maximum and minimum plausible values for the decision horizon and the yearly number of people to benefit from the decision and then the user would look for combination of these two parameters that result in a high probability for a cost-effective study. These results are based on a fixed willingness-to-pay and study sample size. Figure 4.7 demonstrates that the study is unlikely to be cost-effective if the expected decision horizon is less than 2 years, but if the decision horizon is near 10 years then the study can go ahead, provided the number of patients affected by the decision exceeds 15,000 per year.

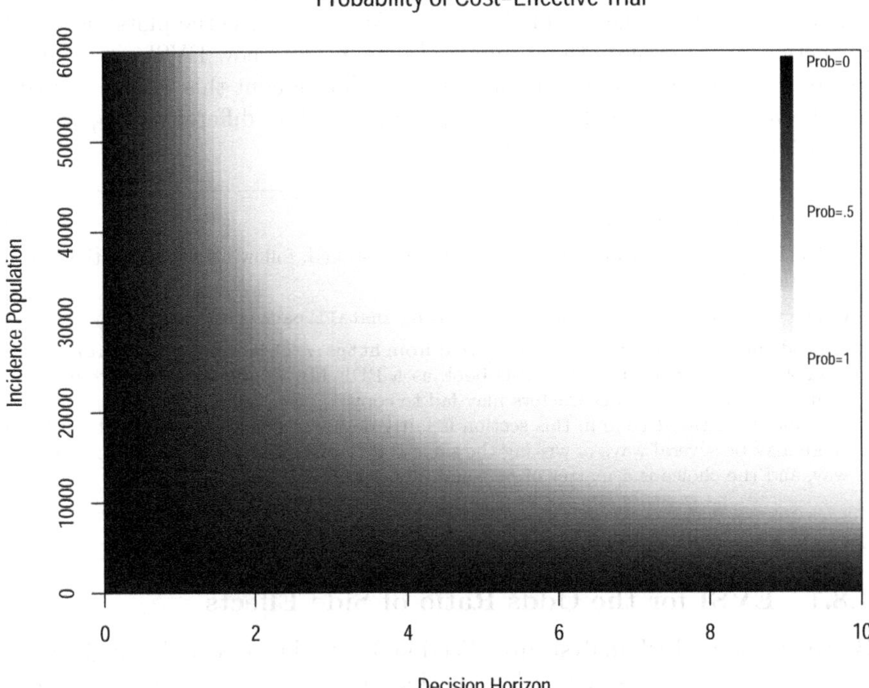

FIGURE 4.7
The probability of a cost-effective trial for different lengths for the time horizon and levels of incidence population. The probability represents the uncertainty in the study costs and EVSI estimates.

Overall, no single plot is able to explore all the key aspects of an ENBS analysis, and these plots should be used in combination to support decision-making around study design. Furthermore, these plots can be explored across different key values; for example, the plot of the probability of a cost-effective trial could be explored for different sample sizes, study follow-up times, outcomes collected and sub-groups targeted, to eliminate study sample designs where the majority of decision horizons and patient population sizes result in a study with negative value. We will explore how these plots can be used in practical study design in the next section, which focuses on calculating EVSI for our case study.

4.8 An EVSI Anaysis: The Chemotherapy Example

In this section, we demonstrate an EVSI analysis using the Chemotherapy case study from Chapter 2. We will present five different study designs to target a range of different model parameters and evaluate which of these designs could provide the highest value for money. We will utilise the plots discussed in Section 4.7 to explore the results and demonstrate how EVSI can be used for research prioritisation and study design. Throughout this section, we will use the `voi` package to calculate EVSI, and clarify how different study designs can be evaluated within the package.

To run the code in this section:

- The Chemotherapy example model should be installed, following the instructions in Section 2.3.5.
- The `voi` R package should be installed, using `install.packages("voi")`.

All code in the book is provided and linked from `https://chjackson.github.io/voi/articles/book.html`. If reading this book as a PDF file, don't paste code from the PDF, since some special characters may fail to copy.

Note that the R code in this section is fairly detailed. Due to the flexibility of R, there may be several ways of writing the same code – often there is no single "correct" way, and the choice is a matter of personal preference.

4.8.1 EVSI for the Odds Ratio of Side Effects

Based on the EVPPI analysis from Section 3.4, we know that the single parameter with the highest EVPPI is the log odds ratio of side effects for the novel treatment compared to standard care. Thus, we begin by considering EVSI for a RCT that aims to update information for this log odds ratio. As in Section 4.5, we plan to undertake a two-arm RCT with 1:1 randomisation between standard care and the novel intervention. The primary outcome will be whether the individual experienced side effects within 1 year of initiating chemotherapy. EVSI for this study could be computed using the built-in study design `"trial_binary"`, which updates information on the log-odds ratio and the baseline probability of side effects.

However, trial data will often only be used to provide information about the relative effect of a treatment. Therefore, we calculate the EVSI of an RCT that only updates information for the log-odds ratio, by using a bespoke study design in `voi`, and the regression-based computation method. The data-generating function is:

```
> OR_datagen_fn <- function(inputs, n = 500){
+   p_side_effects_t1 <- inputs[, "p_side_effects_t1"]
+   p_side_effects_t2 <- inputs[, "p_side_effects_t2"]
+   X1 <- rbinom(length(p_side_effects_t1), n, p_side_effects_t1)
+   X2 <- rbinom(length(p_side_effects_t2), n, p_side_effects_t2)
+   # Create odds ratio as summary statistic
+   OR <- (n - X2) / X2 / ((n - X1) / X1)
+   data_save <- data.frame(OR = OR)
+   return(data_save)
+ }
```

In this case, the data are generated from the probability of side effects for the two interventions, `p_side_effects_t1` and `p_side_effects_t2`, and then summarised by computing the odds ratio. Finally, the odds ratio itself is saved for each of the simulated datasets.

EVSI can then be computed using the GAM regression method in the `evsi` function:

```
> # EVSI calculation using GAM regression.
> evsi_OR <- evsi(outputs = chemotherapy_output,
+                 inputs = m_params,
+                 pars = c("p_side_effects_t1", "p_side_effects_t2"),
+                 n = seq(50, 1500, by = 50),
+                 method = "gam",
+                 datagen_fn = OR_datagen_fn,
+                 par_fn = generate_psa_parameters)
```

This code computes EVSI for sample sizes between 50 and 1500, in intervals of 50, allowing us to evaluate EVSI across different sample sizes and optimise the sample size of the proposed RCT. As we are using a single value to summarise the data, EVSI can be calculated using GAM regression with minimal computational cost. As with all analyses for the Chemotherapy example, EVSI is also computed across a grid of willingness-to-pay thresholds from £0 to £50000 per QALY gained.

Figure 4.8 displays the resulting EVSI estimates, plotted against willingness-to-pay threshold, for the 29 sample sizes considered in this analysis. We can see that EVSI quickly approaches the maximum possible value of EVPPI as sample size increases and that for all willingness-to-pay values between £20000 to £30000 per QALY gained, EVSI is positive for all studies we have considered.

4.8.2 ENBS for the Odds Ratio of Side Effects

To use EVSI for study design, however, we need to compute ENBS, and therefore require an estimate of the fixed and per-person costs of this analysis (c.f., Section 4.6). In this analysis, we assume that the research results will not affect patient care until the study has been completed, thus, we do not include EVSI for the studied individuals in the calculation of research costs. We will also discount the initial patient population to account for the fact that the study will take an estimated 3 years to complete (a more realistic assumption could relate the sample size to the recruitment time for the study and discount initial population deferentially for smaller studies).

The fixed costs for an RCT include staff costs to screen individuals for recruitment, analyst time, protocol and statistical analysis plan development, monitoring, database development and maintenance, and knowledge transfer costs. To make reasonable assumptions about these costs, we assume that this RCT will recruit patients at ten different sites, and each site will need a full-time staff member to support recruitment; assuming these staff are paid

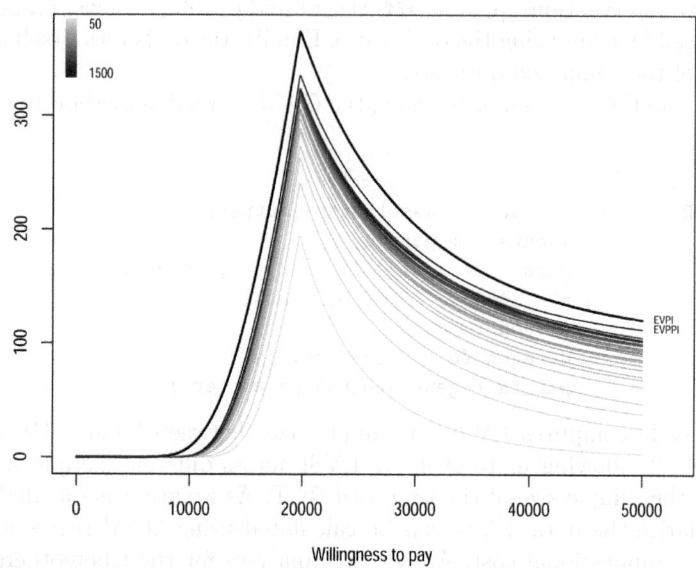

FIGURE 4.8
EVPI, EVPPI and EVSI (grey lines) against willingness-to-pay threshold for a RCT that updates information on the log odds ratio of side effects for the Chemotherapy example. EVSI with smaller sample sizes are plotted in lighter colours.

£40,000, this requires a total staff budget of £1.2 million. In addition to this, analyst and database manager time is budgeted at £55,000 a year but is only required at the beginning and end of the project, which we will assume requires a year of funding. Adding another £5,000 to account for knowledge translation requires a total budget of £1.26 million. To account for potential underestimation in these costs, we specify an upper limit of potential fixed costs as £1.4 million.

For per-person costs, we have already accounted for staff time collecting data and recruiting patients in the fixed costs so these will be smaller. It is common to provide sites with payments per patient to incentivise recruitment and account for additional burden on healthcare providers, which we assume will be £500. In addition, the drug costs should be included, and are either £120 for the standard care, or £1,975 for the novel treatment. This represents an average per person drug cost of £1,047.50. Finally, we need to account for the opportunity cost of randomising an individual to the non-optimal treatment; measured as the expected incremental net benefit between the optimal intervention given current information and the non-optimal intervention. At £20,000 per QALY gained, the incremental net benefit is £26.10, which is £13.05 per patient enrolled in the study, as half the patients receive the novel

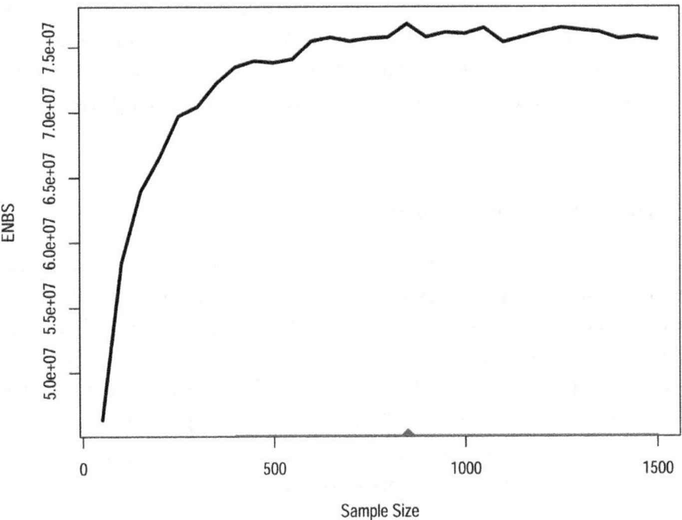

FIGURE 4.9
The Expected Net Benefit of Sampling (ENBS) against sample size for an RCT that updates information on the log odds ratio of side effects for the Chemotherapy example. The curve is not smooth as EVSI is estimated separately for each sample size resulting in Monte Carlo error. The solid line on the axis represents the interval in which ENBS is within 5% of the maximum, highlighted by a triangle.

intervention, which is optimal. Thus, the total assumed per-patient cost is £1,560.55. To account for potential under-estimation, we will specify a maximum potential per person cost of £1,600.

Based on these costs, Figure 4.9 represents the ENBS across different sample sizes for a current population 46,000, discounted at 3.5%, with a decision horizon of 10 years. The optimal sample size for this study is 806 patients per arm, which generates an ENBS of £77 million. The ENBS curve is very flat, seen by a wide interval for the ENBS within 5% of the optimal; 400 to 1,500 patients per arm. Note that this curve for ENBS is not completely smooth. This is because the regression-based method estimates EVSI independently for each sample size. Thus, the regression-based method is subject to Monte Carlo error, which varies between each simulation. The variation in EVSI for high value studies that are close to EVPPI is due to this Monte Carlo simulation error. Note that the moment-matching and Gaussian approximation methods estimate a smooth curve for EVSI and will, thus, result in a smooth curve for ENBS (as seen in Figure 4.5). The estimates of EVSI from the regression-based method could also be smoothly interpolated for sample sizes that are not part of the original estimation set – this procedure is implemented in the `voi` package.

This proposed study is easy to conceptualise, and has a value close to EVPI, making it an efficient study for reducing model uncertainty. However, clinical trials will often collect information on a range of outcomes of interest, not just the primary outcome. Thus, we now consider an RCT that provides information on multiple outcomes, that will be used to inform our decision-analytic model.

4.8.3 Side Effects and Follow-Up

As presented in Section 3.4, one potential RCT would collect information on the follow-up trajectory for all patients that experienced side effects to update information for the log odds ratios of side effects and the transition probabilities in the Markov model. The data-generation process for this proposed study is more complex than for the initial RCT in the previous section. Firstly, we need to generate the number of patients who experience side effects, assuming a recruitment of n patients per arm, using the same data-generation model as in the previous section, based on probabilities π_0 and π_1;

$$X^0_{SE} \sim Binomial(n, \pi_0), \; X^1_{SE} \sim Binomial(n, \pi_1).$$

From this, we define $Y_{SE} = X^0_{SE} + X^1_{SE}$ as the total number of patients experiencing side effects. We then need to simulate the number of patients who are hospitalised, conditionally on the 1-year probability of hospitalisation, and the number of patients who die, conditionally on the 1-year probability of death, within the first year after receiving the intervention. These simulations depend on the 1-year probability of hospitalisation (Γ_1) and death (Γ_2), respectively;

$$Y_{Hosp} \sim Binomial(Y_{SE}, \Gamma_1), \; Y_{dead} \sim Binomial(Y_{Hosp}, \Gamma_2).$$

Finally, as discussed in Section 4.2, the probabilities of transition to recovery from home and in-hospital care are updated by information on the recovery time for patients. In this case, conditionally on the transition probabilities from home and hospital care to recovery, transformed into rates, we generate data points for the number of patients who recovered from home $R_{Home} = X^0_{SE} + X^1_{SE} - Y_{Hosp}$ and the number of patients who recovered from in-hospital care $R_{Hosp} = Y_{Hosp} - Y_{Dead}$ from an exponential distribution.

To generate the data at the individual level, we use the following data-generation function:

```
> # Data generation function
> full_datagen_fn <- function(inputs, n = 500){
+   p_side_effects_t1 <- inputs[, "p_side_effects_t1"]
+   logor_side_effects <- inputs[, "logor_side_effects"]
+   # Odds for side effects for treatment 1
+   odds_side_effects_t1 <- p_side_effects_t1 / (1 - p_side_effects_t1)
+   # Odds for side effects on treatment 2
+   odds_side_effects_t2 <- odds_side_effects_t1 * exp(logor_side_effects)
```

```
+    # Probability of side effects under treatment 2
+    p_side_effects_t2 <- odds_side_effects_t2 / (1 + odds_side_effects_t2)
+    p_hospitalised_total <- inputs[, "p_hospitalised_total"]
+    p_died <- inputs[, "p_died"]
+    lambda_home <- inputs[, "lambda_home"]
+    lambda_hosp <- inputs[, "lambda_hosp"]
+    rate_recover_hosp <- -log(1 -lambda_hosp )
+    rate_recover_home <- -log(1 - lambda_home)
+
+    X1 <- X2 <- X_hosp <- X_dead <- N_recover_home <-
+      N_recover_hospital <- vector("numeric", length = dim(inputs)[1])
+    T_home <- T_hosp <- matrix(NA, nrow = dim(inputs)[1], ncol = 2 * n)
+    for(i in 1:dim(inputs)[1]){
+      # Simulate the number of patients with side effects
+      X1[i] <- rbinom(1, n, p_side_effects_t1[i])
+      X2[i] <- rbinom(1, n, p_side_effects_t2[i])
+
+      # Simulate the number of patients hospitalised
+      X_hosp[i] <- rbinom(1, X1[i] + X2[i], p_hospitalised_total[i])
+      # Simulate the number of patients die
+      X_dead[i] <- rbinom(1, X_hosp[i], p_died[i])
+
+      ## Simulate recovery times for patients
+      N_recover_home[i] <- X1[i] + X2[i] - X_hosp[i]
+      if(N_recover_home[i] > 0){
+        T_home[i, 1:N_recover_home[i]] <- rexp(N_recover_home[i],
+                                               rate_recover_home[i])
+      }
+      N_recover_hospital[i] <- X_hosp[i] - X_dead[i]
+      if(N_recover_hospital[i] > 0){
+        T_hosp[i, 1:N_recover_hospital[i]] <- rexp(N_recover_hospital[i],
+                                                   rate_recover_hosp[i])
+      }
+
+    }
+
+    data_save_dat <- data.frame(cbind(X1 = X1, X2 = X2,
+                                      X_hosp = X_hosp, X_dead = X_dead,
+                                      N_recover_home = N_recover_home,
+                                      N_recover_hospital =
+                                        N_recover_hospital,
+                                      T_home = T_home, T_hosp = T_hosp))
+    return(data_save_dat)
+ }
```

In this function, the data for each simulation are saved as a single row, meaning that the final output from this function has S rows (the number of PA simulations) and $4n+6$ columns. This allows for the possibility of all individuals having a recovery time from hospital or at home. In all simulations, there will be some columns of this data frame with missing values, which correspond to the individuals who did not require a simulated recovery time because either they (i) did not experience side effects or (ii) did not recover from the state of interest.

Alternatively, for the regression-based method, the data must be summarised into low-dimensional summary statistics and saved in a data frame with a relatively small number of columns. In this example, we reduce the number of outcomes to five, the odds ratio, the proportion of patients hospitalised and who died and the sum of the recovery times for patients receiving care at home and in hospital. Thus, the data-generation function for the regression-based method can be provided as follows;

```
> full_datagen_fn_RB <- function(inputs, n = 500){
+   # Generate the data
+   data_sim <- full_datagen_fn(inputs, n)
+   T_home <- data_sim[, (1:(2 * n)) + 6]
+   T_hosp <- data_sim[, (6 + 2 * n) + (1:n)]
+   OR <- (n - data_sim$X2) / data_sim$X2 /
+     ((n - data_sim$X1) / data_sim$X1)
+   p_hosp <- data_sim$X_hosp / (data_sim$X1 + data_sim$X2)
+   p_dead <- data_sim$X_dead / data_sim$X_hosp
+   T_home_sum <- rowSums(T_home, na.rm = TRUE)
+   T_hosp_sum <- rowSums(T_hosp, na.rm = TRUE)
+
+   data_save_dat <- data.frame(cbind(OR = OR,
+                                     p_hosp = p_hosp, p_dead = p_dead,
+                                     T_home_sum = T_home_sum,
+                                     T_hosp_sum = T_hosp_sum))
+   return(data_save_dat)
+ }
```

In this function, we use the individual level data-generation function to generate the data before summarising. The recovery times for the patients who recover at home and in hospital are saved in $2n$ columns so these must be extracted before the sum of the recovery times is computed.

4.8.3.1 EVSI Calculation Using the Regression-Based Method

As there are five outcomes in the study, the regression-based method for calculating EVSI involves a regression on five summary statistics. The default `gam` method will be inefficient in that case (c.f., Section 3.3.5). Instead, we calculate EVSI using the MARS method for constructing a GAM, by setting `method = "earth"` in the call to `evsi()`;

```
> # EVSI calculation using MARS regression - large number of parameters.
> evsi_OR_allout <- evsi(outputs = chemotherapy_output,
+                        inputs = m_params,
+                        pars = c("p_side_effects_t1",
+                                 "logor_side_effects",
+                                 "p_hospitalised_total", "p_died",
+                                 "lambda_home", "lambda_hosp"),
+                        n = seq(50, 1500, by = 50),
+                        method = "earth",
+                        datagen_fn = full_datagen_fn_RB,
+                        par_fn = generate_psa_parameters)
```

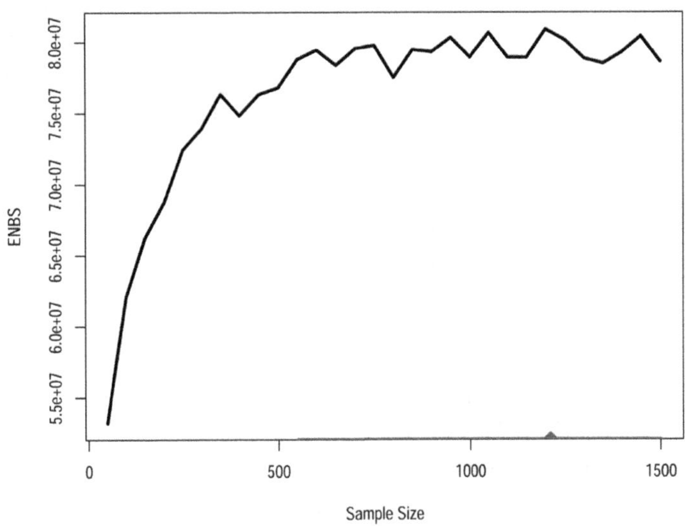

FIGURE 4.10
The Expected Net Benefit of Sampling for against sample size for an RCT that updates information on the log odds ratio of side effects and all transition probabilities for the Chemotherapy example. The curve is not smooth as EVSI is estimated separately for each sample size resulting in Monte Carlo error. The solid line on the axis represents the interval in which ENBS is within 5% of the maximum, highlighted by a triangle.

Assuming that the cost of this study is the same as for the previous study, as we provided the costs for running a clinical trial and the additional cost of the additional outcomes are minimal, Figure 4.10 displays the ENBS across sample size for this RCT. This study has a higher optimal sample size compared to the study that only collects data on the log odds ratio at 1211. This is because additional value is generated from the information for the transition probabilities and thus larger sample sizes are needed before the additional per-patient cost exceeds the value generated from each study participant. The range of sample sizes within 5% of this optimal remains similar at 550 to 1,500. The ENBS at this sample size increases from the single study to is £81 million, representing an increase of approximately £4 million or 5% higher.

Based on the analysis from the regression-based method, we can also plot the COSS (Figure 4.11). In this instance, we see that the optimal sample size remains relatively consistent for willingess to pay thresholds above £20,000 per QALY gained. This is particularly true for the interval that represents sample sizes that are within 5% of the optimal value. The curve in Figure 4.11 is not smooth like the one presented in Figure 4.6. This is again due to the Monte

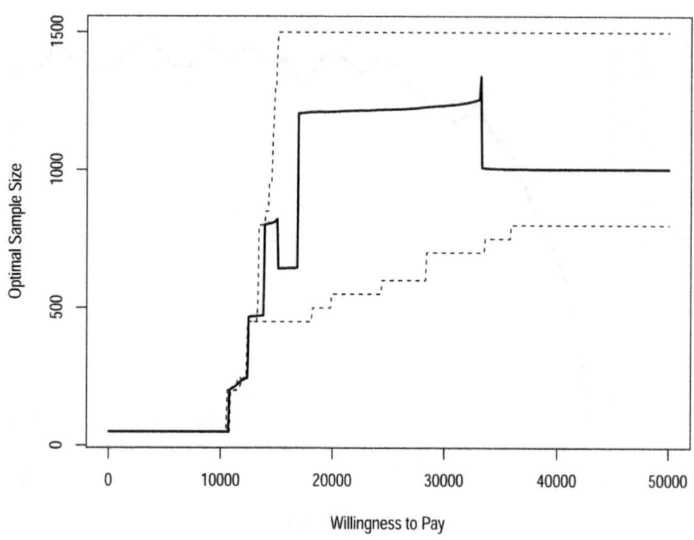

FIGURE 4.11
The Curve of Optimal Sample Size for a study collecting information on the log odds ratio of side effects and follow up information for those experiencing side effects for the Chemotherapy example. The dashed lines represents the sample size that maintains the ENBS within 5% of the maximum value.

Carlo error in the regression-based method EVSI estimation. As the "optimal" sample size is highly sensitive to the exact value estimated for EVSI, the COSS selects the sample sizes that are associated with the higher EVSI values. To avoid these issues, we can consider using the moment-matching method for EVSI estimation, which would produce a smoother fitted function of sample size.

EVSI Calculation Using the Moment-Matching Method

Note the code for this calculation is more advanced, and some knowledge of Bayesian inference and JAGS is assumed. A brief introduction to JAGS is provided in Section 4.5.2.3.

To calculate EVSI using the moment-matching method for this example, we need to create an analysis function that updates all the parameters based on the observed data. In this example, we use the Bayesian software JAGS to simulate from the posterior for the model parameters.

```
> full_datagen_fn <- function(inputs, n = 500){
+   p_side_effects_t1 <- inputs[, "p_side_effects_t1"]
```

```
+   logor_side_effects <- inputs[, "logor_side_effects"]
+   # Odds for side effects for treatment 1
+   odds_side_effects_t1 <- p_side_effects_t1 / (1 - p_side_effects_t1)
+   # Odds for side effects on treatment 2
+   odds_side_effects_t2 <- odds_side_effects_t1 * exp(logor_side_effects)
+   # Probability of side effects under treatment 2
+   p_side_effects_t2 <- odds_side_effects_t2 / (1 + odds_side_effects_t2)
+   p_hospitalised_total <- inputs[, "p_hospitalised_total"]
+   p_died <- inputs[, "p_died"]
+   lambda_home <- inputs[, "lambda_home"]
+   lambda_hosp <- inputs[, "lambda_hosp"]
+   rate_recover_hosp <- -log(1 -lambda_hosp )
+   rate_recover_home <- -log(1 - lambda_home)
+
+   X1 <- X2 <- X_hosp <- X_dead <- N_recover_home <-
+     N_recover_hospital <- vector("numeric", length = dim(inputs)[1])
+   T_home <- T_hosp <- matrix(NA, nrow = dim(inputs)[1], ncol = 2 * n)
+   for(i in 1:dim(inputs)[1]){
+     # Simulate the number of patients with side effects
+     X1[i] <- rbinom(1, n, p_side_effects_t1[i])
+     X2[i] <- rbinom(1, n, p_side_effects_t2[i])
+
+     # Simulate the number of patients hospitalised
+     X_hosp[i] <- rbinom(1, X1[i] + X2[i], p_hospitalised_total[i])
+     # Simulate the number of patients die
+     X_dead[i] <- rbinom(1, X_hosp[i], p_died[i])
+
+     ## Simulate recovery times for patients
+     N_recover_home[i] <- X1[i] + X2[i] - X_hosp[i]
+     if(N_recover_home[i] > 0){
+       T_home[i, 1:N_recover_home[i]] <- rexp(N_recover_home[i],
+                                              rate_recover_home[i])
+     }
+     N_recover_hospital[i] <- X_hosp[i] - X_dead[i]
+     if(N_recover_hospital[i] > 0){
+       T_hosp[i, 1:N_recover_hospital[i]] <- rexp(N_recover_hospital[i],
+                                                  rate_recover_hosp[i])
+     }
+
+   }
+
+   data_save_dat <- data.frame(cbind(X1 = X1, X2 = X2,
+                                     X_hosp = X_hosp, X_dead = X_dead,
+                                     N_recover_home = N_recover_home,
+                                     N_recover_hospital =
+                                       N_recover_hospital,
+                                     T_home = T_home, T_hosp = T_hosp))
+   return(data_save_dat)
+ }
> # Analysis function based on JAGS
> full_analysis_fn <- function(data, args, pars){
+   ## Format Data - Adjust for 0 recovery times
+   T_home <- NA
+   if(data$N_recover_home > 0){
```

```
+     T_home <- as.numeric(as.matrix(data[, (1:data$N_recover_home) + 6]))
+   }
+
+   T_hosp <- NA
+   if(data$N_recover_hospital > 0){
+     T_hosp <- as.vector(as.matrix(data[,
+                                        (6 + 2 * args$n) +
+                                          (1:data$N_recover_hospital)]))
+   }
+
+   # Create the data list for JAGS
+   data_jags <- list(X1 = data$X1,
+                     X2 = data$X2,
+                     X_hosp = data$X_hosp,
+                     X_dead = data$X_dead,
+                     T_home = T_home,
+                     T_hosp = T_hosp,
+                     N_recover_home = ifelse(data$N_recover_home > 0,
+                                            data$N_recover_home,
+                                            1),
+                     N_recover_hosp = ifelse(data$N_recover_hospital > 0,
+                                            data$N_recover_hospital,
+                                            1),
+                     n = args$n,
+                     n_side_effects = args$n_side_effects,
+                     n_patients = args$n_patients,
+                     logor_side_effects_mu = args$logor_side_effects_mu,
+                     logor_side_effects_sd = args$logor_side_effects_sd,
+                     p_recovery_home_alpha =
+                       betaPar(args$p_recovery_home_mu,
+                               args$p_recovery_home_sd)$alpha,
+                     p_recovery_home_beta =
+                       betaPar(args$p_recovery_home_mu,
+                               args$p_recovery_home_sd)$beta,
+                     p_recovery_hosp_alpha =
+                       betaPar(args$p_recovery_hosp_mu,
+                               args$p_recovery_hosp_sd)$alpha,
+                     p_recovery_hosp_beta =
+                       betaPar(args$p_recovery_hosp_mu,
+                               args$p_recovery_hosp_sd)$beta,
+                     n_died = args$n_died,
+                     n_hospitalised = args$n_hospitalised)
+
+   LogOR_addoutcomes_trial <-
+     "model{
+
+     ## Models for the data
+     X1 ~ dbin(p_side_effects_t1, n)
+     X2 ~ dbin(p_side_effects_t2, n)
+
+     X_hosp ~ dbinom(p_hospitalised_total, X1 + X2)
+     X_dead ~ dbin(p_died, X_hosp)
+
+     rate_recover_home <- -log(1 - lambda_home)
```

```
+     rate_recover_hosp <- -log(1 - lambda_hosp)
+
+     for(i in 1:N_recover_home){
+       T_home[i] ~ dexp(rate_recover_home)
+     }
+
+     for(i in 1:N_recover_hosp){
+       T_hosp[i] ~ dexp(rate_recover_hosp)
+     }
+
+     # Probability of side effects under treatment 1
+     p_side_effects_t1 ~ dbeta(1 + n_side_effects,
+                               1 + n_patients - n_side_effects)
+
+     # Log odds of side effects on treatment 2
+     logor_side_effects ~ dnorm(logor_side_effects_mu,
+       logor_side_effects_sd)
+     # Odds of side effects on treatment 1
+     odds_side_effects_t1 <- p_side_effects_t1 / (1 - p_side_effects_t1)
+     # Odds for side effects on treatment 2
+     odds_side_effects_t2 <- odds_side_effects_t1 *
+       exp(logor_side_effects)
+
+     # Probability of side effects under treatment 2
+     p_side_effects_t2    <- odds_side_effects_t2 /
+       (1 + odds_side_effects_t2)
+
+     ## Variables to define transition probabilities
+     # Probability that a patient is hospitalised over the time horizon
+     p_hospitalised_total ~ dbeta(1 + n_hospitalised,
+                                  1 + n_side_effects - n_hospitalised)
+     # Probability that a patient dies over the time horizon given they
+     # were hospitalised
+     p_died ~ dbeta(1 + n_died, 1 + n_hospitalised - n_died)
+     # Lambda_home: Conditional probability that a patient recovers
+     # considering that they are not hospitalised
+     lambda_home ~ dbeta(p_recovery_home_alpha, p_recovery_home_beta)
+     # Lambda_hosp: Conditional probability that a patient recovers
+     # considering that they do not die
+     lambda_hosp ~ dbeta(p_recovery_hosp_alpha, p_recovery_hosp_beta)
+   }
+   "
+
+   filein <- file.path(tempdir(),fileext="datmodel.txt")
+   cat(LogOR_addoutcomes_trial, file=filein)
+
+   # Perform the MCMC simulation with JAGS.
+   bugs.data <- jags(
+     data =  data_jags,
+     parameters.to.save = pars,
+     model.file = filein,
+     n.chains = 1,
+     n.iter = args$n.iter,
+     n.thin = 1,
```

```
+     n.burnin = 250, progress.bar = "none")
+
+   data_extract <- bugs.data$BUGSoutput$sims.matrix
+   logor_side_effects <- data_extract[, "logor_side_effects"]
+   p_hosp_total <- data_extract[, "p_hospitalised_total"]
+   p_died <- data_extract[, "p_died"]
+   lambda_home <- data_extract[, "lambda_home"]
+   lambda_hosp <- data_extract[, "lambda_hosp"]
+
+   return(data.frame(logor_side_effects = logor_side_effects,
+                     p_hospitalised_total= p_hosp_total,
+                     p_died = p_died,
+                     lambda_home = lambda_home,
+                     lambda_hosp = lambda_hosp))
+ }
```

In this function, we load all the generated data and the values that define the priors for the parameters in a data list `data_jags`. The values to define the priors for the parameters should be loaded in the `args` function of the analysis function, for example `args$n_side_effects` loads the number of side effects in the original study, which defined the PA distribution for our decision-analytic model. We then use the `jags` `R` function to sample from the posterior distribution that is defined by the JAGS model code. Note that we reload the PA distribution for the probability of side effects for the standard care, which ensures that the data from the study do not update the baseline risk of side effects as these are not updated through the Bayesian model.

Once the analysis function has been defined, EVSI can be computed using the moment-matching method. Firstly, we must define a list that contains all of the inputs that are required for the analysis function:

```
> analysis_args <- list(n_side_effects = n_side_effects,
+                       n_patients = n_patients,
+                       n = 50,
+                       logor_side_effects_mu = logor_side_effects_mu,
+                       logor_side_effects_sd = logor_side_effects_sd,
+                       betaPar = betaPar,
+                       p_recovery_home_mu = p_recovery_home_mu,
+                       p_recovery_home_sd = p_recovery_home_sd,
+                       p_recovery_hosp_mu = p_recovery_hosp_mu,
+                       p_recovery_hosp_sd = p_recovery_hosp_sd,
+                       n.iter = 5250,
+                       n_died = n_died,
+                       n_hospitalised = n_hospitalised,
+                       p_side_effects_t1 = m_params$p_side_effects_t1)
```

Based on this list and the key functions defined above, EVSI can be computed:

```
> # EVSI calculation using the momemt matching method.
> evsi_OR_allout_MM <- evsi(outputs = chemotherapy_output,
+                           inputs = m_params,
```

```
+                           pars = c("logor_side_effects",
+                                    "p_hospitalised_total", "p_died",
+                                    "lambda_home", "lambda_hosp"),
+                           pars_datagen = c("p_side_effects_t1",
+                                            "logor_side_effects",
+                                            "p_hospitalised_total",
+                                            "p_died",
+                                            "lambda_home", "lambda_hosp"),
+                           n = seq(50, 1500, by = 50),
+                           method = "mm",
+                           datagen_fn = full_datagen_fn,
+                           model_fn = calculate_costs_effects,
+                           analysis_args = analysis_args,
+                           analysis_fn = full_analysis_fn,
+                           par_fn = generate_psa_parameters,
+                           npreg_method = "earth")
```

Within this code, note that the parameters used to generate the data (`pars_datagen`) and the parameters updated by the data (`pars`) are different as the probability of side effects under the standard care is not updated. Additionally, note that to use the moment-matching method, the EVPPI for the parameters of interest must be computed. As the number of key parameters is greater than 4, it can be challenging to compute EVPPI using the default `gam` procedure, and so `npreg_method` argument is supplied to indicate that EVPPI should be computed using MARS regression instead.

Figure 4.12 displays the ENBS across different sample sizes and the COSS for the side effects and follow up RCT calculated using the moment-matching method. As expected, the ENBS curve is smooth with respect to sample size, which leads to an alternative optimal sample size of 1,397 patients. This is higher than the 1,211 patients identified using the regression-based method. However, the range of values within 5% of this optimal value is 550 to 1,500, which is the same range observed for the regression-based method. This demonstrates that the exact optimal sample size can be dependent on the method used to compute EVSI but the range of valuable sample sizes is likely to be more stable. Thus, sample size determination should also consider external considerations including feasibility of recruitment and other stakeholder perspectives.

For comparison, we have included the ENBS for the regression-based method in Figure 4.12 as a dotted line. From this, we can see that the regression-based method estimates higher values for the ENBS compared to the moment-matching method, leading to an optimal ENBS of £77 million. As both methods are approximate, it is unclear which of these estimates are closer to the true value of EVSI. In this case, this disagreement would probably not affect any trial design decision resulting from the ENBS analysis, so it is not useful to determine which method gives the more accurate ENBS estimate. However, in settings where the decision would be different, i.e., where one method provides positive ENBS and the other negative, we

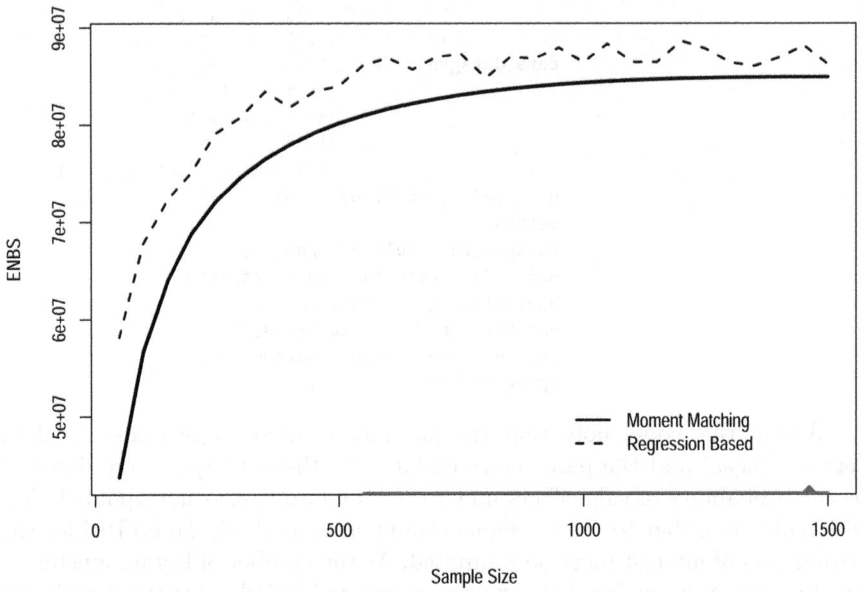

FIGURE 4.12
The ENBS plotted across different sample sizes for an RCT investigating the probability of side effects and transition probabilities for individuals who experience side effects, calculated using the regression based method (dashed line) and the moment matching method (solid line). The solid line on the axis represents the interval in which ENBS is within 5% of the maximum, highlighted by a triangle.

might want to obtain more precise EVSI estimates, for example, by using more samples in the PA dataset for the regression-based method, or by increasing the number of nested simulations or the number of posterior simulations for the moment-matching method.

Finally, the COSS calculated with the moment-matching method (Figure 4.13) is smoother than that calculated using the regression-based method and indicates that the optimal sample size is 1,500 patients for willingness-to-pay thresholds over £25,000 and 0 for willingness-to-pay below around £11,000. The lower bound of the 5% interval broadly increases as the willingness-to-pay threshold increases.

As a final sensitivity analysis, we consider the probability of a cost-effective trial for a willingness-to-pay of £20,000 per QALY gained and a sample size of 1,020 (Figure 4.14), calculated using the regression-based (a) and moment-matching method (b). These two plots are very similar and demonstrate that provided the decision horizon is above two and the population to benefit is

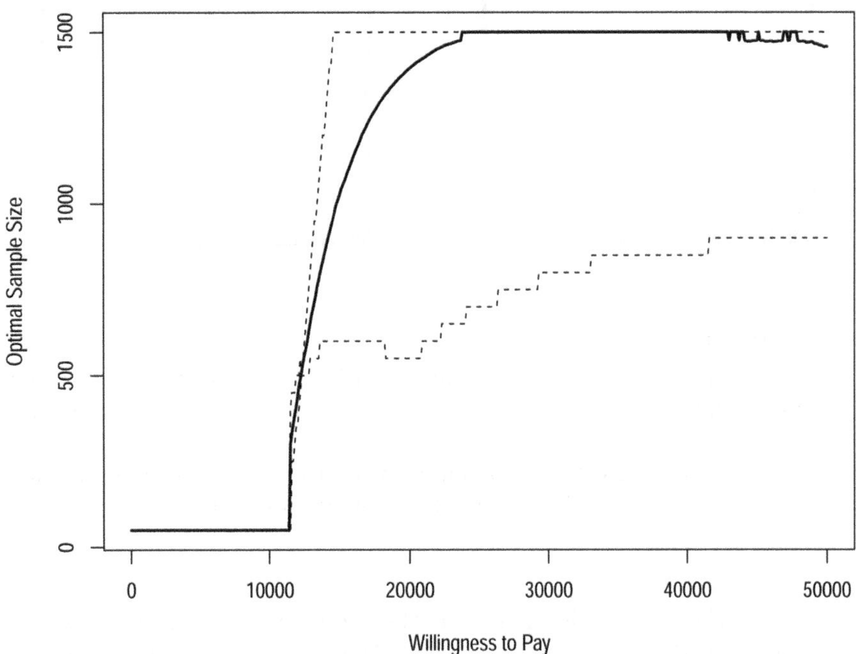

FIGURE 4.13
The Curve of Optimal Sample Size for an RCT investigating the probability of side effects and transition probabilities for individuals who experience side effects, calculated using the moment-matching method. The dashed lines represents the sample size that maintains the ENBS within 5% of the maximum value.

above 10,000, an RCT focusing on the odds ratio of side effects with the two treatments and the long-term follow-up is likely to be cost-effective.

4.8.4 Utilities

Having identified that an RCT is a potentially valuable study, particularly with sample sizes between 550 and 1,500 per arm, we now consider whether alternative study designs have the potential to offer value for decision makers. While the log odds ratio of side effects was associated with the highest value of information, the utilities (i.e., the parameters measuring individuals' preferences between health states) were also associated with a high EVPPI, at approximately a third of the value of the log odds ratio. Thus, we now consider the potential value of a survey enrolling participants to understand the utility of the three key health states, home care, hospital care and recovery.

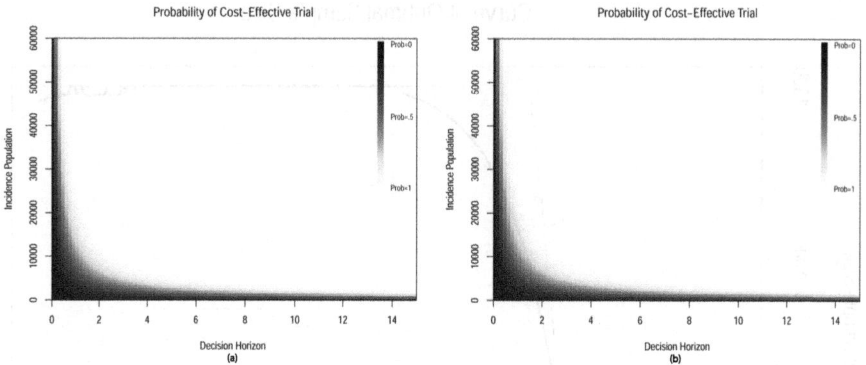

FIGURE 4.14
The probability of a cost-effective trial for an RCT investigating the probability of side effects and transition probabilities for individuals who experience side effects for a willingness-to-pay of £20,000 per QALY gained and a sample size of 1,080. Sub-figure (a) is computed using the regression based method and sub-figure (b) is calculated using the moment-matching method.

> Again, the code for the analyses in this section is fairly advanced: some knowledge of Bayesian inference and JAGS is assumed! A brief introduction to JAGS is provided in Section 4.5.2.3.

Section 4.5.2 presents the code to compute EVSI for a study that updates information for the in-hospital utility. Thus, we extend this code to include utility measurements for the three health states. We assume that study participants will be asked to value a single state that they are currently experiencing, i.e., participants who are experiencing side effects at home will be asked to provide their current utility. This means that we can generate independent utility values when accounting for individual level variation using normal distributions, truncated above by 1. An alternative assumption would be that participants will value multiple health states, in which case, we would need to account for correlation between the responses for a specific individual. For simplicity, we will also assume that the investigators are targeting the same number of utility measurements for each state, resulting in a sample size of $3n$, where n is the number of participants that value each health state.

As in Section 4.5.2, we require an estimate of the standard deviation that quantifies individual-level variation in the utility outcomes. Similar to above, we account for uncertainty in this parameter by including a sample from its uncertainty distribution in the `m_params` matrix. To create the distribution for the individual-level data, we assume that the mean of the individual-level utility distribution is then equal to the mean utility for the three states of interest. Thus, the full data-generating function for the moment-matching method is

```
> # Data generation function - individual data (for other EVSI methods)
> utility_datagen_fn_indiv <- function(inputs, n = 500){
+   # Load the data
+   X_home_care <- X_hospital <-
+     X_recovery <- matrix(NA, nrow = dim(inputs)[1], ncol = n[1])
+   for(i in 1:nrow(inputs)){
+     set.seed(123 + i)
+     m_recovery <- inputs[i, "u_recovery"]
+     m_home_care <- inputs[i, "u_home_care"]
+     m_hospital <- inputs[i, "u_hospital"]
+     sd_recovery <- inputs[i, "sd_iid_recovery"]
+     sd_home_care <- inputs[i, "sd_iid_home_care"]
+     sd_hospital <- inputs[i, "sd_iid_hospital"]
+     # Simulate the costs
+     X_recovery[i, ] <- truncnorm::rtruncnorm(n[1], mean = m_recovery,
+                                              sd = sd_recovery,
+                                              a = -Inf, b = 1)
+     X_home_care[i, ] <- truncnorm::rtruncnorm(n[1], mean = m_home_care,
+                                               sd = sd_home_care,
+                                               a = -Inf, b = 1)
+     X_hospital[i, ] <- truncnorm::rtruncnorm(n[1], mean = m_hospital,
+                                              sd = sd_hospital,
+                                              a = -Inf, b = 1)
+   }
+   data_save_dat <- data.frame(cbind(X_recovery = X_recovery,
+                                     X_home_care = X_home_care,
+                                     X_hospital = X_hospital))
+   return(data_save_dat)
+ }
```

This data-generating function results in a matrix with $3n$ columns, one datapoint generated for each participant in the study. To calculate EVSI, we also need to define an analysis function for the data:

```
> # Analysis function based on JAGS
> utility_analysis_fn <- function(data, args, pars){
+   # Create the data list for JAGS
+   recovery_idx <- (1:args$n)
+   home_care_idx <- args$n + (1:args$n)
+   hospital_idx <- 2*args$n + (1:args$n)
+
+   data_jags <- list(X_recovery =
+                       as.vector(as.matrix(data[, recovery_idx])),
+                     X_home_care =
+                       as.vector(as.matrix(data[, home_care_idx])),
+                     X_hospital =
+                       as.vector(as.matrix(data[, hospital_idx])),
+                     n = args$n,
+                     alpha_recovery = args$betaPar(
+                       args$u_recovery_mu,
+                       args$u_recovery_sd
```

```
+                     )$alpha,
+                     beta_recovery = args$betaPar(
+                       args$u_recovery_mu,
+                       args$u_recovery_sd
+                     )$beta,
+                     alpha_home_care = args$betaPar(
+                       args$u_home_care_mu,
+                       args$u_home_care_sd
+                     )$alpha,
+                     beta_home_care = args$betaPar(
+                       args$u_home_care_mu,
+                       args$u_home_care_sd
+                     )$beta,
+                     alpha_hospital = args$betaPar(
+                       args$u_hospital_mu,
+                       args$u_hospital_sd
+                     )$alpha,
+                     beta_hospital = args$betaPar(
+                       args$u_hospital_mu,
+                       args$u_hospital_sd
+                     )$beta)
+
+   trial <- "model{
+     for(i in 1:n){
+       X_recovery[i] ~ dnorm(u_recovery, tau_recovery) T(, 1)
+       X_home_care[i] ~ dnorm(u_home_care, tau_home_care) T(, 1)
+       X_hospital[i] ~ dnorm(u_hospital, tau_hospital) T(, 1)
+     }
+
+     u_recovery ~ dbeta(alpha_recovery, beta_recovery)
+     u_home_care ~ dbeta(alpha_home_care, beta_home_care)
+     u_hospital ~ dbeta(alpha_hospital, beta_hospital)
+
+     sd_recovery ~ dunif(0.000001, 0.15)
+     sd_home_care ~ dunif(0.00001, 0.6)
+     sd_hospital ~ dunif(0.00001, 0.4)
+
+     tau_recovery <- 1 / sd_recovery ^ 2
+     tau_home_care <- 1 / sd_home_care ^ 2
+     tau_hospital <- 1 / sd_hospital ^ 2
+   }
+   "
+
+   filein <- file.path(tempdir(),fileext="datmodel.txt")
+   cat(trial, file=filein)
+
+   # Perform the MCMC simulation with OpenBUGS.
+   # Close OpenBUGS once it has finished (if debug is set to TRUE)
+   bugs.data <- jags(
+     data =  data_jags,
+     parameters.to.save = pars,
+     model.file = filein,
+     n.chains = 1,
+     n.iter = args$n.iter,
```

```
+       n.thin = 1,
+       n.burnin = 250, progress.bar = "none")
+
+   u_recovery <- bugs.data$BUGSoutput$sims.matrix[, "u_recovery"]
+   u_home_care <- bugs.data$BUGSoutput$sims.matrix[, "u_home_care"]
+   u_hospital <- bugs.data$BUGSoutput$sims.matrix[, "u_hospital"]
+
+   return(data.frame(u_recovery = u_recovery,
+                     u_home_care = u_home_care,
+                     u_hospital = u_hospital))
+ }
```

In this function, the data for the JAGS model are loaded by creating a `data_jags` list and contain all the parameters required to define the prior distributions for the mean utilities. Secondly, the truncated normal distribution in JAGS is specified using the `T(, 1)` function, which truncates the normal distribution below 1. Uncertainty distributions are then defined for the individual-level utility standard deviations, and added as columns of the `m_params` data frame.

```
> # Add individual level variation parameters
> m_params$sd_iid_recovery <- runif(nrow(m_params), 0.000001, 0.15)
> m_params$sd_iid_home_care <- runif(nrow(m_params), 0.000001, 0.6)
> m_params$sd_iid_hospital <- runif(nrow(m_params), 0.000001, 0.4)
```

In this example, we use uniform distributions with limits 0.000001, 0.15, 0.00001, 0.6 and 0.00001, 0.4 for recovery, home care and hospital care, respectively. Finally, EVSI can be computed using the `evsi` function:

```
> # EVSI calculation using the momemt matching method.
> evsi_utility <- evsi(outputs = chemotherapy_output,
+                      inputs = m_params,
+                      pars = c("u_recovery", "u_home_care", "u_hospital"),
+                      pars_datagen = c("u_recovery", "u_home_care",
+                                       "u_hospital",
+                                       "sd_iid_recovery",
+                                       "sd_iid_home_care",
+                                       "sd_iid_hospital"),
+                      n = seq(20, 300, by = 10),
+                      method = "mm",
+                      datagen_fn = utility_datagen_fn_indiv,
+                      model_fn = calculate_costs_effects,
+                      analysis_args = list(n = 20,
+                                           betaPar = betaPar,
+                                           u_recovery_mu = u_recovery_mu,
+                                           u_recovery_sd = u_recovery_sd,
+                                          u_home_care_mu = u_home_care_mu,
+                                          u_home_care_sd = u_home_care_sd,
+                                           u_hospital_mu = u_hospital_mu,
+                                           u_hospital_sd = u_hospital_sd,
+                                           n.iter = 5000),
+                      analysis_fn = utility_analysis_fn,
+                      par_fn = generate_psa_parameters,
+                      Q = 50)
```

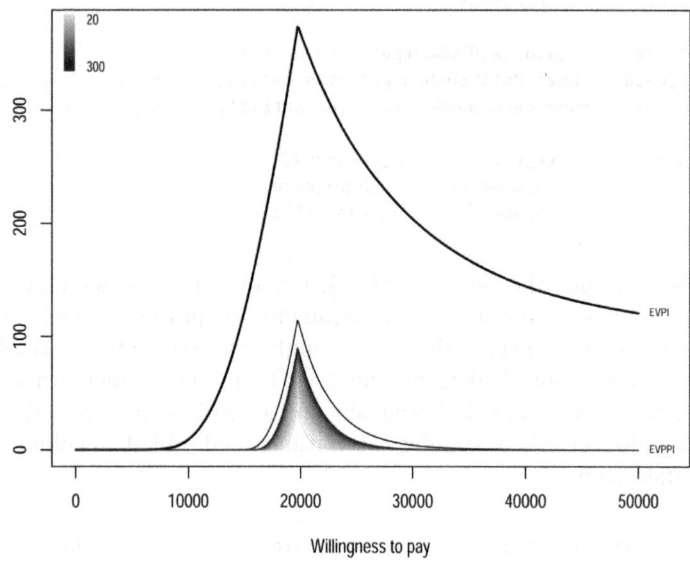

FIGURE 4.15
The EVSI plotted across willingness-to-pay for a study investigating the utilities of the three key health states in the Chemotherapy example.

where `betaPar` is a function that translates the mean and standard deviation values that define the PA distributions for the mean utilities into the parameters for the beta distribution. In this code, we compute EVSI for sample sizes between 20 and 300 in increments of 10. This is because we assume 300 patients per state is the maximum potential recruitment for this study. Figure 4.15 displays EVSI plotted in terms of the willingness-to-pay for the proposed study investigating utilities. The maximum sample size of 300 patients provides approximately the maximum value for the study, which is substantially below EVPPI. In general, the study is unlikely to provide any value at willingness-to-pay thresholds above £25,000 per QALY gained.

4.8.4.1 Expected Net Benefit of Sampling for a Survey to Measure Utilities

To determine the potential cost of undertaking this proposed study to explore ENBS, we consider, as usual, the fixed and per-person costs of undertaking this study. For the fixed cost, we include designing the survey, monitoring for eligible participants and disseminating the results. We assume that the survey will take one person 6 months to design, and the study will enrol for 1 year, with the same individual monitoring for participants, before the analysis takes the remaining 6 months. Based on the clinical researcher salary discussed

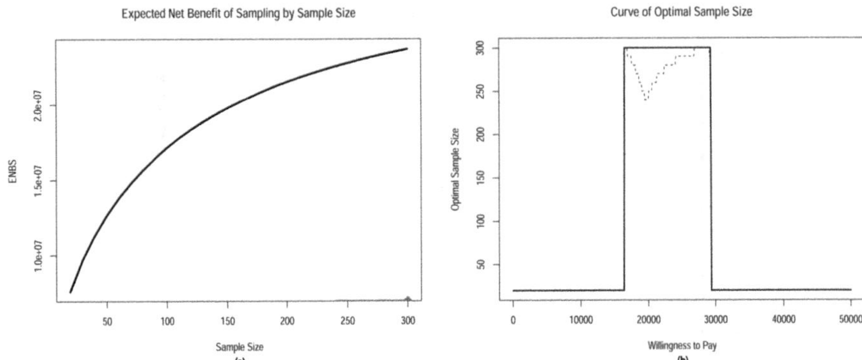

FIGURE 4.16
(a) The ENBS plotted across different sample sizes for a survey collecting information on the utility of the three health states in the Chemotherapy example. (b) The Curve of Optimal Sample Size for the same study. For both sub-figures, EVSI is calculated using the moment-matching method. The dashed line represents the sample size that maintains the ENBS within 5% of the maximum value.

previously, this could be around £40,000 per year to a total of £80,000. To account for dissemination costs, we add another £10,000 to a total of £90,000. For the per-person cost, we assume that the survey will take 10 minutes to complete and will use an EQ-5D questionnaire, which will then be mapped to a utility score. This is substantially faster than other potential utility tasks such as time trade off. It will also exert a lower burden on the patients who may be experiencing severe side effects in hospital. In addition to this, participants will be given a gift to thank them for participation in the study, and the travel costs will be reimbursed for those who have recovered and are receiving care at home. Thus, we assume that 10 minutes of recruitment time costs approximately £220, and participants will receive a £100 gift, with a maximum allowance for travel of £50. Thus, our total per-person cost is £370. This compares favourably with the cost of undertaking an RCT. Finally, we adjust the population to account for the fact that the study will take 2 years and use a decision horizon of 8 years, which matches with the decision horizon chosen for the RCT, accounting for the shorter study time.

Figure 4.16 displays the ENBS across different sample sizes for a willingness-to-pay of £20,000 per QALY gained, a population size of 46,000, decision horizon of 8 years and a discounting factor of 3.5%. It also displays the COSS. We can see that, due to the low cost of the study and the large population size, the optimal sample size for this study is the maximum recruitment level of 300 patients per state. ENBS for this sample size is £25 million, which is substantially below ENBS of the proposed RCT. The COSS demonstrates that the optimal sample size is 300 for all studies that have positive value and 0 for those studies with no potential value. This means that

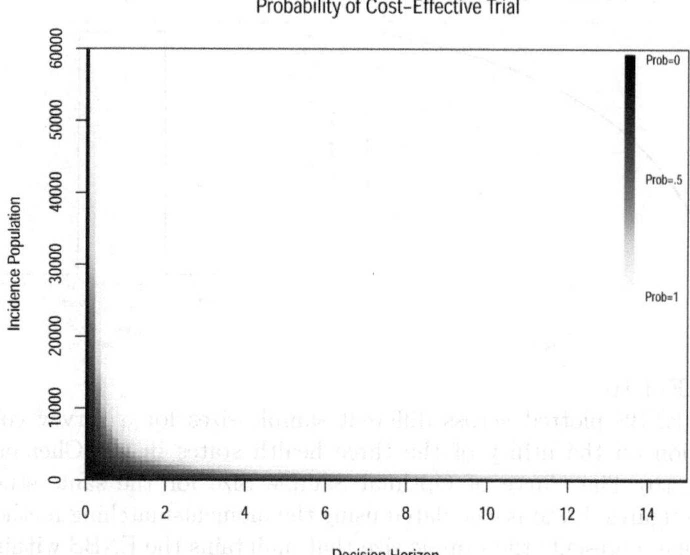

FIGURE 4.17
The probability of a cost-effective trial at a willingness-to-pay of £20,000 per QALY gained and sample size of 300 per utility state for a study investigating the utilities of the three key health states in the Chemotherapy example.

the population size and decision horizon are key factors in determining the optimal sample size, as any value results in a valuable study.

Figure 4.17 demonstrates that the conclusion of a positive ENBS for this study design, i.e., willingness-to-pay of £20,000 per QALY gained and sample size of 300, is robust to the decision horizon and population size. The probability of a cost-effective trial is 1 for almost all combinations of these two parameters.

4.8.5 Long-Term Survival

In addition to a RCT and a survey to investigate utilities, we can also estimate the value of obtaining more information on the long-term survival rate, as it was associated with a positive population-level EVPPI, and can be informed by an observational study. Observational studies will typically be cheaper and larger than RCTs and surveys, as they primarily require fixed costs, i.e., the analyst time will be the primary cost, which will not change irrespective of the sample size. In a retrospective data analysis, specifically, the data will have already been collected and the sample size known, so there is no need to compare the value of studies with different sample sizes. Instead, we could compare the expected value of a retrospective study to the value of the proposed RCT and survey. Another crucial element of the proposed retrospective analysis would be that the results of the study will be quicker to obtain, meaning that a larger population can benefit from the improved information. Particularly, the study could take as little as 6 months to complete, making the comparable decision horizon 9.5 years.

To generate long-term survival data, we assume that we have access to data from a population-level database that records the diagnosis and death date for all individuals. For simplicity, we assume that we will be taking a single year cohort from 20 years ago, representing a study size of 46,000 individuals. We assume that the follow-up time for this cohort is sufficient to avoid considering censoring in our data simulation procedure, although censoring for survival data could also have been simulated (Heath et al. 2022). Note that in making this assumption, we are assuming that the baseline risk of death has remained the same in this disease for 20 years. Depending on the disease area, this may be an unrealistic assumption and adjustments to the proposed study and accompanying data-generating procedure would be needed. Our decision-analytic model defines a constant rate of death over the whole lifetime time horizon. While this is likely an unrealistic assumption, it makes the choice of data-generating process simple, as this will be an exponential distribution. To simplify the data generation, we note that the sum of exponentially distributed times with the same rate parameter r can be generated from a Gamma distribution with shape parameter $2n$ and scale parameter $\frac{n}{r}$, where n is the sample size of the study. Thus, the data-generating function for all methods is:

```
> #### STUDY 3: Long-term Survival ####
> longterm_datagen_fn <- function(inputs, n = 46000){
+   rate_longterm <- inputs[, "rate_longterm"]
+   sum_of_surv <- rgamma(dim(inputs)[1], shape = 2 * n,
+                         scale = n / rate_longterm)
+   return(data.frame(surv_sum = sum_of_surv))
+ }
```

EVSI can be calculated using the regression-based method with the following code, which is very efficient for a single sample size of 46000.

```
> evsi_longterm <- evsi(outputs = chemotherapy_output,
+                       inputs = m_params,
+                       pars = c("rate_longterm"),
+                       n = 46000,
+                       method = "gam",
+                       datagen_fn = longterm_datagen_fn,
+                       par_fn = generate_psa_parameters)
```

As EVSI is not estimated for different sample sizes, and only updates a single model parameter, the regression-based method is the best choice, as it can efficiently estimate EVSI for a single sample size. Figure 4.18 displays EVSI for this study in terms of the willingness-to-pay threshold, calculated using the regression-based method. We can clearly see that the proposed study essentially reaches EVPPI for all willingness-to-pay thresholds, meaning that

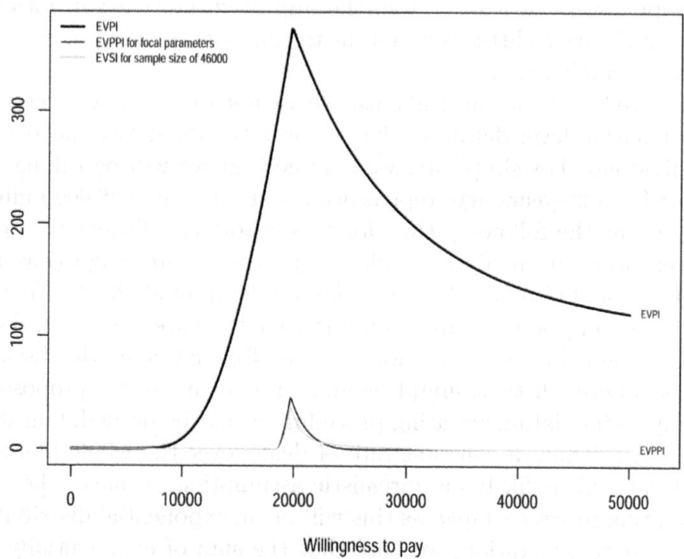

FIGURE 4.18
The EVSI against willingness-to-pay for a study investigating the long-term survival for the Chemotherapy example.

the study attains its maximum potential value. However, it is clear that the maximum expected value of this study is substantially lower than that of the proposed RCT. The costs, however, are also substantially lower, requiring analyst time and dissemination costs. Based on the estimates above, we determine that 1 year for an analyst is £55,000 and add an extra £5,000 for dissemination costs. Thus, for willingness-to-pay thresholds of £20,000, 25,000 and 30,000 per QALY gained, respectively, ENBS is £10.7 million, £970,000 and £255,000. Therefore, this study provides value for money, but an RCT and the utility survey will provide greater value, even if this study is faster and therefore increases the size of the population who would benefit from the better information. Finally, Figure 4.19 displays the probability of a cost-effective study for a decision horizon between 0 and 15, and populations between 0 and double our current estimates. This clearly displays that there is positive value in this study almost irrespectively of the decision horizon and population size, but this value is still dwarfed by the value provided by the RCT.

4.8.6 Overall Study Design

Based on these analyses, it is clear that many different studies have the potential to provide benefit for decision makers. The value of the information

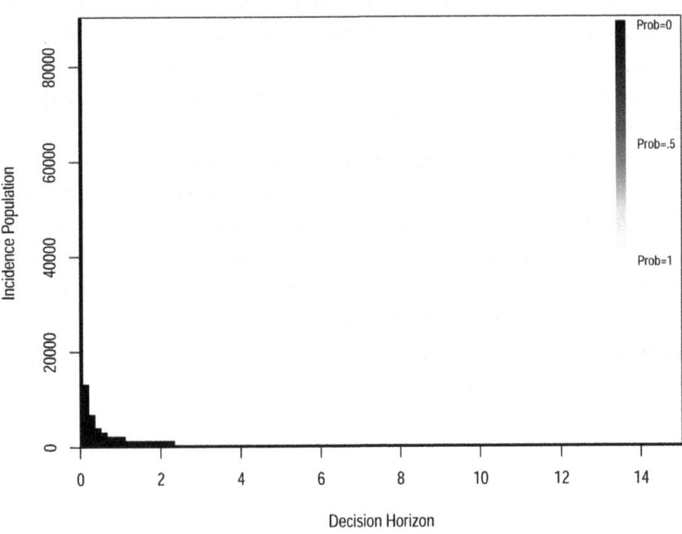

FIGURE 4.19
The probability of a cost-effective trial for a study investigating the long-term survival for the Chemotherapy example. The white area of the plot represents combinations of incidence population and decision horizon that would result in a cost-effective study. The small black area in the bottom left corner represents the small number of combinations of incidence population and decision horizon for which the study would not be cost-effective.

that could be obtained with an RCT primarily investigating the odds ratio of experiencing side effects with the novel chemotherapy (compared to the standard care) is substantial, and exceeds the potential value from other proposed studies. The ENBS analysis highlights that for a willingness-to-pay threshold of £20,000 per QALY gained, a sample size of between 1,000 and 1,200 patients is optimal, but sample sizes as small as 550 could provide similar value. The exact optimal sample size depends on the EVSI estimation method, but broadly the Curve of Optimal Sample Size indicates that for commonly used UK thresholds of between £20,000 and £30,000 the optimal sample size is over 500.

Note that it is possible to consider funding more than one of these studies, and that additional value could potentially be generated by adding studies together in a portfolio of research. This approach can have impact on the optimal study design(s) and is discussed in Chapter 10. It is also important to consider that, similarly to EVPPI, the total value of studies targeting different parameters will not necessarily be additive, so funding two or more studies will not always provide substantial additional value. Conversely, funding two studies may provide substantially higher value than funding each study separately. This would occur when each study itself is not able to improve decision-making, but together they could make substantial difference. For example, a study of the transition probabilities and a study of the health-related utility values would provide information to almost completely define the consequences of side effects.

In summary, this chapter has presented common methods for calculating EVSI, and discussed the theory of using EVSI to optimise study design using the ENBS. We have presented key graphical presentations for exploring the results of EVSI and ENBS analyses, and presented how to perform EVSI computations in practice through the `voi` package. Finally, we have presented an applied example of an EVSI analysis using our running case study. In the next chapter, we discuss a range of important considerations for reporting the results of a VOI analysis to key stakeholders.

5

Reporting and Presenting Value of Information Analyses

Anna Heath, Natalia Kunst, and Sabine Grimm

In the previous two chapters, we have presented methods for computing Value of Information (VOI) measures and provided an example of how VOI can be used in a range of healthcare decision-making situations for sensitivity analysis and to inform the prioritisation and design of research. The chapters have primarily focused on introducing to key concepts for VOI analyses and supporting readers to undertake a VOI analysis through the presentation of computation methods. However, once a VOI analysis has been completed, these results must be presented to key stakeholders in a way that ensures confidence in the results, and helps them to use VOI results to inform health policy and research prioritisation decisions.

Thus, in this chapter, we will discuss two key elements of using VOI in practice; reporting the results of VOI analysis in peer-reviewed publications and technical reports and communicating the implications of the VOI analysis to potentially non-technical stakeholders. Specifically, in Section 5.1, we will present a checklist that highlights the key elements that should be included in a manuscript or report that outlines the results of a VOI analysis. These reporting guidelines ensure that sufficient details are available in the report to critique the VOI results.

Then, in Section 5.2, we highlight some specific graphics and quantities that can aid in communicating VOI results to stakeholders such as health technology assessment (HTA) bodies. The goal of representing VOI using these methods is to support the engagement of individuals who engage with the HTA process but are not specialists in the interpretation and use of VOI measures. Through these two sections, we hope to improve the presentation of the results of VOI analyses, and increase their applicability and accessibility.

DOI: 10.1201/9781003156109-5

5.1 Reporting the Results of a Value of Information Analysis

To improve the reporting of different types of analyses, standardised reporting guidelines have been developed for many different types of studies, including randomised trials (Schulz, Altman and Moher 2010), observational studies (Von Elm et al. 2007), systematic reviews (Page et al. 2021) and economic evaluations (Husereau et al. 2022), which are crucial for VOI reporting. Standardised reporting guidelines are simple, structured tools that detail what information should be reported in a manuscript or report (Simera et al. 2010). They specify a minimum set of information that should be included when reporting the outcomes of different types of studies. In addition to reporting guidelines for broadly defined types of studies, e.g., economic evaluations, extension reporting guidelines have also been developed to clarify additional items that should be reported when undertaking specific subtypes of studies, e.g., adaptive trials (Dimairo et al. 2020).

The most often used reporting guideline for economic evaluations is the Consolidated Health Economic Evaluation Reporting Standards (CHEERS) statement, which was recently updated to a "CHEERS 2022" statement, and provides a minimum set of items required for reporting economic evaluations (Husereau et al. 2022). As VOI analyses are typically performed alongside an economic evaluation, an extension checklist has been developed to specify the minimum, essential information for reporting a VOI analysis, the CHEERS VOI checklist (Kunst et al. 2023a). The CHEERS checklist should be used instead of CHEERS when a VOI analysis is performed alongside an economic evaluation. The overall aim of the CHEERS-VOI checklist is to improve how easily studies can be interpreted and compared, and ensure the analyses provide accurate research recommendations by standardising the reporting of VOI analyses.

The CHEERS VOI checklist was developed through standard methodology for guideline development (Moher et al. 2010; Husereau et al. 2022). An initial set of candidate reporting items was extracted by updating a previous literature review (Koffijberg et al. 2018). These items were then reviewed, edited and confirmed through discussion with experts in VOI. Items were classified as being relevant to VOI analysis in general or only relevant to specific VOI measures, e.g., EVPI, EVPPI, EVSI and ENBS. To determine whether these items were crucial for the minimal reporting of VOI, they were reviewed in an international Delphi study (Hsu and Sandford 2007) by individuals who had a self-reported familiarity or strong expertise with VOI. The final checklist was developed through two 4-hour virtual consensus meetings based on the results of the Delphi study.

5.1.1 Using the CHEERS-VOI Checklist

The CHEERS-VOI checklist can be used by individuals reporting a VOI analysis to ensure they report the key information as part of their study write-up, which should improve the transparency, reproduciblity and rigour of VOI reporting. It can also be used by individuals evaluating reports of VOI methods to determine whether all appropriate information has been provided. It is important to note that individuals can report additional information they believe is relevant to the interpretation of their VOI analysis as the checklist is not an *exhaustive* list of all the elements that could be included. In contrast, it is the *minimum* set of information that should be included to ensure reproducibility and transparency.

Table 5.1 presents the CHEERS-VOI checklist, which contains items included in the CHEERS statement and new items specific to VOI analyses. Items that were included in the checklist provided by the CHEERS statement are numbered, although some of these items required additional elaboration when considering VOI analysis, and so they differ from the original CHEERS items. Further, the new items which are specific to VOI analyses are numbered starting with 1 and preceded with the letter 'S.' Each item is placed in the order that was assessed to be the most logical in the flow of a typical manuscript, but these orderings do not necessarily have to be respected when reporting. Finally, when reporting a VOI analysis, the section heading and paragraph number of the article where each item can be found should be reported.

5.1.2 Explaining the Items in the CHEERS-VOI Checklist

This section provides a more in-depth description of each of the items in the CHEERS-VOI checklist to support the reporting of VOI analyses in practice. Explanations were either extracted from the original publication of the CHEERS statement (Husereau et al. 2022) or the publication of the CHEERS-VOI checklist (Kunst et al. 2023a).

Item 1: Title

Identify the study as an economic evaluation and specify the interventions being compared.

Explanation: The title should enable economic evaluations and value of information analyses to be easily identified through a literature search. Authors should use the terms "economic evaluation" and "value of information" in the title and specify the interventions compared and the setting under investigation to ensure appropriate indexing. Alternatively, authors could use specific terms to define the form of analysis (e.g., "cost-effectiveness" for economic evaluation and "value of further research" for VOI analysis).

TABLE 5.1
Items included in the CHEERS-VOI checklist (Kunst et al. 2023a).

Topic	Item no.	Guidance for Reporting	Reported in Section
Title	1	Identify the study as an economic evaluation and as a VOI analysis, and specify the interventions being compared.	
Abstract	2	Provide a structured summary that highlights context, key methods, results and alternative analyses.	
		Introduction	
Background and objectives	3	Give the context for the study, the study question and its practical relevance for decision-making in policy or practice.	
		Methods	
Health economic analysis plan	4	Indicate whether a health economic analysis plan was developed and where available.	
Study population	5	Describe characteristics of the study population (such as age range, demographics, socioeconomic, or clinical characteristics). If population-level VOI measures are estimated, describe and justify how the population who benefits from the research was defined, how the incidence and/or prevalence were derived, and state and justify the decision time horizon used.	
Setting and location	6	Provide relevant contextual information that may influence findings.	
Comparators	7	Describe the interventions or strategies being compared and why chosen.	
Perspective	8	State the perspective(s) adopted by the study and why chosen.	
Time horizon	9	State the time horizon for the study assumed in economic evaluation and why appropriate.	

(*Continued on next page*)

TABLE 5.1
(Continued).

Topic	Item no.	Guidance for Reporting	Reported in Section
Discount rate	10	Report the discount rate(s) used in economic evaluation and for the population-level VOI analysis, and reason chosen.	
Selection of outcomes	11	Describe what outcomes were used as the measure(s) of benefit(s) and harm(s).	
Measurement of outcomes	12	Describe how outcomes used to capture benefit(s) and harm(s) were measured.	
Valuation of outcomes	13	Describe the population and methods used to measure and value outcomes.	
Measurement and valuation of resources and costs	14	Describe how costs were valued.	
Currency, price date and conversion	15	Report the dates of the estimated resource quantities and unit costs, plus the currency and year of conversion.	
Rationale and description of model	16	If modelling is used, describe in detail and why used. Describe the model structure and justify structural assumptions that have been made. Report if the model is publicly available and where it can be accessed.	
VOI estimation methods	S1	(a) Describe the method used to estimate EVPI and any checks used to determine its accuracy. (b) Describe the method used to estimate EVPPI and any checks used to determine its accuracy. (c) Describe the method(s) used to estimate EVSI and any checks used to determine accuracy.	

(*Continued on next page*)

TABLE 5.1
(Continued).

Topic	**Item no.**	**Guidance for Reporting**	**Reported in Section**
Analytics and assumptions	17	Describe any methods for analysing or statistically transforming data, any extrapolation methods, and approaches for validating any model used. Make sure to: (a) State and justify the cost-effectiveness threshold(s) chosen. (b) Describe and justify the statistical and methodological choices that were made to estimate parameters (e.g., methods of data synthesis, calibration).	
Evidence base	S2	Discuss the potential risk of bias and heterogeneity in the evidence base and describe whether these were adjusted for in the model.	
Characterising heterogeneity	18	Describe any methods used for estimating how the results of the study vary for subgroups.	
Characterising distributional effects	19	Describe how impacts are distributed across different individuals or adjustments made to reflect priority populations.	
Characterising uncertainty	20	Describe methods to characterise any sources of uncertainty in the analysis. Discuss the key structural uncertainties and how these have been addressed. For parameters subject to a probabilistic analysis, describe how the distributions reflecting uncertainty were derived, including any dependencies between parameters. For parameters that were not subject to a probabilistic analysis, explain why they are considered fixed and known with certainty.	

(*Continued on next page*)

TABLE 5.1
(Continued).

Topic	**Item no.**	**Guidance for Reporting**	**Reported in Section**
Parameters of interest in VOI analysis	S3	(a) Specify the individual and/or combinations of parameters for which EVPPI has been computed. (b) Specify the individual and/or combinations of parameters for which EVSI has been computed.	
Study design(s) proposed in VOI analysis	S4	Describe and justify the design of all proposed future research studies and indicate the model parameters that will be updated by these studies.	
Data generation for EVSI	S5	Describe the assumptions used to generate the study data.	
Costs of research studies for ENBS	S6	Provide the costs of research studies proposed in VOI analysis (e.g., fixed costs, variable costs, opportunity costs of allocation), and describe how they were calculated.	
Approach to engagement with patients and others affected by the study	21	Describe any approaches to engage patients or service recipients, the general public, communities, or stakeholders (such as clinicians or payers) in the design of the study.	
		Results	
Study parameters	22	Report all analytic inputs (such as values, ranges, references) including uncertainty or distributional assumptions.	
Summary of main results	23	Report the mean values for the main categories of costs and outcomes of interest and summarise them in the most appropriate overall measure. Report the values for the VOI measures considered.	
Effect of uncertainty	24	Describe how uncertainty about analytic judgments, inputs, or projections affect findings. Report the effect of choice of discount rate and time horizon, if applicable.	

(*Continued on next page*)

TABLE 5.1
(Continued).

Topic	Item no.	Guidance for Reporting	Reported in Section
Effect of engagement with patients and others affected by the study	25	Report on any difference patient/service recipient, general public, community, or stakeholder involvement made to the approach or findings of the study.	
Discussion			
Study findings, limitations, generalisability, and current knowledge	26	Report key findings, limitations, ethical or equity considerations not captured and how these could affect patients, policy or practice.	
Other relevant information			
Source of funding	27	Describe how the study was funded and any role of the funder in the identification, design, conduct and reporting of the analysis.	
Conflicts of interest	28	Report authors, conflicts of interest according to journal or International Committee of Medical Journal Editors requirements.	

Item 2: Abstract

Provide a structured summary that highlights context, key methods, results and alternative analyses.

Explanation: Authors should include a structured summary that includes:

1. The aim or objectives of the study.
2. Key methods, including study population, comparators, input parameters and setting, such as country, time horizon, perspective, currency year and discount rate.
3. Results (mean values of costs and outcomes), including base case and key alternative analyses, VOI analyses and conclusions.
4. Conclusions, which should indicate any potential impact on patients, the public or application in policy or patient care and describe the impact of relevant analyses of uncertainty.

Item 3: Background and objectives

Give the context for the study, the study question and its practical relevance for decision-making in policy or practice.

Explanation: Authors should provide an explicit statement of the motivation for the study, present the study question (i.e., decision problem), explain its relevance for health policy or practice decisions, and describe its importance to patients and the general population. The motivation should not simply state that "the study aims to assess the cost-effectiveness of treatment X," rather it should state the study population and subgroups, the setting and location, the study perspective and the interventions or strategies being compared.

Item 4: Health economic analysis plan

Indicate whether a health economic analysis plan was developed and where available.

Explanation: Health economics analysis plans (HEAPs) outline the proposed health economic analysis. If the HEAP is available, the user should indicate where it can be found and are encouraged to upload it as supplementary material to the submission.

Item 5: Study population

Describe characteristics of the study population (such as age range, demographics, socioeconomic or clinical characteristics). If population-level VOI measures are estimated, describe and justify how the population who benefits from the research was defined, how the incidence and/or prevalence were derived and state and justify the decision time horizon used.

Explanation: Population features (such as age range, gender, sex, income level and ethnic groups) and clinical characteristics (severity, subtypes of illness, histology, etc.) should be reported. These will clarify the relevance of the results to populations of interest. For population-level VOI analysis, the population may be different from the population assumed in the health economic model. As such the characteristics of the population used in VOI analysis must be defined alongside the methods used to determine its size. As seen in Section 4.5, one common method for estimating the population size is to define the yearly number of individuals affected by the decision (yearly incidence) and multiply by the number of years before the decision is reassessed (decision time horizon), discounting appropriately. Thus, the decision time horizon must be explained and justified.

Item 6: Setting and location

Provide relevant contextual information that may influence findings.

Explanation: This should include a clear description of the location, setting or other relevant aspects of the system in which the intervention is provided. It is also suggested that authors subsequently interpret findings in light of system-specific factors in the "Discussion" section.

Item 7: Comparators

Describe the interventions or strategies being compared and why chosen.

Explanation: A detailed description of the elements of each intervention compared in the economic evaluation, particularly for complex interventions, should be provided. Note that comparators may include "do nothing" or "current practice" and these interventions should also be clearly defined. Finally, authors should describe why particular comparators were chosen, including if any potentially relevant comparators were excluded.

Item 8: Study perspectives

State the perspectives adopted by the study and why chosen.

Explanation: The study perspective is the viewpoint from which costs and consequences associated with the comparators are evaluated. This may be a patient perspective, an institutional perspective (e.g., hospital), a healthcare payer's perspective or a societal perspective. As these perspectives lack standard definitions, authors should describe the perspective in terms of which costs and effects have been included, and how this relates to the decision problem and the key stakeholders.

Item 9: Time horizon

State the time horizon for the study and why appropriate.

Explanation: The time horizon is the length of time over which costs and consequences of the interventions are evaluated and reported. This should be long enough to capture the most important differences in costs and consequences and should be justified.

Item 10: Discount rate

Report the discount rate(s) used in economic evaluation and for the population-level VOI analysis, and reason chosen.

Explanation: Both the health economic analysis and the value of information generated by research should be discounted over the assumed time horizon to put less weight on the decisions informed in the more distant future and more weight on the decisions informed in the near future and this discounting rate should be clarified.

Item 11: Selection of outcomes

Describe what outcomes were used as the measures of benefits and harms.

Explanation: As discussed in Chapter 1, the consequences of interventions in an economic evaluation can be evaluated in a range of ways, for example, in terms of QALYs. The outcomes should be clearly defined, especially if they are composite outcomes, in which case the definition of the composite outcome and its constituent parts should be clear. The rationale for the chosen outcome(s) should also be discussed.

Item 12: Measurement of outcomes

Describe how outcomes used to capture benefits and harms were measured.

Explanation: The outcomes selected in item 11 must be measured to inform the model, i.e., estimated from available data or elicited from experts. It is crucial to clarify how all outcomes were measured, and what data were available to inform them.

Item 13: Valuation of outcomes

Describe the population and methods used to measure and value outcomes.

Explanation: In economic evaluation, preference-based values must be valued in order to be compared across different decision options. The methods used to value outcomes should be described, e.g., time-trade off, or use of value sets.

Item 14: Measurement and valuation of resources and costs

Describe how costs were valued.

Explanation: The data used to determine the costs in the study should be described, alongside any additional information required to explain how these data were processed in order to obtain the costs used in the study.

Item 15: Currency, price date, and conversion

Report the dates of the estimated resource quantities and unit costs, plus the currency and year of conversion.

Explanation: It is crucial to report dates and currencies of the analysis, as conversions are often required to ensure that all costs are relevant to the population of interest.

Item 16: Rationale and description of model

If modelling is used, describe in detail and justify why used. Describe the model structure and justify structural assumptions that have been made. Report if the model is publicly available and where it can be accessed.

Explanation: The article should describe the model structure used for analysis and explain why it is appropriate for use in the study. In most cases, a figure illustrating the model structure and patient flows through the model is recommended. Furthermore, the report should indicate the modelling technique used, the model structure (e.g., health states and allowed transitions), and all associated assumptions. The provided information is recommended to be sufficiently detailed to ensure transparency and credibility of the model.

Item S1: VOI estimation methods

(a) *Describe the method used to estimate EVPI and any checks used to determine its accuracy.*

(b) *Describe the method used to estimate EVPPI and any checks used to determine its accuracy.*

(c) *Describe the method(s) used to estimate EVSI and any checks used to determine accuracy.*

Explanation: This book has detailed a range of methods for computing EVPI, EVPPI and EVSI. The chosen method used for each VOI measure should be clarified when reporting a VOI analysis, including simulation size, and any checks for accuracy should be detailed, e.g., calculation of standard errors.

Item 17: Analytics and assumptions

Describe any methods for analysing or statistically transforming data, any extrapolation methods and approaches for validating any model used. Make sure to:

(a) *State and justify the cost-effectiveness threshold(s) chosen.*

(b) *Describe and justify the statistical and methodological choices that were made to estimate parameters (e.g., methods of data synthesis, calibration).*

Explanation: As discussed in Chapter 1, the input values for the model usually require a set of assumptions and calculations to transform the available data into relevant measures of benefits, harms and costs. Assumptions may include information about the characteristics of the modelled population, disease natural history and disease management patterns, including the choice of comparators and treatment pathways. Analytic methods may also include approaches to transform or extrapolate data beyond observed values. There is a range of statistical methodologies that can estimate model parameters, from calculating means and variances to complex methodologies that account for incomplete or potentially biased evidence, such as network meta-analysis (Lumley 2002), digitisation of survival curves (Guyot et al. 2012), matched-adjusted indirect comparison (Signorovitch et al. 2010) and model calibration (Vanni et al. 2011; Enns et al. 2015; Shewmaker et al. 2022). It is important to justify why specific estimation methods were used, and how they provide estimates for the parameters and the associated uncertainty. Furthermore, VOI analyses require a univariate measure of treatment benefit, usually the net monetary benefit. Thus, justification for the specific value or range of values for the cost-effectiveness thresholds used to provide a univariate measure of benefit should be provided.

Item S2: Evidence base

Discuss the potential risk of bias or heterogeneity in the evidence base and describe whether such risks were adjusted for in the model.

Explanation: VOI analyses assume that the updated information is added to the current evidence base, meaning that the current evidence directly impacts the VOI. It is important to clarify (i) which studies are included in the evidence base and the limitations of these studies and/or (ii) how expert opinion was formally extracted. Heterogeneity and bias in the previous studies or elicitation will impact the VOI and should be described.

Item 18: Characterising heterogeneity

Describe any methods used for estimating how the results of the study vary between subgroups.

Explanation: As discussed in Chapter 1, heterogeneity in outcomes between subgroups of the population is separate from parameter uncertainty, and should be explored separately. Thus, authors should report the effect of different types of heterogeneity, including heterogeneity in treatment effects and baseline risks.

Item 19: Characterising distributional effects

Describe how impacts are distributed across different individuals or adjustments made to reflect priority populations.

Explanation: "Distributional" impacts relate to the differential effects across a population, and are crucial to investigate if decision makers are interested in the equity impacts of the interventions. Authors should describe any methods used address distributional concerns through the use of population-specific parameters.

Item 20: Characterising uncertainty

Describe methods to characterise any sources of uncertainty in the analysis. Discuss key structural uncertainties and how these have been addressed. For parameters subject to a probabilistic analysis, describe how the distributions reflecting uncertainty were derived, including any dependencies between parameters. For parameters that were not subject to a probabilistic analysis, explain why they are considered fixed and known with certainty.

Explanation: Statistical uncertainty due to the use of individual patient data can be extracted from the statistical analysis. Parameter uncertainty can be also extracted from experts or the literature. To undertake a VOI analysis, probabilistic analysis must be performed. In a probabilistic analysis, the model inputs can be either subject to parameter uncertainty or fixed. It is crucial to report how the uncertainty distributions are derived for model inputs that are uncertain. This will link to the reporting of item 17. Secondly, if some of the model inputs are fixed, it is important to justify why those inputs are considered be known with certainty. Finally, sensitivity analyses should be performed to evaluate the impact of the assumptions relating to model structure on the reported results.

Item S3: Parameter of Interest in VOI analysis

(a) Specify the individual and/or combinations of parameters for which EVPPI has been computed.

(b) Specify the individual and/or combinations of parameters for which EVSI has been computed.

Explanation: It is crucial to specify which individual parameters or subsets of parameters have been considered in EVPPI analyses and which parameters are being updated by the proposed study for EVSI. Justification for why certain parameter subsets have been considered should also be provided.

Item S4: Study Design(s) proposed in VOI analysis

Describe and justify the design of all proposed future research studies and indicate the model parameters that will be updated by these studies.

Explanation: For EVSI, it is crucial to clearly describe the design of the studies that are considered, as seen in Chapter 4. The description of the study design should include the research question (e.g., population, interventions, control, outcome), and an explanation of how particular model parameter(s) will be learnt from the study. The study type(s), e.g., randomised controlled trial, observational study, administrative data study, questionnaire, should also be clarified, alongside other key design considerations, e.g., sample size.

Item S5: Data generation

Describe the assumptions used to generate the study data.

Explanation: As seen in Section 4.2, EVSI computations usually require simulated future data. Thus, the distributional assumptions required to simulate such data should be reported.

Item S6: Costs of research studies for ENBS

Provide the costs of research studies proposed in VOI analysis (e.g., fixed costs, variable costs, opportunity costs of allocation), and describe how they were calculated.

Explanation: It is critical to specify the assumed cost of research used to compute ENBS. Thus, it is important to report which research costs have been considered and how these costs have been estimated. Section 4.6 clarifies the assumptions that can be made to estimate research costs.

Item 21: Approach to engagement with patients and others affected by the study

Describe any approaches to engage patients or service recipients, the general public, communities or stakeholders (e.g., clinicians or payers) in the design of the study.

Explanation: Patient engagement, wider community engagement and stakeholder involvement are important to enhance the relevance, acceptability and appropriateness of research, ultimately improving its quality. This item asks authors to report any approaches they use to support wider engagement in research.

Item 22: Study parameters

Report all analytic inputs (e.g., values, ranges, references) including uncertainty or distributional assumptions.

Explanation: This item should report the results for the analytic model inputs, i.e., summary values of the assumed distributions for the model inputs.

Item 23: Summary of main results

Report the mean values for the main categories of costs and outcomes of interest and summarise them in the most appropriate overall measure. Report the values for the VOI measures considered.

Explanation: The results of the study should be reported and summarised numerically and graphically. As the study includes both the health economic evaluation and VOI analysis, both should be reported in the results section.

Item 24: Effect of uncertainty

Describe how uncertainty about analytic judgments, inputs or projections affect findings. Report the effect of the choice of discount rate and time horizon, if applicable.

Explanation: The results of the probabilistic analysis and the sensitivity analyses for key model inputs should be reported using appropriate graphics and numerical summaries. Note that when VOI analyses are considered as sensitivity analyses, the reporting of VOI results could address this item.

Item 25: Effect of engagement with patients and others affected by the study

Report on any difference that the involvement of stakeholders such as patient/service recipients, the general public or community made to the approach or findings of the study.

Explanation: This item reports whether stakeholder engagement changed the approach or the findings of the study.

Item 26: Study findings, limitations, generalisability, and current knowledge

Report key findings, limitations, ethical or equity considerations not captured, and how these could affect patients, policy or practice.

Explanation: The discussion should contextualise the results, and help the reader interpret and review the study finding. It should include a discussion of the potential impact of the research on patients, policy or practice, highlight how the study advances knowledge, and relate the study findings back to the original decision question. Remaining or open questions can also be highlighted.

Item 27: Source of funding

Describe how the study was funded, and any role of the founder in the identification, design, conduct and reporting of the analysis.

Explanation: The funding relationships for the project should be clearly articulated.

Item 28: Conflicts of interest

Report authors' conflicts of interest according to the journal or International Committee of Medical Journal Editors (ICMJE) requirements.

Explanation: All potential conflicts of interest should be disclosed.

5.1.3 The Importance of the CHEERS-VOI Checklist

The CHEERS-VOI checklist is intended to standardise reporting of VOI analyses performed alongside economic evaluation to improve the interpretation, transparency and comparison of studies to ensure that VOI analyses provide accurate research recommendations (Kunst et al. 2023a). The CHEERS-VOI checklist should be used instead of the CHEERS statement when economic evaluations include a VOI analysis. The CHEERS-VOI checklist is intended for researchers, analysts and other stakeholders who either report or assess an economic evaluation with a VOI analysis which is aimed to be published or submitted for HTA. Thus, the CHEERS-VOI checklist is a vital tool for individuals who have undertaken a VOI analysis and wish to communicate the results of their analysis to researchers and stakeholders.

5.2 Communicating Value of Information to HTA and Reimbursement Bodies

5.2.1 Introduction

Having discussed how VOI analyses should be reported in manuscripts and reports, we now turn our attention to how to communicate VOI analyses so they can reach their potential in health technology assessment (HTA). In applied HTA processes, decision/policy makers are aiming to determine whether or not to recommend a new technology for reimbursement in their health system. As we know, VOI methods can be particularly useful when the evidence base is immature and decisions need to be made about (Fenwick et al. 2020):

1. Whether the new health technology should be adopted and approved for reimbursement.
2. Whether additional evidence is required before the reimbursement or adoption decision is made.

In fact, VOI analysis allows decision makers to weigh up key factors including (i) the current level of uncertainty, (ii) the expected net benefit of the new technology and (iii) the expected opportunity loss associated with (not) recommending the technology for reimbursement at a given time point. Thus, it is a critical tool for decision-making. This is particularly true in jurisdictions where reimbursement/approval can be granted subject to certain conditions, which can include requiring future research or different financial schemes such as "pay for performance." These are known as *Managed Entry Agreements* (MEA) (McCabe et al. 2010; Dabbous et al. 2020) and can be thought of as adding a third option of "Yes but ..." to the standard responses from the HTA process of "Yes" and "No."

However, to use VOI in HTA decision-making, both as a methodology to assess the sensitivity of decisions to parameter uncertainty (Coyle, Buxton and O'Brien 2003) and to develop MEAs, requires the communication of VOI methods and their results to HTA reviewers and committees. The key challenge is that these committees often consist of a diverse set of stakeholders, including health economists, clinicians, patient representatives, public health experts and payers. This is because committee members tend to find the interpretation of VOI measures difficult (Grimm et al. 2021a), acting as a barrier to the use of VOI in practice. In particular, the following barriers to use of VOI in the context of HTA reimbursement decisions have been identified:

1. A perception that VOI is difficult.

2. A lack of clarity about how VOI analysis is helpful in supporting decision-making.
3. Challenges in interpreting common VOI outputs.

This section presents an overview of methods for communicating VOI results in the context of HTA reimbursement decision-making, with a focus on improving the communication of these results to address these key concerns.

5.2.2 The EVPI Curve

As we discuss in Chapter 3, EVPI results can be presented graphically for a range of values of the cost-effectiveness threshold (Figure 3.2). This graphical presentation is recommended by the ISPOR task force reports on VOI (Fenwick et al. 2020; Rothery et al. 2020). It is important to clarify to stakeholders that VOI depends on the valuation of health opportunity cost, i.e., the willingness-to-pay threshold, and that as this threshold changes, the VOI changes. While the EVPI curve is only relevant for EVPI and displays the maximum potential cost of parameter uncertainty, EVPPI and EVSI can also be presented in this fashion (see Figures 3.6 and 4.3, respectively).

5.2.3 Expected Loss Curves

Another relevant graphic that can support the interpretation of EVPI consists of a curve of the expected loss against the willingness-to-pay threshold (as previously mentioned in Section 1.6.5). These expected loss curves (ELC) aggregate the EVPI and the expected net benefit forgone because of recommending a (sub-optimal) technology and plot the sum as a function of the cost-effectiveness threshold (Eckermann, Briggs and Willan 2008; Alarid-Escudero et al. 2019a). Thus, the expected loss curves quantify the consequences of parameteric uncertainty and choosing a suboptimal intervention in terms of expected foregone benefits and uncertainty.

At a given threshold, the expected loss is equal to the EVPI for the most cost-effective technology in the decision problem; while for the other technologies, the expected loss is greater than the EVPI. ELCs can communicate which technology is optimal given current evidence, the extent to which the other technologies are suboptimal and the consequences of decision uncertainty, in a single graphical representation. This is a unique advantage when compared with the commonly used cost-effectiveness acceptability curve or frontier (see Section 1.6.5), which crucially do not present information about how much better one strategy is over another, nor about the potential value of reducing parameter uncertainty. The ELC for the chemotherapy example is shown in Figure 5.1.

In Figure 5.1, we can see that at £50,000 per QALY, the expected loss due to parameter uncertainty in our model is relatively small, but if the decision maker were to continue to implement the standard care, there would be a very

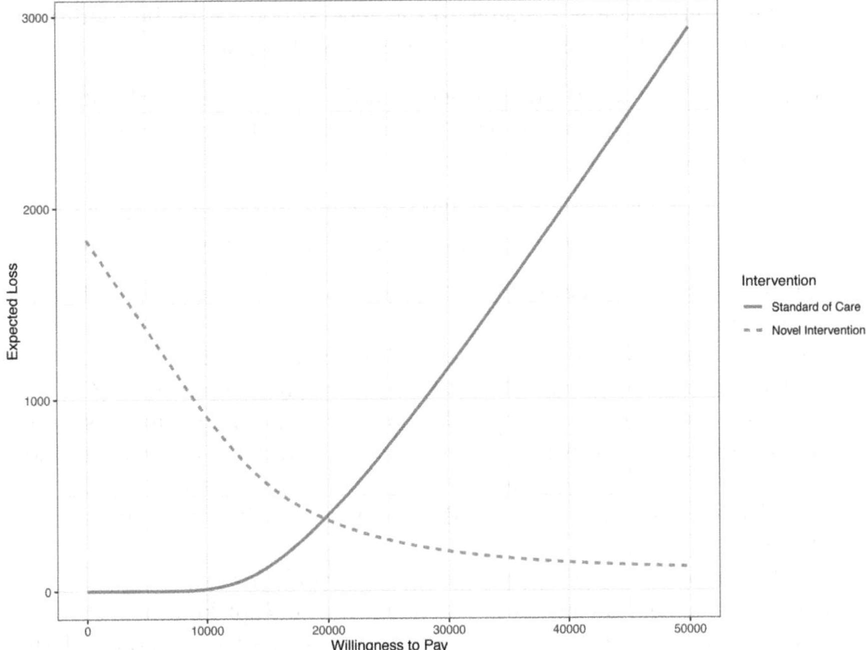

FIGURE 5.1
The expected opportunity loss associated with implementing the standard of care (solid line) and the novel intervention (dashed line) plotted across willingness-to-pay for Chemotherapy example.

large expected loss as this intervention is non-optimal. Conversely, at £20,000 per QALY, the parameter uncertainty drives the expected opportunity loss for both decisions. In general, we can see that when the decision between the two interventions is relatively certain, the expected loss of selecting the incorrect decision option is large when compared to the expected loss when the decision is uncertain.

5.2.4 HTA Risk Analysis Chart

Another useful graphical display when communicating with diverse stakeholders is the HTA risk analysis chart. This chart aims to facilitate decision-making about reimbursement of health technologies by illustrating the value of MEAs (Grimm et al. 2017). As discussed previously, HTA reimbursement decisions typically aim to determine whether an intervention should be implemented, i.e., a “Yes” or “No” decision. However, MEAs provide a range of alternative potential decisions, including

- The novel intervention is available to the general population but additional information should be collected and this approval decision may be revoked.
- The novel intervention is not currently available to the general public but can be made available for research purposes and a future decision will be made based on additional evidence (McKenna et al. 2015).
- The novel intervention is made available to the general population but at a reduced price, which may be increased depending on future evidence.

Thus, it is relevant for decision makers to see how VOI changes with different prices and evidence.

A HTA risk analysis plots the expected loss (defined by Eckermann, Briggs and Willan (2008) and discussed above). However, in contrast to ELCs, each HTA risk analysis presents the expected loss for a specific willingness-to-pay threshold. In jurisdictions where this threshold is externally determined, these charts can be produced once and presented to stakeholders but in other settings, multiple charts may be needed to perform sensitivity analysis to the thresholds.

Apart from presenting the expected loss separately for different thresholds, the expected loss concept has been extended to a concept known as the *payer strategy and uncertainty burden* (P-SUB). The use of the P-SUB nomenclature is to clarify the concept that expected loss imposes a burden onto society. In addition, the P-SUB explicitly highlights the two components of the expected loss:

- The loss that is due to imperfect information, which is known as the *payer uncertainty burden* or PUB in this framework and is equal to the EVPI.
- The loss that is incurred by recommending a sub-optimal strategy or technology, the *payer strategy burden* or PSB.

Thus, the P-SUB is equal to the total expected loss, the PUB equals the EVPI, and the PSB is equal to the absolute value of the incremental net benefit of a given strategy versus the most cost-effective strategy in the decision problem at a given willingness-to-pay threshold. The HTA risk analysis chart can be drawn multiple times under different assumptions to explore the impact of different potential MEAs. Firstly, we can illustrate the effect of a price reduction on PUB and PSB, which would impact the net monetary benefit and change PUB and PSB. Secondly, we can illustrate the effect of data collection on PUB and PSB by subtracting EVSI from the PUB. Overall, the goal of the decision maker is to lower the P-SUB as much as possible based on two potential tools: price reductions and data collection.

As an example, Figure 5.2 shows the individual-level P-SUB calculated for the worked chemotherapy example assuming a threshold of £20,000. Note that as with standard VOI analyses, the P-SUB can also be presented at the population level (see Section 3.2.1). Figure 5.2 displays the following three sub-plots:

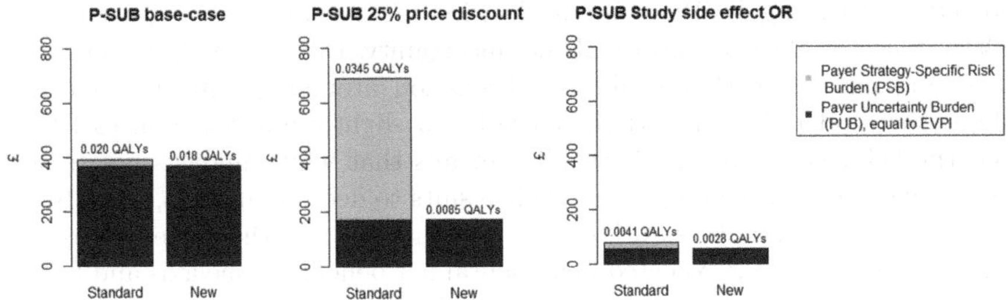

FIGURE 5.2
P-SUB calculated for three scenarios in chemotherapy example.

- Left-hand figure; the P-SUB for the base-case.
- Centre figure; the P-SUB recalculated when a price discount of 25% is offered for the new intervention.
- Right-hand side figure; the P-SUB recalculated assuming that a study with sample size $n = 900$ was performed to investigate the OR of side effects.

Overall, the P-SUB (or expected loss) increases for the standard care when the new intervention is offered at a price discount, as the novel intervention becomes more cost-effective when the price of the novel intervention decreases. The PUB (or EVPI) is also reduced by approximately half, as the decision to recommend the new intervention becomes more certain. Finally, with the planned study, the PUB (or EVPI) is reduced the most as the remaining uncertainty after the study is substantially reduced (Section 4.8).

Thus, the HTA risk analysis chart can help the decision maker provide recommendations for MEAs. It can also illustrate when a simple recommendation to either approve or reject the novel intervention can be appropriate. However, as EVPI calculates the value of eliminating all parameter uncertainty, it can be challenging for decision makers to understand when values of EVPI/PUB require a request for additional research. Further work will focus on translating these concepts to improve the understanding of EVPPI and EVSI. This is especially critical as EVPPI can be used to help decision makers understand which parameters are driving decision uncertainty (Coyle, Buxton and O'Brien 2003) to potentially reduce their concerns over the level of uncertainty about certain parameters, particularly when EVPPI for those parameters separately and in combination is low.

5.2.5 Appraisal of Risk Chart

Finally, the Appraisal of Risk Chart (ARCH) was developed based on interviews and a workshop with HTA stakeholders, mostly from the Netherlands (Grimm et al. 2021b). It was developed to address decision makers' objections to using EVPI to guide decision-making as they perceived that EVPI does not always capture all the uncertainty. Decision makers were also searching for a tool that facilitates discussion around qualitative concerns about the model. The ARCH therefore has a slightly different aim to ELCs and the HTA risk analysis chart. This means that while analysts create the two previous graphics and present their results to decision makers, the ARCH is used by decision makers during their deliberations. Furthermore, it was observed that presenting expected incremental net benefit on one axis and EVPI on the other supported decision makers in deriving appropriate actions.

Thus, the ARCH was developed as a plot with two axes (e.g., Figure 5.3):

- The x-axis relates to the *risk* associated with decision-making, which can be subjectively assessed on a scale from low to high.
- The y-axis relates to the incremental net benefit of the new intervention versus its comparators, which can either be positive, when the novel intervention results in a positive change in net benefit, or negative, when the comparators are optimal.

The challenge of developing the ARCH is the subjective assessment of risk. The easiest way to achieve this is to base it on the EVPI (or relevant VOI measure) alone. Alternatively, it can be informed by the decision makers in addition to the EVPI, which can incorporate model uncertainties that were not quantified in the probabilistic analysis (PA) and are therefore not reflected in EVPI. As an example, consider that the model required an extrapolation of overall survival beyond the trial follow-up period. An appropriate distribution must be chosen to achieve this extrapolation and the different distributions can lead to vastly different model outputs. However, it is unlikely that the chosen distribution truly reflects the long-term survival for patients and a weighted average of the different candidate extrapolations would be more appropriate. However, model averaging is rarely used in HTA submissions and the resulting PA is therefore likely under-estimating the uncertainty surrounding the overall survival of patients. This can be qualitatively considered using the ARCH. Thus, we recommend that before the ARCH is used, an inventory of all uncertainties be made and that it be indicated whether the uncertainty is appropriately reflected in the PA or not.

When viewed in a simplified, non-gradient, manner, ARCH has four quadrants representing the following combinations (Figure 5.3):

- Positive net benefit (of the intervention versus comparators) and low risk.
- Positive net benefit and high risk.
- Negative net benefit or net loss and low risk.
- Net loss and high risk.

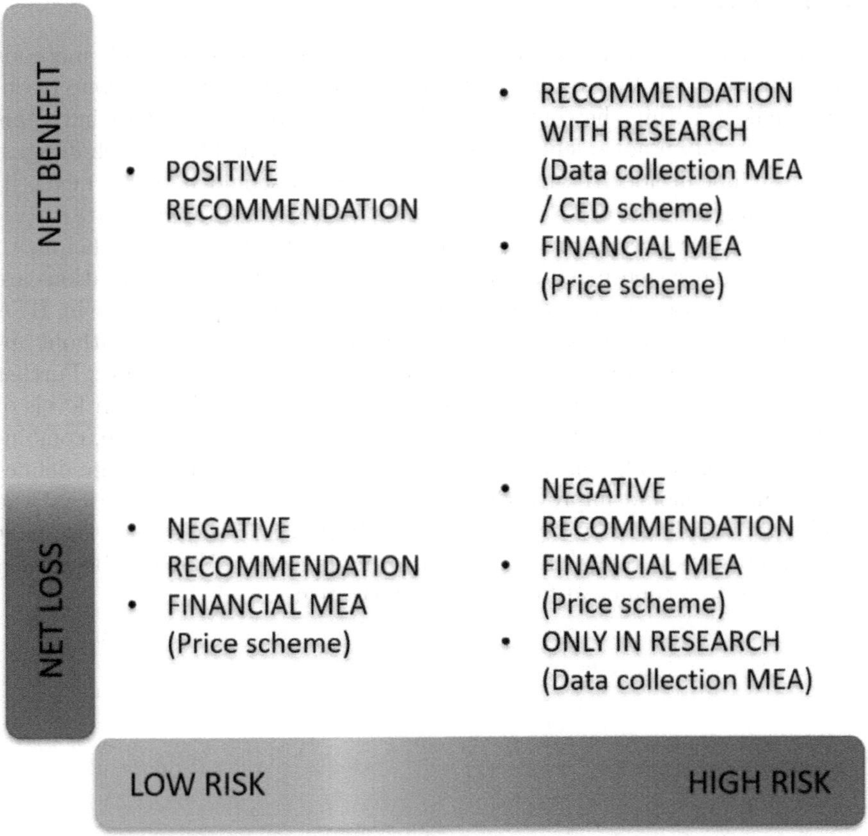

NET BENEFIT

- POSITIVE RECOMMENDATION

- RECOMMENDATION WITH RESEARCH (Data collection MEA / CED scheme)
- FINANCIAL MEA (Price scheme)

NET LOSS

- NEGATIVE RECOMMENDATION
- FINANCIAL MEA (Price scheme)

- NEGATIVE RECOMMENDATION
- FINANCIAL MEA (Price scheme)
- ONLY IN RESEARCH (Data collection MEA)

LOW RISK HIGH RISK

FIGURE 5.3
The appraisal of risk chart (ARCH) displaying the potential actions that could be taken by decision makers for each quadrant. MEA: Managed Entry Agreement, CED: Coverage with Evidence Development.

These four states of the world require different actions and decisions from the decision makers, represented in Figure 5.3. Decision makers can deliberate which quadrant they are in and then pick the appropriate action(s) or request further analyses or data collection to improve their decision-making. For example, in the negative incremental net benefit/high risk quadrant, decision makers may investigate whether a price reduction may lead to an acceptable risk. Alternatively, they may request EVSI and ENBS analyses for study designs that would reduce decision uncertainty. Following this, they could display the post-study (or post-price reduction) risk and net benefit in a new ELC, HTA risk analysis chart and/or ARCH to support their final decision.

5.2.6 Discussion

In conclusion, the second half of this chapter has discussed available methods for communicating VOI results and concepts to a heterogeneous group of decision makers within the context of HTA and reimbursement. Crucially, this audience are often not solely interested in the most valuable research designs, but must first decide (i) whether further research should be performed, (ii) whether simple "Yes" or "No" recommendations are most appropriate or (iii) whether price reductions are indicated. The presented methods can help with this task and decision-making. However, these methods and presentations are not currently used in practical HTA as VOI is used sporadically in HTA decision-making. This means that MEAs are being recommended without appropriate consideration of the risk reduction that they would achieve. Further research should focus on decision makers' information needs and on levels of EVPI that are deemed acceptable by decision makers. Furthermore, combining these recommendations and graphical tools with the reporting guidelines in Section 5.1 should improve the practical use of VOI and ensure that the results are understood by key stakeholders. Building on these ideas, the next chapter presents a range of real-world examples where VOI has been used successfully to recommend future research strategies.

6

Value of Information: Success Stories

Natalia Kunst, Edward C. F. Wilson, Haitham Tuffaha, Christopher Jackson, Nicky J. Welton, Hawre Jalal, and Fernando Alarid-Escudero

Value of Information (VOI) methods are not just of theoretical interest. They have been widely used to inform decision-making about research to inform health policy, in a variety of clinical areas. This chapter tells a series of such "success stories."

6.1 Population-Based Newborn Screening for Germline TP53 Variants: Clinical Benefits, Cost-Effectiveness, and Value of Further Research

Natalia Kunst

The following story is based on Kunst et al. (2022).

Although paediatric cancer is rare, it represents the leading cause of disease-related death in children and adolescents (Cunningham, Walton and Carter 2018). The estimated number of individuals in the United States diagnosed with cancer before turning 20 years old exceeded 15,000, and almost 1,800 died due to cancer, in 2018 (Siegel, Miller and Jemal 2018; National Cancer Institute 2021). Despite the treatment advances, which have substantially improved survival of some childhood cancers, survival outcomes in children with cancer remain poor (Jemal et al. 2017; National Cancer Institute 2021).

In recent years, an increasing number of genetic variants associated with childhood cancers have been identified (Therrell Jr 2001; Genetic Alliance; District of Columbia Department of Health 2010). Thus, population-based newborn genetic screening for paediatric cancer predisposition may help improve survival outcomes by providing an opportunity for early detection and prevention of advanced disease. However, newborn genetic screening may also

DOI: 10.1201/9781003156109-6

lead to several disadvantages, including unnecessarily increased parental anxiety and surveillance costs (Yeh et al. 2021).

The tumor suppressor gene, TP53, is one gene that is often altered in cancer patients, with variants being highly associated with carriers' predisposition to several paediatric malignancies (Wunder et al. 2005; Petitjean et al. 2007; Tabori et al. 2010; Seidinger et al. 2011; Alderfer et al. 2015; Wasserman et al. 2015; Welch et al. 2016). In many cancer types, TP53 variants are associated with worse health outcomes including poorer survival and more frequent disease recurrence (Marchetti et al. 1993; Bardeesy et al. 1994; Børresen et al. 1995; Aas et al. 1996; Birch et al. 1998; Erber et al. 1998; Bernard et al. 2020). Given that children with identified TP53 variants have a high risk of cancer, population-based newborn genetic screening for TP53 can help identify individuals that would benefit from surveillance and improve their health outcomes. However, the clinical benefits and cost-effectiveness of newborn genetic screening for TP53 at a population level remain unclear.

Before implementing newborn genetic screening for TP53 population-wide, there is a need to establish clinical utility and cost-effectiveness of this preventive measure and evaluate whether the underlying evidence is sufficient to justify implementation of universal screening (Robson et al. 2010). For some interventions, a randomized controlled trial, which is often considered the gold standard to establish the impact of various interventions, is not a feasible approach. Newborn genetic screening for TP53 represents this type of intervention, because establishing its impact would require a long time horizon, as well as a large sample size, given the rarity of childhood cancers. Instead, decision-analytic methods can be used to leverage available data and estimate the expected clinical outcomes, including benefits and harms, and costs associated with screening. Furthermore, these methods can be applied to evaluate the potential need and focus of additional data collection, and determine the optimal design of research studies that should be prioritized to efficiently reduce decision uncertainty.

We developed the Precision Medicine Prevention and Treatment (PreEMPT) model (Yeh et al. 2021), a decision-analytic microsimulation model, to assess the clinical benefits of newborn genetic screening for TP53 and determine whether implementing screening population-wide would be cost-effective in the United States. Further, we used this model and applied a value of information analysis to evaluate the need for additional research to resolve uncertainty about implementation of newborn genetic screening for TP53 population-wide.

This study focused on the TP53-associated paediatric cancer predisposition syndromes and modelled the benefits of identifying TP53 variants early. Specifically, we considered a subset of cancer conditions diagnosed under 20 years of age, including adrenocortical carcinoma, choroid plexus carcinoma, osteosarcoma and rhabdomyosarcoma. More specifically, we used the PreEMPT model to simulate a cohort of 4 million newborns, which is representative of an annual birth population in the United States, and considered two strategies: usual care and newborn genetic screening for TP53. Under the screening strategy, newborns with an identified TP53 variant undergo cancer surveillance in line with established guidelines (Kratz et al. 2017). Using VOI methods, we designed and assessed the potential benefits of five different research studies. More specifically we proposed four separate data collection designs for populations with each Li-Fraumeni syndrome-associated cancer, and a data collection design for a general population without any of the cancers. Further, in our VOI analyses, we assumed a willingness-to-pay threshold of $100,000 per life year gained and a decision time horizon of 10 years.

Our findings suggest that newborn screening for TP53 would be expected to identify 894 individuals with P/LP TP53 variants who would undergo cancer surveillance. Further, it would lead to a 7.2% decrease in the number of cancer-related deaths. While our study suggests that newborn screening could be cost-effective, there is considerable decision uncertainty. We found that additional research to estimate the probability of TP53 variants given rhabdomyosarcoma would efficiently reduce the identified decision uncertainty and would result in a high expected benefit of 349 life years gained. Specifically, this would be achieved by a retrospective registry study with a relatively small sample of 200 individuals with rhabdomyosarcoma.

We applied decision-analytic methods to design and prioritize the most efficient research study designs for newborn genetic screening for TP53. Our findings indicate that a retrospective observational study in rhabdomyosarcoma patients should be prioritized, due to its highest expected benefit. Further, we predicted that there would be value in performing a study on the prevalence of TP53 variants in the general population. Although the general population study was associated with lower expected benefits, we estimated that this data collection would also be worthwhile, leading to a gain of 75 life years with a sample size of 4,000. Note that, similar to estimates of cost-effectiveness, the expected value of information depends on decision makers' willingness to pay. Given that there are no clear recommendations for willingness-to-pay thresholds, for each proposed research study we estimated its expected value across a wide range of thresholds, in addition to a base-case ($100,000 per life year gained).

The results of this study highlight the need for additional research to establish health and cost outcomes associated with newborn genetic screening for TP53. We identified research studies that would efficiently improve decision-making and help develop best practices with respect to newborn genetic screening. Importantly, the research studies identified as optimal are feasible and inexpensive, and could reduce enough uncertainty to pave the way to a change in universal newborn screening and reduce burden of disease in children.

6.2 A Value of Information Analysis of Research on the 21-Gene Assay for Breast Cancer Management

Natalia Kunst

The presented story is based on Kunst et al. (2019).

The benefits of adjuvant (i.e., post-surgical) chemotherapy in women with node-negative, estrogen receptor (ER)-positive, human epidermal growth factor 2 (HER2)-negative breast cancer remain uncertain (Hayes 2012; Burstein, Hayes and Vora 2018). Although adjuvant chemotherapy reduces disease recurrence risk and improves patients' survival (Early Breast Cancer Trialists' Collaborative Group 2005), in some patients it leads to adverse events and toxicities, including congestive heart failure and acute myeloid leukaemia (Crump et al. 2003; Bowles et al. 2012; Calip et al. 2015). A number of risk-stratification models have been developed to inform decisions about the use of chemotherapy in women with early breast cancer, e.g., the prediction tools Adjuvant! Online (Ravdin et al. 2001) and PREDICT (Wishart et al. 2010; National Health Service 2018). These models stratify patients into different risk groups based on clinicopathologic information. However, the predictions of these risk stratification models may not be accurate because the literature suggests that tumors may have different metastatic potential despite their similar pathologic characteristics (Tang et al. 2011).

Tumor-specific gene expression profile testing can be used to identify women that are unlikely to benefit from chemotherapy, and may help spare them chemotherapy toxicity and other chemotherapy-related risks (Hassett et al. 2012). The 21-gene assay Oncotype DX (Genomic Health, Redwood City, CA) is one of the available gene expression profile tests, which has been widely used in women with early breast cancer in the United States (Orucevic, Heidel and Bell 2016). The 21-gene assay performs better than clinicopathologic risk stratification models (Hornberger et al. 2012; Rutter et al. 2016), but it is costly, and, thus, its unrestricted use may cause an economic burden on society.

We applied the iterative decision-making framework in health and medicine (Kunst et al. 2023b) to evaluate the cost-effectiveness of the use of the 21-gene assay and guide decisions about the use of adjuvant chemotherapy in women with early breast cancer. More specifically, our study carefully conceptualised the decision problem and decision-analytic model, and addressed all assumptions from the previous cost-effectiveness analyses of the 21-gene assay that were deemed questionable. Specifically, we applied existing knowledge about patients' distant recurrence risk and response to chemotherapy, and evaluated the cost-effectiveness of the 21-gene assay in women stratified by their risk of disease recurrence using settings that accurately represented clinical practice. Further, we used VOI methods to examine the potential benefits of conducting additional research on the use of the 21-gene assay to support treatment decisions in women with early-stage breast cancer.

Our study found that the cost-effectiveness of the 21-gene assay varied depending on patients' risk. While our findings indicated that the effectiveness and cost-effectiveness of the 21-gene assay in women with high and intermediate risk is well-established and robust in the face of essentially all parameter uncertainty, the cost-effectiveness of the 21-gene assay in women with low risk is more uncertain. Consequently, for this subpopulation, we continued with the evaluation and performed a full VOI analysis. The VOI results suggested that a randomized controlled trial (RCT) to examine the predictive value of the 21-gene assay had the greatest potential to efficiently reduce the current decision uncertainty about the use of the 21-gene assay in women with low risk of distant recurrence. This study was predicted to be valuable, after accounting for its expected costs, and considering a large number of different sample sizes and willingness-to-pay thresholds. There was also expected value in the other two research studies considered: the prospective observational study for the prevalence of chemotherapy use stratified by clinical risk with and without the 21-gene assay, and the retrospective observational study for the joint distribution of the 21-gene assay and the PREDICT clinical risk. Although the marginal expected benefits of these two observational studies were lower compared to the RCT, these research studies may be considered if the RCT would not be feasible. As recommended, we reported VOI results across a wide range of willingness-to-pay thresholds to provide information useful to many decision makers.

The majority of previous cost-effectiveness studies did not evaluate the use of the 21-gene assay separately for women with low, intermediate, and high clinical risk. This was probably due to lack of evidence on all risk-stratified outcomes required to perform a risk-stratified cost-effectiveness analysis. However, omitting this important feature could provide decision makers with misleading information indicating that the 21-gene assay is cost-effective in all women with early breast cancer, while it may be cost-effective only for subgroups of patients with a specific clinical risk. Our study used recently collected evidence (Connecticut Tumor Registry 2018) to inform the decision-analytic model about the 21-gene assay-PREDICT joint distribution, and addressed all the other questionable assumptions identified by the recent literature review (Wang et al. 2018). While a number of previous cost-effectiveness studies combined women with low, intermediate and high risk, and concluded that the 21-gene assay is cost-effective (Rouzier et al. 2013; Ward et al. 2013; Marrone, Stewart and Dotson 2015), our results indicated that the evidence strongly supports the cost-effectiveness of the 21-gene assay only for women with high and intermediate clinical risk. For women with low risk, uncertainty regarding the cost-effectiveness of the 21-gene assay persists. Our findings further indicated substantial decision uncertainty surrounding the 21-gene assay cost-effectiveness in this patient subpopulation, which could be efficiently resolved with the proposed RCT to determine the predictive value of the 21-gene assay.

Ours was the first study to perform a full VOI analysis, including estimates of the expected value of perfect information (EVPI), partial perfect information (EVPPI) and sample information (EVSI) and expected net benefit of sampling (ENBS), for the 21-gene assay use in women with early breast cancer. The results of this work provided important insights on the 21-gene assay decision-making, and highlighted the important role of applying the iterative decision-making framework (Kunst et al. 2023b), which includes a VOI analysis, to inform decision-making, improve disease management and inform efficient research study design and prioritisation.

6.3 Late Pregnancy Ultrasound to Screen for and Manage Potential Birth Complications in Nulliparous Women

Edward C. F. Wilson

In the NHS in England, current practice during pregnancy is for the mother to undergo two ultrasound scans. The first "dating scan" (around 12 weeks) is to confirm the number of foetuses and estimated date of delivery. This scan can also include detection of a small number of abnormalities such as spina bifida or Down's syndrome. The second "anomaly scan" (around 20 weeks) checks for a larger number of abnormalities. A later, third trimester scan (at around 36 weeks), is only offered in higher risk pregnancies or where anomalies have been detected. However, offering a routine late pregnancy scan to all mothers could detect late developing anomalies such as large or small for gestational age (LGA/SGA) or breech presentation. This would allow intervention to reduce the risk at delivery, in particular stillbirth.

As stillbirth is thankfully a rare outcome from pregnancy in the UK (around 0.4% of all births) (National Statistics 2019), a randomised controlled trial powered to detect a statistically significant difference in this would be unfeasibly large (enrolling in excess of 50,000 mothers). However, there may be value in exploring whether other aspects of a late pregnancy ultrasound scan could be a cost-effective approach allowing intervention to reduce perinatal morbidity or mortality.

The National Institute of Health and Care Research (NIHR) commissioned a study to conduct a systematic review, economic evaluation and value of information analysis to (i) identify what measurements could be taken as part of a third-trimester scan, (ii) their sensitivity and specificity at predicting adverse outcomes, (iii) whether a scan including "promising" measurements would represent a cost-effective use of healthcare resources and (iv) whether there was value in further research to reduce uncertainty, and if so where (Smith et al. 2021).

The systematic review identified several measurements including umbilical artery Doppler flow, cerebroplacental ratio and assessment of oligohydramnios, all of which were only weakly predictive of adverse outcomes. However, foetal biometry to assess small and large for gestational age (SGA and LGA respectively), and presentation (breech or cephalic) appeared to show some potential. These were therefore explored within a decision model, exploring various strategies including a simple, rapid presentation scan that could be performed by a midwife with a hand-held scanner at a routine antenatal visit, to a more detailed scan assessing whether the foetus is predicted to be small or large for gestational age. Following a positive assessment for breech, the model assumed a mother would be offered external cephalic version (ECV). Following detection of SGA, the mother would be offered induction of labour, and following detection of LGA, the model compared induction of labour and expectant management (Wilson et al. 2021).

Based on current information, the most cost-effective strategy was to offer a simple presentation-only scan to all mothers, yielding an expected incremental net benefit compared with current practice of £87.36 per mother (95% credible interval £4.88 to £205.68, INB calculated at £20,000 per QALY). Across England, this equates to an expected INB of £17.1 million or 857 QALYs per annual birth cohort. However, due to uncertainties in the evidence base (that is, parameter uncertainty), there is only a 44% probability that this conclusion is correct: there is a 56% probability that one of the other strategies or status quo could be the most cost-effective. The expected loss associated with this uncertainty (i.e., the EVPI) is £31.56 per pregnancy, or £53.3 million to England as a whole (assuming a ten-year time horizon for the value of information analysis).

Exploring the value of information further, the parameter with the biggest impact on decision uncertainty was the cost of induction of labour, or specifically, the difference in cost between an induced delivery and expectant management. This was somewhat surprising and could not have been predicted beforehand. However, as modelled, the cost does not relate simply to the cost of the induction procedure, but also uncertainty in the timing of the procedure and the impact on antenatal visits, as well as the cost of the delivery itself. Subsequent modelling would benefit from breaking down this area in more detail than the approximations used in the analysis.

Using the moment-matching approach (Heath, Manolopoulou and Baio 2018), we estimated that a study of "reasonable size" to explore the cost of induction vs expectant management would yield a positive return on investment. For example, the EVSI of a study enrolling 1,000 mothers per arm is in excess of £11 million. If this were delivered for a cost of, say, £1 million, it would yield a greater than 10-fold return on investment. Of equal (or perhaps greater) importance is the finding that studies on the outcomes of small or large for gestational age did not yield a positive expected return on investment.

6.4 VOI Applications in the Research Portfolio of the Australian National Centre of Research Excellence in Nursing

Haitham Tuffaha

The National Centre of Research Excellence in Nursing (NCREN) in Queensland (2012-2018), was Australia's first centre of research excellence in nursing funded by the National Health and Medical Research Council (NHMRC) (Tuffaha, Gordon and Scuffham 2016). A major stream of research within the NCREN was on interventions supporting skin integrity in hospitalised patients (e.g., intravascular devices). Those interventions could benefit a wide population of patients, but there was a paucity of published evidence to support a decision on their implementation. NCREN researchers conducted systematic reviews and pilot studies to inform the need for and feasibility of larger clinical trials. Although most NCREN researchers were familiar with the role of economic evaluation in assessing the value for money of new interventions, VOI analysis was totally new to them. NCREN health economists presented to the team about the VOI principles, and explained the applications of VOI methods in informing funding decisions, trial design and research prioritisation. Senior NCREN researchers were interested in exploring the potential of VOI in informing research prioritisation and efficient trial design. They were also interested in using VOI analysis to demonstrate the value for money of their investigator-initiated research projects for research funding organisations (e.g., NHMRC and Medical Research Future Fund).

Four interventions were identified by the NCREN for VOI analysis:

1. Clinically indicated peripheral intravenous catheter replacement compared with routine replacement every 72–96 hours.
2. Tissue adhesive for securing arterial catheters compared with standard securement options (e.g., polyurethane dressing).
3. Negative pressure wound therapy (NPWT) compared with standard dressing to prevent surgical site infections in Caesarean section.
4. Nutritional support compared with standard hospital diet in preventing pressure injury in high-risk patients.

The team chose these interventions for VOI analysis because they targeted a wide population of hospitalised patients, the evidence to support their adoption was limited, and there were NCREN studies or systematic reviews on their effectiveness. For each intervention, an economic evaluation was performed, decision uncertainty characterised and VOI measures including EVPI, EVPPI, EVSI and ENBS were calculated using Monte Carlo simulations. The benefits and costs of additional research were considered together with the costs and consequences of implementing the intervention now versus waiting for more information. The optimal trial design would be the one that maximised the expected net research benefit. Finally, the future research studies were ranked according to their expected net monetary benefits (Tuffaha, Gordon and Scuffham 2016).

All interventions were deemed to be cost-effective, but with various levels of decision uncertainty (Tuffaha et al. 2014a; Tuffaha et al. 2014b; Tuffaha et al. 2015a; Tuffaha et al. 2015; Tuffaha et al. 2016; Tuffaha, Gordon and Scuffham 2016). Negligible uncertainty regarding the clinically indicated catheter replacement intervention suggested that current evidence was sufficient for implementation (Tuffaha et al. 2014a). For the tissue adhesive intervention, the results indicated that an additional research study before implementation could be worthwhile: a four-arm trial of 220 patients in each arm, collecting data on the cost and efficacy of this intervention compared to other securement devices (Tuffaha et al. 2014b). Additional research on NPWT before implementation could be also worthwhile, consisting of a two-arm trial of 200 patients per arm, investigating the relative effectiveness of NPWT for preventing surgical site infections in Caesarean section patients compared with standard dressings (Tuffaha et al. 2015a; Tuffaha et al. 2015). The results also indicated that nutritional support should be implemented concurrently with a two-arm trial of 1,200 patients in each arm, evaluating the relative effectiveness of nutritional support in preventing pressure injury compared with standard hospital diet (Tuffaha et al. 2015; Tuffaha et al. 2016). Based on their expected net monetary benefits, the proposed future studies were ranked as: (1) NPWT (1.2 million Australian dollars [AUD]), (2) tissue adhesive (AUD 0.3 million) and (3) nutritional support (AUD 0.1 million) (Tuffaha et al. 2014a; Tuffaha et al. 2014b; Tuffaha et al. 2015a; Tuffaha et al. 2015; Tuffaha et al. 2016; Tuffaha, Gordon and Scuffham 2016).

Regarding catheter replacement, recommendations towards the clinically indicated approach have been implemented in hospitals across the US, UK and Australia. This is now part of international guidelines, including the Michigan Appropriateness Guide for Intravenous Catheters (MAGIC) (Moureau and Chopra 2016). The optimal choice of a device to secure catheters has been evaluated in a large trial by the NCREN, which found that there was no difference in the safety and effectiveness among the devices (Tuuli et al. 2020). Using the standard polyurethane dressing to secure catheters could save $10 billion/year of hospital budgets worldwide (Tuuli et al. 2020). A recent large trial, evaluating the efficacy of NPWT in obese women undergoing Caesarean delivery showed that, compared with standard wound dressing, NPWT did not significantly reduce the risk of surgical-site infection (National Pressure Ulcer Advisory Panel, European Pressure Ulcer Advisory Panel, Pan Pacific Pressure Injury Alliance 2019). Nutritional support in high-risk hospitalised patients has been recommended in the clinical guidelines by the European Pressure Injury Advisory Panel, the National Pressure Injury Advisory Panel and the Pan Pacific Pressure Injury Alliance (Rickard et al. 2018).

6.5 Strategies for Wound Dressing to Reduce Surgical Site Infection

Christopher H. Jackson[1] *and Nicky J. Welton*[2]

Reeves et al. (2019) compared three strategies for dressing wounds after surgery to reduce surgical site infection (SSI). The cost-effectiveness of these strategies was considered from the perspective of the National Health Service in England and Wales. They used EVPPI and EVSI methods to estimate the potential benefits of further research about the effectiveness of different strategies.

The key evidence that informed the analysis was obtained from a systematic review and network meta-analysis of 29 randomised controlled trials (Dumville et al. 2016). This compared a range of dressing strategies, that were categorised as "no dressing," "simple dressing," "complex dressing" and "glue" (or tissue adhesive used as a dressing). This indicated substantial uncertainty, with an odds ratio of infection of 1.05 (95% credible interval 0.37–2.41) for glue compared to simple dressings, and 0.98 (0.56–1.55) for no dressing compared to simple.

These estimates informed a decision model that also incorporated data on the baseline risks of SSI, the costs of dressing, management of SSIs and other healthcare, and the impact of SSI on health-related quality of life. The model estimated the expected costs, QALY and net benefit associated with each dressing strategy.

The expected value of partial perfect information (per person) was estimated for each model parameter, using algebraic methods. These estimates were converted to population-level EVPPI, using estimates of the number of surgical wounds in England and Wales over an assumed period of 5 years where the decision question is still relevant. The parameters with the greatest EVPPI were those describing the relative effectiveness of the dressing strategies, which had EVPPI of around £2,000 million. As this is much greater than the cost of a trial, this result showed that a further trial to gain more precise estimates of these effects would probably be cost-effective.

Therefore, an analysis of the expected value of sample information was conducted, to investigate the potential benefits of a randomised controlled trial of the three strategies, with different sample sizes and randomisation ratios. At the same time, a "pilot" trial ("Bluebelle") was conducted, where 394 patients were randomised to no dressing, simple dressing or glue (Blazeby 2020), both to obtain a small amount of further evidence, and to gain practical experience in running a study in this area that would be helpful to inform the feasibility of a larger study. Including this study in the network meta-analysis reduced the

1. Told the "success story".
2. Conducted the original VOI analysis.

uncertainty in the estimates of relative effectiveness of the different strategies, with odds ratios 0.98 (0.48–1.75) for glue compared to simple dressings, and 1.02 (0.64–1.49) for exposed compared to simple dressings.

The EVSI was calculated with and without the data observed from this study. Without the pilot data, the population EVSIs for even larger studies of 3,000 patients were around £2,000 million, and afterwards around £1,500 million. This indicates that such a large study would have benefits that vastly exceed the cost of conducting it. The optimal design might also be chosen to maximise the ENBS, though this was not considered in the published example.

In conclusion, a Value of Information analysis provided evidence that a large trial to provide better information about dressing effectiveness would lead to substantially better-informed policy on use of health service resources, hence to population health benefits that would be at least worth the cost of the trial.

6.6 Research Prioritisation for Gouty Arthritis

Hawre Jalal and Fernando Alarid-Escudero

Jutkowitz et al. (2017) studied research prioritisation for gouty arthritis, and developed the Curve of Optimal Sample Size (COSS, Jutkowitz et al. 2019).

Gouty arthritis is a common inflammatory arthritis affecting more than 8 million people in the US. The disease is associated with increased uric acid and results in sudden and debilitating attacks of severe pain in the affected joints. Clinical guidelines recommend urate-lowering therapy for established gout cases. Both allopurinol and febuxostat are recommended as first-line options. Allopurinol is the standard of care, as it is effective, well tolerated and inexpensive. Febuxastat is an alternative therapy that is more expensive and effective in lowering uric acid in the blood. Studies have shown allopurinol to be cost-saving relative to no treatment and allopurinol-febuxastat sequential therapy to be cost-effective. However, several parameters in the cost-effectiveness analysis model are uncertain, including the efficacy of allopurinol and the effectiveness of the allopurinol-febuxastat strategy. Value of Information analysis allowed for examining the opportunity loss due to uncertainty, hence informing future research prioritization, and the optimal sample size for collecting additional information.

Jutkowitz et al. (2017) conducted a comprehensive VOI analysis to determine the most important sources of uncertainty, and determine opportunities for future data collection. Specifically, the analysis involved computing the EVPI, the EVPPI for individual parameters and groups of parameters that were likely to be included in the same clinical trial or data collection approach, the EVSI and the expected net benefit of sampling (ENBS), and estimating the optimal sample size. Computations were performed using the Gaussian approximation method. For EVPPI and EVSI, the study examined four realistic data collection study designs:

1. A clinical trial evaluating the efficacy of allopurinol dose escalation.
2. A clinical trial evaluating the efficacy of febuxostat dose escalation.
3. A prospective observational study evaluating the health utility of gout patients.
4. A comprehensive clinical trial evaluating the efficacy of allopurinol dose escalation, febuxostat dose escalation, ancillary outcomes and the health utilities of gout patients.

The analyses revealed several areas for which future research and resource allocation may be beneficial. First, the EVPI as a function of the willingness-to-pay (WTP) revealed two ranges of WTP where the EVPI was highest: around \$60,000 and \$340,000/QALY. These were the areas where the strategies were most competitive, the choice among them was less certain, hence the value of information to inform this choice was higher. In addition, design 4, which involved a comprehensive study design comprising elements from the other designs, had the highest EVPPI over the entire WTP range tested, followed by designs 1, 2 and 3. All of these showed that EVPPI was highest at WTP values of around \$50,000 and \$340,000 as before.

In addition, the study computed the cost of research and population EVSI for each proposed study design. It also produced the ENBS curve over various sample sizes, which enabled the optimal sample size to be estimated. The optimal sample sizes for the designs were 735, 710, 2,095 and 4,000 for designs 1 to 4, respectively. Determining the optimal sample size using EVSI has advantages compared to power calculations in clinical trials, which are based on achieving particular error rates for hypothesis tests of treatment effects. Power calculations do not explicitly take into account the expected benefits and costs of treatment decisions made based on the trial data, whereas these are explicitly accounted for by decision modelling and VOI analysis.

The study's main conclusion was that additional research to collect information on the effectiveness of allopurinol and febuxostat dose escalation is expected to be valuable, and should be prioritised. A unique feature of this study is that it evaluated several realistic data collection experiments on a preexisting model and evaluated VOI for various combinations of parameters, designs and sample sizes. This study was feasible because the analysis used the Gaussian approximation method that is specifically designed to be computationally efficient, enabling many VOI calculations to be performed in a range of examples.

Furthermore, Jutkowitz et al. (2019) was the first manuscript to develop the COSS, which examines the optimal sample size at various WTP thresholds. This requires the computation of EVSI at various WTP thresholds and sample sizes, which can be challenging. However, using the Gaussian approximation method helped reduce the computation time dramatically and aided the development of the COSS (Jutkowitz et al. 2019). The COSS also allows for examining the impact of both parameter uncertainty and scenario analysis on the WTP threshold.

Part II

Ongoing Research in Value of Information

7

Value of Information for Estimation Instead of Decision-Making

Christopher Jackson

In most of this book, Value of Information (VOI) methods are described in contexts where a decision maker has to choose between a finite number of options, to minimise a quantifiable expected loss (or maximise an expected utility). As we describe in Chapter 1, in typical health economic models, there are a number of treatments, each with a net benefit that is estimated as a function of uncertain parameters. The optimal treatment is the one that maximises the expected net benefit.

However, in many fields of application, models that combine evidence from different sources are developed for purposes other than formally making decisions between discrete actions. Mathematical models are often used to represent complex real-world processes, based on evidence from different sources that is subject to uncertainty. The published results of the models may indirectly inform policy. Even if there is no single, clearly defined policy decision motivating the model, it is still important to determine which model inputs are most uncertain, and where specific further data would be most valuable – since more precise estimates eventually lead to better-informed decisions. Furthermore, we may want to estimate the optimal design of a study to collect specific data.

Value of Information methods can still be used to address these questions. This involves considering *estimation*, in itself, as an example of a decision problem. While explaining how this works, we also explain how VOI is defined and computed for very general classes of decision problems, beyond the discrete choices discussed in the book. These ideas are connected to the field of *sensitivity analysis* in mathematical models, and also connect to the ideas of *design of experiments* from a Bayesian perspective.

A brief review of the main ideas is given here. For a more broadly accessible tutorial, see Jackson et al. (2021), for a review of the literature from a statistical perspective, see Jackson et al. (2022), and for the application to multi-parameter evidence synthesis, see Jackson et al. (2019).

DOI: 10.1201/9781003156109-7

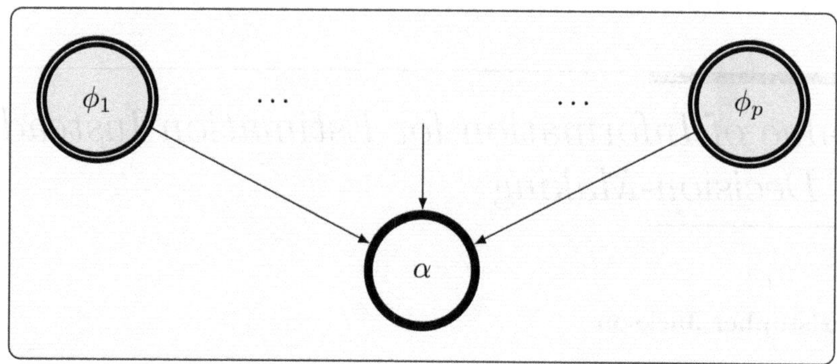

FIGURE 7.1
Directed acyclic graph representing a known deterministic model. (From Jackson et al. (2019), used under the terms of the Creative Commons CC BY license.)

7.1 Model Framework

As in the preceding chapters of this book, we have a *model* that defines how a quantity of interest α (or a vector of quantities) is related to a set of uncertain "input" parameters ϕ (Figure 7.1).

In standard health economic models, α would typically be the incremental net benefit of a new treatment compared to the standard of care, and the model is a *known deterministic function* $\alpha = f(\phi)$. However in some mathematical models that inform policy, there is no explicit decision – a policy maker simply requires an estimate of some quantity α. For example, a transport policy maker may want to know how a reduction in the use of cars might affect public health (through changes in air pollution exposure, physical activity and risk of road injuries) before considering any particular policy to achieve that change (Jackson et al. 2021).

An even more general class of models is known as *Bayesian multi-parameter evidence synthesis* (MPES) models (Jackson et al. 2019) in which α and ϕ are related to each other and to observed data y by a network of statistical models and deterministic functions. An example in medicine is a model for estimating prevalence of HIV infection in the UK based on a combination of indirect data sources (Presanis et al. 2010; De Angelis et al. 2014). Estimates are published annually and have informed public campaigns and policies around testing and treatment. The difference from standard health economic models is that it cannot be expressed as a deterministic function – given specific values for all parameters ϕ, we cannot compute a corresponding value of α. However, we can calculate the *joint distribution* representing our beliefs about α and ϕ, given prior distributions on ϕ, data y and a set of statistical models relating them,

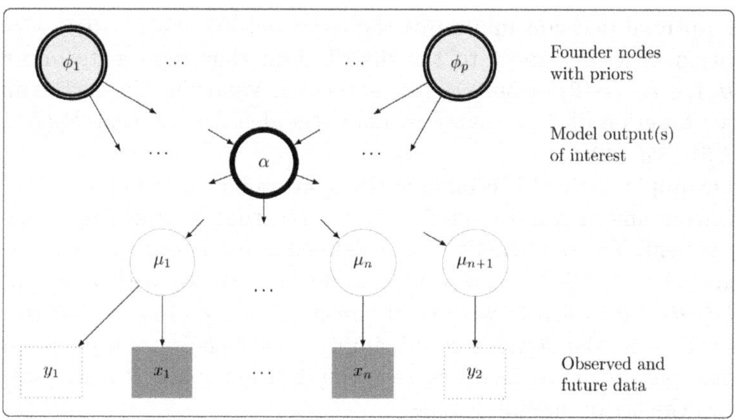

FIGURE 7.2
Directed acyclic graph for Bayesian multi-parameter evidence synthesis. (From Jackson et al. (2019), used under the terms of the Creative Commons CC BY license.)

by Bayesian inference, typically using Markov Chain Monte Carlo (MCMC) simulation.

In either case, we want to learn which parameters ϕ contribute most to the uncertainty about α, hence where further research should be targeted to reduce this uncertainty. Potentially we also want to estimate the expected value of a specific study to get information on one or more of the ϕ. Depending on what is feasible in practice, this study might directly inform ϕ or a function of the ϕ.

7.2 General Definitions of Value of Information

In Chapters 3 and 4, the expected value of information was defined in terms of decisions between finite actions. Here we give a broader definition in terms of a subtly more general decision problem (and in terms of losses instead of net benefits or utilities). To define a decision problem we need

1. A set of d chosen from a space of possible decisions $\mathcal{D}$. Unlike earlier in the book, this space is not necessarily a finite, discrete set of actions. For example, it could be a continuous space representing the possible estimates for a parameter of interest.
2. A set of uncertain parameters $\boldsymbol{\theta}$ in a model of the problem of interest.
3. A loss function $L(d, \boldsymbol{\theta})$ associated with each action d if the parameters take values $\boldsymbol{\theta}$.

The optimal decision minimises the expected loss $E_{\boldsymbol{\theta}}(L(d,\boldsymbol{\theta}))$, where the expectation is with respect to the distribution that represents uncertainty about $\boldsymbol{\theta}$. Let $\boldsymbol{\alpha} = \boldsymbol{\alpha}(\boldsymbol{\theta})$ be our *quantity of interest.* We define this as the minimal subset or function of $\boldsymbol{\theta}$ necessary to make the decision, so that $E_{\boldsymbol{\theta}}(L(d,\boldsymbol{\theta})) = E_{\boldsymbol{\alpha}}(L(d,\boldsymbol{\theta}))$, $\forall d \in \mathcal{D}$.

For example, in health economic decision problems, we choose the treatment d, from among a finite set $\mathcal{D} = \{1,\ldots,D\}$, that is expected to maximise the net benefit. The net benefit NB_d is defined as a function of the parameters $\boldsymbol{\theta}$, so that the quantity of interest $\boldsymbol{\alpha}$ would be a vector with D components $\alpha_d = NB_d(\boldsymbol{\theta})$. Equivalently we are minimising a loss $L(d,\boldsymbol{\theta}) = -NB_d(\boldsymbol{\theta})$.

Alternatively, the decision could simply be the choice of a point estimate $\hat{\boldsymbol{\alpha}}$ of some parameter $\boldsymbol{\alpha}$, in which case the decision space $\mathcal{D}$ is the *support* of $\boldsymbol{\alpha}$, that is the set of possible values.

Alongside making a decision, we wish to also determine the expected value of further information to reduce decision uncertainty. The value of perfect, partial perfect and sample information are defined respectively as follows.

$$\begin{aligned} EVPI &= E_{\boldsymbol{\theta}}(L(d^*,\boldsymbol{\theta})) - E_{\boldsymbol{\theta}}(L(d^*_{\boldsymbol{\alpha}},\boldsymbol{\theta})) \\ EVPPI &= E_{\boldsymbol{\theta}}(L(d^*,\boldsymbol{\theta})) - E_{\boldsymbol{\phi}}[E_{\boldsymbol{\theta}|\boldsymbol{\phi}}(L(d^*_{\boldsymbol{\phi}},\boldsymbol{\theta}))] \\ EVSI &= E_{\boldsymbol{\theta}}(L(d^*,\boldsymbol{\theta})) - E_{\mathbf{y}}\left[E_{\boldsymbol{\theta}|\mathbf{y}}(L(d^*_{\mathbf{y}},\boldsymbol{\theta}))\right] \end{aligned}$$

Note that the definitions of EVPI and EVPPI given in Chapter 3 and EVSI in Chapter 4 are special cases of this more general definition, where $L(d,\boldsymbol{\theta}) = -NB_d(\boldsymbol{\theta})$ and d^* is the decision that maximises $E_{\boldsymbol{\theta}}(NB_d(\boldsymbol{\theta}))$.

7.3 Point Estimation as a Decision Problem

Here we consider the choice of the most appropriate point estimate $\hat{\alpha}$ of an unknown parameter α as a decision problem. We would generally define the loss to be higher for estimates $\hat{\alpha}$ that are further from the true value α, for example

- squared error loss:

$$L(\hat{\alpha},\alpha) = (\hat{\alpha}-\alpha)^2$$

- absolute error loss:

$$L(\hat{\alpha},\alpha) = |\hat{\alpha}-\alpha|$$

The mathematical difference from the typical health economic decision problems considered in this book is that the decision space is *continuous* rather than discrete.

The optimal decision is the one that minimises the expected loss. The expectation in this expected loss is taken with respect to a distribution that represents the current state of knowledge about α. It can be shown, by minimising $E_\alpha(L(\hat{\alpha}, \alpha))$ as a function of $\hat{\alpha}$, that the optimal point estimate is $E_\alpha(\alpha)$ under squared error loss, that is, the *mean* of the uncertainty distribution. Likewise the *median* of this distribution is the optimal estimate under absolute error loss, and the *mode* is the optimal estimate under a loss defined to equal one where $\hat{\alpha} = \alpha$ and zero otherwise. See, e.g., Bernardo and Smith (2009, 2nd ed. pp. 257–258), for proofs.

For squared error loss, the expected loss under current information is $var(\alpha)$, the variance of the uncertainty distribution. Gains from new information are expressed as reductions in variance: the EVPPI for a parameter ϕ is

$$var(\alpha) - E_{\phi}\left[var_{\alpha|\phi}(\alpha|\phi)\right] \tag{7.1}$$

and the EVSI for new data y is

$$EVSI(\mathbf{y}) = var(\alpha) - E_{\mathbf{y}}\left[var_{\alpha|\mathbf{y}}(\alpha|\mathbf{y})\right] \tag{7.2}$$

Thus the value of infomation is the *reduction in variance* expected in the quantity of interest after learning new information.

The quantity termed the EVPPI here measures the sensitivity of the model output α to the uncertain input ϕ. This quantity is well known in the field of *sensitivity analysis* for computer models, and is sometimes termed the "main effect" of ϕ (see, e.g., Saltelli, Chan and Scott (2000), Oakley and O'Hagan (2004)). See also Borgonovo, Hazen and Plischke (2016) and Borgonovo and Plischke (2016) for a detailed review of these and other probabilistic measures of model sensitivity.

The idea of valuing information about a single, scalar parameter α by reductions in its variance can be generalised to vectors $\boldsymbol{\alpha}$ of multiple parameters. These ideas are central to the field of Bayesian design of experiments. For more information of how to construct measures of the value of information in these cases see, e.g., Jackson et al. (2019) and Jackson et al. (2022). This can be done using different functions of the covariance matrix of $\boldsymbol{\alpha}$, which correspond to different views about how reductions in variance of different parameters are valued by the decision maker. For example, is a decision maker concerned with *absolute* reductions or *relative* reductions in variance? In the former case, they would pay the same amount for information that reduces the variance from 200 to 100, as for information that reduces the variance from 2,000 to 1,900. In the latter case, they would value a variance reduction from 200 to 100 the same as a reduction from 2,000 to 1,000.

7.4 Computation of VOI for Point Estimation

The value of information in point estimation problems is easy to compute using variants of the non-parametric regression methods described in Section 3.3.4 in

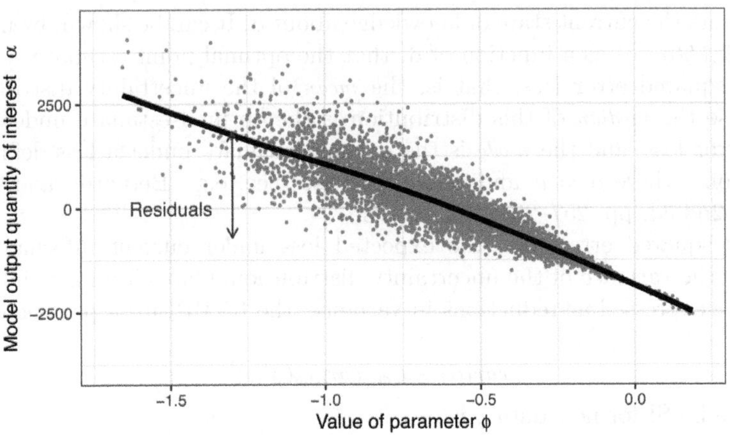

FIGURE 7.3
Illustration of using regression to compute EVPPI of a scalar model parameter ϕ, as the expected reduction in variance of a model output α after learning the true value of ϕ, using a sample from the joint distribution of α and ϕ. The fitted value at $\phi^{(s)}$ is $E(\alpha|\phi = \phi^{(s)})$. The variance of the fitted values is the EVPPI, and the variance of the residuals is the variance remaining after learning ϕ, $\text{Var}(E(\alpha|\phi = \phi^{(s)}))$.

Chapter 3. This technique is also discussed from the perspective of computer model sensitivity analysis by Stanfill et al. (2015). This works as follows, and is illustrated in Figure 7.3 (a variant of Figure 3.3 in Chapter 3).

- As before, a "probabilistic analysis" sample is generated, that is, a sample $(\alpha^{(s)}, \phi^{(s)}) : s = 1, \ldots, S$ from the uncertainty distribution of the model inputs ϕ, and the corresponding model outputs α.
- A regression model is fitted to the "dataset" comprising the $\alpha^{(s)}$ as the outcome, and the $\phi^{(s)}$ as the predictors: $\alpha^{(s)} = g(\phi^{(s)}) + \epsilon_s$, analogously to Equation (3.6), though with just one regression model instead of one per decision d.
- The fitted values from this regression $\hat{g}(\phi^{(s)})$, at each $\phi^{(s)}$ are extracted for each s. The fitted value is an estimate of $E(\alpha|\phi = \phi^{(s)})$, the expected outcome if we had the "perfect information" that $\phi = \phi^{(s)}$. The variability in the *residuals* $\alpha^{(s)} - \hat{g}(\phi^{(s)})$ represents the variability in the outcome that is not explained by ϕ.
- Since $\text{Var}(\alpha) - E_\phi\left[\text{Var}_{\alpha|\phi}(\alpha)\right] = \text{Var}_\phi\left[E_{\alpha|\phi}(\alpha|\phi)\right]$, the EVPPI (Equation 7.1) can be estimated as the variance of the fitted values, interpreted as the variance explained by ϕ.

- Equivalently we could calculate the EVPPI as the variance of the $\alpha^{(s)}$ minus the residual variance, an estimate of the reduction in variance on learning ϕ.

The `voi` R package has a function, `evppivar`, to calculate the EVPPI using this method. The same regression methods are used as in the `evppi` function that was demonstrated in Chapter 3.

The user supplies a data frame `inputs` with S rows containing the sampled $\phi^{(s)}$, and a vector (or one-column data frame) `outputs` with the corresponding $\alpha^{(s)}$. Suppose the joint EVPPI is wanted for two input variables, labelled as `x1` and `x2` in the columns of `inputs`. This is calculated as:

```
evppivar(outputs, inputs, pars=c("x1","x2"))
```

The same approach can be used for calculating the EVSI (Equation 7.2) describing expected reductions in variance from a proposed study of a specific design and sample size, by regressing on a "summary statistic" of the data (as explained in Chapter 4). A similar function `evsivar` is available in the `voi` package for this.

7.5 VOI in Models Used for Estimation: Further Practical Considerations

In practice, using VOI ideas for sensitivity analysis and research prioritisation in models for estimation is generally a less formal procedure than in models used for explicit decision-making.

One further issue is that models may be used to estimate many different "output" quantities of interest to policy makers. Information about a model "input" may have different value for learning different model outputs. Informal judgements may therefore be required to determine which of these outputs are most important to know, hence where further research should be focused. A useful illustration to aid these judgements is a coloured grid showing the expected reduction in variance of different output quantities, given perfect information about different inputs. Figure 7.4 shows an artificial example of this. The lightest squares correspond to parameters (rows) whose uncertainty has the highest influence on the uncertainty of the model outputs (columns).

Note again that in models for estimation, "value of information" does not represent a monetary value, but a reduction in variance of an uncertain quantity. In this framework, to make research decisions under economic constraints, for example, to choose a study sample size to maximise the expected net benefit of sampling (as in Chapter 4), a decision maker must choose the amount that they are willing to pay for a specific reduction in variance of an important quantity. While this might be difficult to specify, framing in

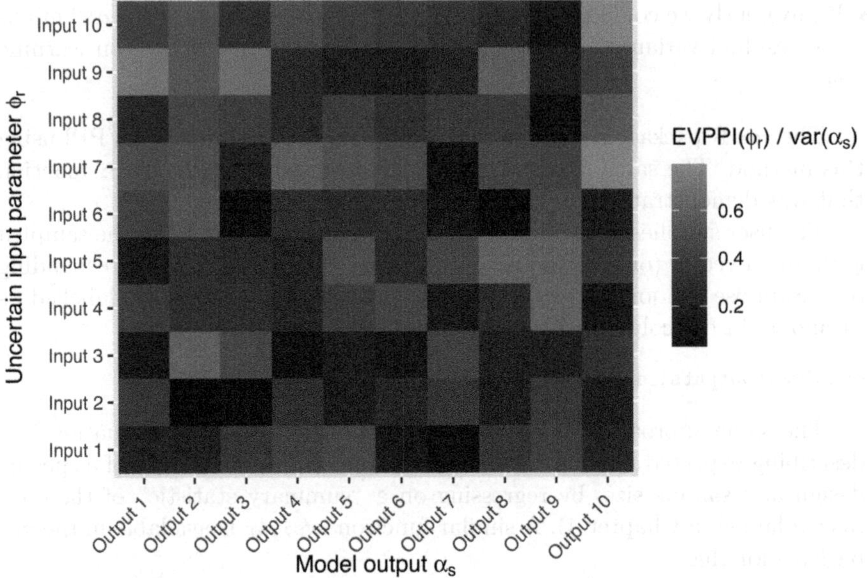

FIGURE 7.4
Example of an EVPPI analysis in a model with multiple outputs used for estimation instead of decision-making. This shows the proportion of the uncertainty variance of a model output α_s explained by the uncertainty around the model input ϕ_r, for ten model outputs s and ten model inputs r. This proportion is equal to the corresponding EVPPI divided by the variance of α_s. Lighter squares represent greater uncertainty.

terms of VOI allows the trade-offs to be made explicit. A worked example is given in Jackson et al. (2019). In health economic decision models, on the other hand, better information can be valued in monetary terms more easily, because better information leads to better health-related decisions, and the resulting improvements in health can often be valued on a generally understood scale such as cost per QALY.

8

Structural Uncertainty and Value of Information

Christopher Jackson

8.1 Structural Uncertainty

Any analysis based on a mathematical model relies on the assumptions of that model, and Value of Information (VOI) analysis is no different. In VOI analysis, the assumptions include the model for how the expected net benefit is defined as a function of parameters $\boldsymbol{\theta}$, and assumptions about the values of $\boldsymbol{\theta}$ and associated uncertainty, as represented by the probability distribution $p(\boldsymbol{\theta})$. EVSI analyses also make assumptions about the sampling distributions of potential observed data. In this section, we refer to these assumptions collectively as "the model."

This dependence on assumptions is both an advantage and a challenge of VOI methods. The advantage is their "formal correctness": if we can specify and implement an appropriate net benefit function and distributions on parameters, we can exactly determine the optimal decision and expected value of further information through Bayesian principles (Chapter 2).

There is a challenge, however, because by their nature, all models are approximations of reality. NB($\boldsymbol{\theta}$) is an approximation of how the costs and effects of decisions arise in the real world, and the $p(\boldsymbol{\theta})$ is an approximation of our current knowledge. There is usually more than one reasonable way to define these. The choice may affect the conclusions of the analysis – an issue generally referred to as *structural uncertainty* or *model uncertainty.*

DOI: 10.1201/9781003156109-8

There are two general approaches to dealing with structural uncertainty:

1. One common approach is to simply determine the results under a series of different assumptions. This is sometimes called "scenario analysis" or "deterministic sensitivity anaysis" (Briggs et al. 2012). We might hope that different reasonable assumptions do not change the conclusions. This can help to understand model assumptions, and help decision-making, together with qualitative judgements about which assumptions are more plausible. However, its dependence on arbitrary choices of "scenario" can be problematic.
2. A more formal approach is based on extending the model to form a more complex model where the structural uncertainty is parameterised. A general overview of this idea is described by Strong, Oakley and Chilcott (2012) and Strong and Oakley (2014). Essentially, an extra parameter, known as a "discrepancy," δ, is inserted into the model in the place where we think the model might be wrong. The original model is defined by a null value of δ, and an uncertainty distribution around δ represents the extent to which the original model might be inaccurate. Similar principles also underlie the ideas of flexible modelling or model averaging to represent uncertainty about statistical models to describe data, and the use of elicitation to enhance weak evidence in decision models (Jackson et al. 2011).

An advantage of the "discrepancy" perspective on structural uncertainty is the connection to VOI analysis. We illustrate this with a simple example.

8.2 Value of Information for a Model Discrepancy or Bias Parameter

In the model of Chapter 2, suppose we suspect that the estimate of the log odds ratio of side effects $\log(\rho)$ used in our model may be biased. This might be because the study that informed the estimate is not representative of the setting of our decision model, or because of flaws in its conduct. In our original "naive" model we used the following distribution:

Model 1. $\theta \sim N(\log(0.52), 0.3)$

To extend this model to account for this structural uncertainty, we could redefine the parameter θ, giving:

Model 2. $\theta = \theta_0 + \delta$, with $\theta_0 \sim N(\log(0.52), 0.3)$.

δ is the "bias correction" parameter, or discrepancy, between the original biased model and our ideal corrected model.

The difficulty with this approach is the choice of distribution for δ. Assuming that only the mean and variance matter, and not the exact distribution family, we could use a normal distribution, $\delta \sim N(\mu_\delta, \sigma^2_\delta)$. Then:

- We might set $\mu_\delta = 0$, indicating that we do not know the *direction* of bias, that is, we are unsure whether the published point estimate of θ is an overestimate or an underestimate.
- The variance σ^2_δ controls the potential amount of bias. To set this, we commonly have to rely on informal judgements, such as *"we are 90% confident that the true θ is within 10% of the biased θ_0."* This is equivalent to arbitrarily inflating the amount of uncertainty around the original θ.

Either Model 1 or Model 2 could be used for decision-making about the optimal intervention. A typical approach might be to use Model 1 to present a "base-case" result, accompanied by a deterministic sensitivity analysis to make sure that different reasonable assumptions about θ do not affect this result.

However, Model 1 does not allow us to determine the value of further information that would reduce this "structural" uncertainty about θ. This must be based on Model 2. Under Model 2, the EVPPI for δ can be seen as the "expected value of model improvement" (Strong and Oakley 2014). In practice, the model might actually be improved through obtaining data on the "unbiased" setting, which would inform θ in Model 2. But either way, we face a similar problem as before: the VOI for δ (or θ in Model 2) will depend on an unverifiable assumption about the potential extent of bias, the choice of variance σ^2_δ. And again, the impact of this assumption can only be assessed by deterministic sensitivity analysis: repeating the VOI calculation for a range of reasonable different values of σ^2_δ, and hoping that this does not affect the conclusions about the potential value of further evidence to improve the model.

This is a simple, artificial illustration, and the details of how this approach might be used in different modelling situations will be different. But the general principle remains: while VOI is a useful tool to guide decision-making and research prioritisation in the presence of uncertainty, it must be used in conjunction with good judgements, given the limitations of data and models in real decision problems. Given weak evidence, we may be unsure whether judgements should be quantified, or whether different evidence should just be presented qualitatively. In these situations, a kind of "reverse" inference might also be useful, e.g., determining how much bias there needs to be for the conclusions of the analysis to change materially. If this amount of bias is obviously implausible, then this will strengthen the conclusions.

9

Rapid Value of Information Using Minimal Modelling

Claire Rothery, David Glynn, and Hendrik Koffijberg

9.1 Introduction

VOI is most commonly applied when a comprehensive decision-analytic model is available to characterise the uncertainty in model inputs and translate this uncertainty into decision uncertainty arising from uncertainty in the model outputs. However, there are a number of barriers to the practical application of VOI in decision-making, including the time and resources available to construct sufficiently realistic decision-analytic models and the computational time required to undertake the analyses. Although important advances have been made in the efficient computation of VOI, and computing time is less likely to represent a barrier today than in the past 10 years, the methods that facilitate efficient computation of VOI metrics from a decision model do not reduce the time and human resources required to construct such models.

Policy decision makers with a responsibility for making research prioritisation and commissioning decisions often operate within rigid timelines, which makes traditional (i.e., full) modelling efforts unsuitable for integration into the research prioritisation process. Within healthcare decision-making, this is further exacerbated by the fact that those institutions with the remit for making reimbursement decisions about new health technologies, which is typically informed by decision-analytic modelling, are often separated from those institutions with the responsibility for prioritising and commissioning new research. Moreover, the models used to support reimbursement decisions may not be publicly available to support research prioritisation and commissioning decisions.

DOI: 10.1201/9781003156109-9

The need for practical and feasible methods within the time and resource constraints of a deliberative process of research prioritisation has called for the development of "rapid" or minimal modelling approaches. These approaches allow for rapid estimation of the benefits of further research without the need for constructing a comprehensive decision-analytic model. This involves simplifying, or omitting components, of the full modelling approach in order to produce sufficient information on the value of research in a timely manner. A rapid approach may also be viewed as offering a transparent, consistent and efficient method for setting research priorities and may provide a workable interface, whereby analysts and relevant decision stakeholders could validate key inputs and assumptions about the existing evidence in real time as part of a deliberative process for determining whether further research is required. The approach may also be relevant to different decision-making contexts, where the decision maker may choose the metric of value that is most relevant to their context.

9.2 Rapid Approaches for Estimating the Value of Research

In healthcare decision-making, minimal modelling has been proposed in the literature as a method for rapid estimation of the value of research (Meltzer et al. 2011). This can be performed when a prior clinical study is available that directly characterises uncertainty in the comprehensive measures of outcome that are sufficient to inform a decision for all relevant decision options. This is possible when: (i) the endpoints of relevance to the decision occur within the timeframe of the study; (ii) the study follows individuals up to the point of death or full recovery and records all relevant outcomes that are important for the decision; and (iii) there are no age-specific competing causes of death or other events after the study ends. Even in circumstances where these criteria are not met, it may be possible to extrapolate intermediate endpoints of a study to a comprehensive metric of value by constructing a relatively simple decision model with few parameters (Bennette et al. 2016).

The rapid approach for estimating the value of research that is proposed in this chapter (known as "rapid VOI") places the focus on a primary outcome of relevance to the decision (Claxton et al. 2015a; McKenna et al. 2016) and uses it to understand the consequences of decision uncertainty. The primary endpoint reported in studies, or a proposed new research study, typically captures the most important aspect of outcome for the decision. In healthcare decision-making, the primary outcome of clinical effectiveness may be used as the starting point to understand the health consequences of decision uncertainty, i.e., what are the consequences, expressed in terms of the primary outcome, if we do not fund the research study based on the current levels of uncertainty about the health technologies? Importantly, starting with a primary outcome does not mean that other outcomes are unimportant, it simply places the focus on a specific outcome of interest as a starting point for deliberation in order to establish the value of reducing decision uncertainty in that outcome. The expected value of research is then expressed as the number of events avoided/gained for a harmful/benefit outcome (or number of individuals experiencing a clinically important improvement for a continuous scale of outcome).

In situations where there are a number of other important aspects of outcome that are not captured in the primary outcome, a minimum clinical difference (MCD) for effect size in the primary outcome may be specified in order to implicitly account for these other unquantified aspects of outcome. The MCD represents a judgement about the improvement, expressed in terms of the primary outcome, that would need to be demonstrated in a new research study for one decision option to be considered superior relative to the alternatives. Requiring that research must demonstrate large differences in effect will tend to reduce the expected value of research because larger differences are less likely to be observed than smaller ones. A plausible MCD may be used as part of the deliberative process to assess whether the proposed research is sufficient to account for these other aspects of outcome. Where the endpoints of a study are not sufficient to capture all valuable aspects of outcome, an alternative approach is to use external evidence to link the endpoints to a comprehensive measure of outcome through simple extrapolation or modelling efforts.

9.3 The Minimum Requirements to Establish the Value of Research

The minimum requirements to conduct a rapid assessment of the value of research are described below for a binary outcome measure:

Primary outcome measure

The primary outcome measure captures the most important aspect of outcome. The value of research is expressed in terms of this outcome measure as "benefits gained" or "harms avoided" depending on whether the outcome is a benefit or harm. Alternative outcomes may also be considered.

Expected outcomes under the control (baseline outcome)

Research typically involves comparing outcomes associated with a new experimental technology and a control. An estimate of the baseline event rate or probability of an outcome in the absence of the new technology, i.e., under the control or current available technology, is required. This should be expressed in terms of the primary outcome measure. The baseline probability of outcome may be informed by the control arm of relevant clinical trials, external evidence, expert opinion or judgements relevant to the decision, and is likely to be subject to uncertainty.

Uncertainty about the relative effect of the experimental technology

An estimate of the relative effect of the experimental technology compared to a control is required for the primary outcome, along with an estimate of its uncertainty. This is usually expressed in terms of an odds ratio or relative risk, with a 95% confidence interval (or standard error) representing the range of plausible values that the quantity can take.

Importantly, some judgement about the uncertainty in this estimate based on what is already known about the quantity must be made in order to determine whether additional evidence is required. This judgement may come from a systematic review and meta-analysis of the available existing evidence, or from alternative sources such as expert elicitation or meta-epidemiological studies. If an estimate is unavailable or considered inadequate, alternative plausible values may be used in a scenario analysis to represent different judgements for deliberation about the uncertain estimate of relative effect.

Incidence per annum

Information generated by research is used to inform decisions for the population of individuals who could benefit from the information. An estimate of the number of individuals affected by the uncertain decision choice each year is required in order to establish the size of the benefits to the target population.

Minimum clinical difference (MCD) in primary outcome

When the primary outcome is not sufficient to capture all valuable aspects of outcome relevant to the decision, an MCD for effect size in the primary outcome may be specified in order to implicitly account for these other unquantified aspects of outcome. For example, the experimental technology may have higher costs associated with it compared to the control that are not captured in the primary measure of outcome. The MCD represents a judgement about the improvement in the primary outcome that would need to be demonstrated in a new research study for the new technology to be considered superior to the control.

Costs of the proposed new study

Some assessment of the likely costs of the proposed new research study is required in order to establish whether the expected value of the research is sufficient to justify the expected costs.

Duration of the proposed new study

A judgement about the duration of time it will take for the proposed research to be conducted and for the results to be reported is required because the value of research declines with the longer it takes for research to be reported. This might be informed by an assessment of study sample size, expected recruitment rates, or historical experience from conducting similar types of studies.

Length of time for which new research is expected to be valuable

The information generated by new research will not be valuable indefinitely because other changes occur over time. For example, over time new and more effective technologies become available, which will eventually make those currently available obsolete. This means that new information about relative effect is only relevant for a specific amount of time. A judgement about the length of time that the evidence from the proposed new study might be valuable is required in order to estimate the expected value of research over an appropriate time horizon. This judgement could be informed by historical evidence or experience about whether a particular research area is likely to see future innovations and/or other evaluative research reporting.

Discount rate

When a time horizon greater than 1 year is considered, discounting should be used to reflect the fact that resources committed today for new research could be invested at a real rate of return to provide more resources in the future.

9.4 Rapid Estimation of the Value of Research

The value of research to resolve decision uncertainty in the primary measure of outcome is estimated by sampling from the uncertainty distributions of relative effect and baseline event rate (as typically determined from the range of plausible values specified by the confidence interval or standard error on these quantities) and multiplying by the number of individuals per annum whose decision choice is to be informed by the research. Each sampled value from the distributions is interpreted as one realisation of how outcomes for the population of interest might turn out in practice, as supported by the existing evidence, i.e., each sampled value represents one possible "true" value of how outcomes could turn out. Repeating this process many times (e.g., 10,000) creates a distribution of the consequences of uncertainty, which is expressed in terms of the primary outcome measure. For example, if the primary outcome is mortality, the consequences of uncertainty are expressed in terms of number of deaths per annum. The distribution of consequences provides the chance of making an "incorrect" decision due to uncertainty in the existing evidence about the optimal decision choice, while the number of individuals affected by the decision provides the scale of the consequences of this uncertainty per annum. The average over this distribution provides an expected upper bound on the value of research to resolve this uncertainty, i.e., the EVPI.

Table 9.1 illustrates the process used for the estimation of the expected value of research for five random samples taken from the independent distributions of relative effect (odds ratio for the experimental technology relative to a control) and baseline risk (control) for a primary outcome of mortality.[1] Each sampled value from the distributions is interpreted as one possible realisation of uncertainty, i.e., one possible "true" value of how individual outcomes might turn out, as supported by the current evidence. The last row of Table 9.1 represents the average across the five sampled values corresponding to the central estimates. Across the five sampled values, the balance of evidence based on what is already known about the experimental technology and control indicates that the experimental technology is expected to reduce the number of deaths per annum by ten compared to the control, i.e., the odds ratio for death of 0.96 is in favour of the experimental technology (column A) and the mean absolute number of deaths for the experimental technology is 302 per annum (column E) compared with the baseline risk for the control of 312 deaths per annum (column D). This means that based on current information (without conducting further research) the balance of evidence is in favour of adopting the experimental technology. However, there is a chance that the experimental technology would increase rather than reduce the number of deaths – this is seen for the sampled realisations 4 and 5, where the odds ratio for the experimental technology is greater than one.

The expected consequences of this uncertainty depend on the likelihood that the experimental technology is less effective than the control, how much less effective it is, and the size of the eligible population. The chance that the experimental technology is less effective is simply the chance of observing an odds ratio greater than one, which is 40%, i.e., in two out of the five samples. The resulting consequences of this uncertainty is the number of additional deaths incurred if the experimental technology is used instead of the control in these instances (i.e., realisations 4 and 5 of column G). The expected value of additional research to resolve this uncertainty is a weighting of the consequences of the uncertainty by the likelihood of them occurring, which is the average of the consequences of uncertainty across the sampled realisations (i.e., the average of the samples in column G). Therefore, the expected upper bound on the value of research to resolve uncertainty in the decision choice between the experimental technology and the control is five deaths averted per annum for this sample.

1. Note that any structural correlation between baseline risk and relative effect should be preserved where possible when sampling from the distributions between these two quantities.

TABLE 9.1
An example of rapid estimation of the value of research for a primary outcome of mortality. MCD: minimum clinical difference.

				Deaths per annum for an incidence of 1,000 eligible patients		**Value of additional evidence (Expected Value of Perfect Information)**		
Sampled realisation of uncertainty	**Odds ratio for death (experimental technology vs. control)**	**Baseline odds of death for control**	**Odds of death for experimental technology ($= AB$, assuming distributions independent)**	**Control ($= \frac{1000B}{1+B}$)**	**Experimental technology ($= \frac{1000C}{1+C}$)**	**Absolute effect for experimental technology in number of deaths per annum ($= E - D$)**	**Consequences of uncertainty for experimental technology ($= F$ if $A > 1$)**	**Consequences of uncertainty for experimental technology when MCD = 10 ($= G$ if $G \geq 10$)**
	[*A*]	[*B*]	[*C*]	[*D*]	[*E*]	[*F*]	[*G*]	[*H*]
1	0.83	0.6	0.5	375	332	−43	0	0
2	0.91	0.54	0.49	351	329	−21	0	0
3	0.95	0.49	0.47	329	318	−11	0	0
4	1.05	0.35	0.37	259	269	9	9	0
5	1.08	0.33	0.36	248	263	15	15	15
Average	0.96	0.46	0.44	312	302	−10	5	3

If a larger difference in outcome between the experimental technology and control (i.e., greater than five deaths averted per annum) is required to be detected in a new research study before the results are considered clinically important, or because there are other important aspects of outcome that are not captured in the endpoint of mortality, then an MCD may be specified. For the sampled data presented in Table 1, an MCD of ten deaths averted per annum means that clinical practice will only change (i.e., revert from the experimental technology based on current information to the control) if the results of the new research study indicate that at least ten additional deaths are averted per year. The last column of Table 9.1 (column H) shows the expected value of research for an MCD of ten additional deaths averted per annum. The chance that the experimental technology is less effective than the control occurs in realisations 4 and 5, but if further research must demonstrate a difference of ten additional deaths averted per year then this only occurs in realisation 5. The need to demonstrate a reduction of ten additional deaths per annum reduces the expected upper bound on the value of research from five to three deaths averted per year.

The value of research over the time horizon for which the research is expected to be of value is the discounted sum of the expected benefits of research for the population of individuals who can benefit from the information up to the point at which the new evidence is no longer considered useful to clinical practice. This provides an expected upper bound on the value of research over the relevant research time horizon. A summary measure of the expected value of conducting the proposed new research study is based on a comparison of the maximum expected value of research to the costs of the research. The costs of the proposed research study may be expressed in terms of the primary outcome measure by using an assessment of the health opportunity costs of research expenditure for that outcome. For example, a study in the UK estimating the relationship between changes in National Health Service (NHS) expenditure and health outcomes suggests that it costs the NHS £114,000 to avert one death (Claxton et al. 2015b). Using this assessment of the health opportunity costs of research expenditure, a new research study costing £2 million would be expected to avoid 17 deaths per annum in the UK NHS. If the expected upper bound on the value of research were less than 17 deaths averted per year then the research would not be considered of value to the UK NHS.

The rapid estimation of the value of research in terms of a primary outcome measure provides a useful starting point for deliberation on the value of the proposed research study. The translation to a more comprehensive and comparable measure of health outcome such as quality-adjusted life years (QALYs) would enable the value of research and its costs to be compared directly across research proposals to determine their relative priority.

9.5 Conclusions

Rapid VOI offers a quick and practical means for estimating the value of research by avoiding the need for developing comprehensive decision-analytic models. One notable limitation associated with taking a minimal approach to modelling is the risk of over-simplification of complex processes; the extent to which the simplified approach adequately addresses the need for further evaluative research is an important consideration and the assumptions underpinning the analysis should be made explicit with an assessment of their likely impact on the results. Nonetheless, within the time and resource constraints of a deliberative process of research prioritisation and commissioning decisions, the rapid estimation of the value of research provides a starting point for deliberation to understand the consequences of uncertainty associated with existing evidence and the value of conducting new research.

10

EVSI Portfolio Optimisation

Michael Fairley and Jeremy Goldhaber-Fiebert

10.1 Introduction

Limited healthcare budgets imply the need for resource allocation and prioritisation. Likewise, limited health research funds imply that clinical research studies must be designed to support better decision-making in the context of an overall study evaluation, prioritisation, and selection process.

While Value of Information (VOI) analyses can be used to estimate expected gains from collecting data to inform health decisions, consider the size of the population that could benefit along with the timing and extent of implementation, optimise study design features (e.g., sample size), and select optimal study designs (Raiffa and Schlaifer 1961; Thompson 1981; Weinstein 1983; Hornberger, Brown and Halpern 1995; Claxton and Posnett 1996; Claxton 1999b; Claxton and Thompson 2001; Ades, Lu and Claxton 2004; Eckermann and Willan 2008; Fenwick, Claxton and Sculpher 2008; Conti and Claxton 2009; Griffin, Welton and Claxton 2010; Willan and Eckermann 2010; McKenna and Claxton 2011; Hall et al. 2012; Wilson 2015), analyses typically evaluate and optimise such outcomes for one decision at a time. They do not consider the research portfolio and how to optimise the overall expected value of information achieved by research investments across all commissioned studies.

Work by Fairley, Cipriano and Goldhaber-Fiebert (2020) develops a framework for analysing research study portfolio investment decisions, where the studies inform different policy decisions and investment in the portfolio is made in the context of a constrained research budget. The key insight from the analysis is more spending on one study (e.g., to conduct it with a

DOI: 10.1201/9781003156109-10

larger sample size) necessarily results in fewer resources available (e.g., a smaller sample size) for one or more other studies. Because of this, there is a necessary trade-off between reducing decision uncertainty for one policy decision and reducing it for other decision(s). The framework employs a budget-constrained optimisation that maximises the sum of the expected net benefit across the portfolios of study investments considered.

10.2 Analytic Framework for the VOI Portfolio Problem

Fairley, Cipriano and Goldhaber-Fiebert (2020) initially assume that each decision problem has only one candidate study under consideration and the outcomes of each study only inform one decision problem but also discuss the consequences of relaxing this assumption.

10.2.1 The Research Funder's Decision Problem

A research funder must allocate a fixed research budget, $B > 0$, selecting research studies to fund from many candidate studies. The candidate studies s in the set $\mathcal{S}$ are each related to a single clinical decision problem. For simplicity, it is assumed that each clinical decision has only one study and that each study will only inform one clinical decision. Extending the general framework of considering research investments for studies for a single decision problem, the subscript $s \in \mathcal{S}$ is added to the decision set, net monetary benefit functions, uncertain parameters, data, sample size, cost function, EVSI and the discounted sum of the number of individuals affected by the decision ($\mathcal{D}_s, u_{sd}(\boldsymbol{\theta}_s), \boldsymbol{\theta}_s, \mathbf{X}_s, n_s, c_s(n_s), v_s(n_s), N_s$, respectively) to obtain the $\text{ENBS}_s(n_s)$ for each decision.

The research funder maximises the expected net present value of research by selecting which studies receive any funding and the sample size for each funded study, subject to the research budget constraint:

$$\begin{aligned} \max_{n_1,\ldots,n_{|S|}} \quad & \sum_{s\in\mathcal{S}} N_s v_s(n_s) - c_s(n_s) && \text{(OPT1)} \\ \text{s.t.} \quad & \sum_{s\in\mathcal{S}} c_s(n_s) \leq B \\ & n_s \in \mathbb{Z}_+ \quad \forall s \end{aligned}$$

The above formulation considers a single-period allocation problem and hence does not consider multi-period budgets or borrowing/lending between periods.

The optimisation problem OPT1 is difficult to solve in general.

Firstly, the EVSI function, $v_s(n_s)$, is generally non-concave. EVSI often increases slowly at small sample sizes because small studies are unlikely to influence the decision, then increases rapidly at sample sizes that are sufficient

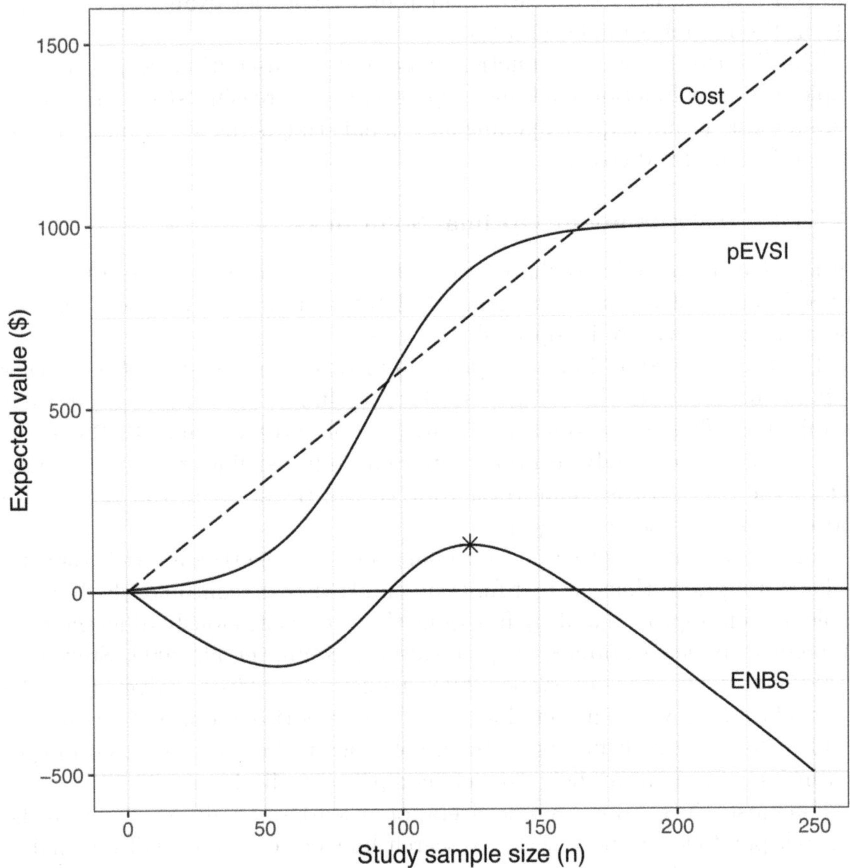

FIGURE 10.1
Example of common shapes of Population EVSI and ENBS where ENBS rises and falls and below a certain sample size it is negative.

to influence the decision, and, finally, asymptotically approaches the expected value of (partial) perfect information (Raiffa and Schlaifer 1961; Frazier and Powell 2010). $\text{ENBS}_s(n_s)$ often falls and then rises, decreasing into negative ENBS at for small values of n_s when the expected value of the study does not exceed the study's costs, then increasing to a point where the marginal value of information equals the marginal cost of information (i.e., $N_s \frac{\partial v_s(n_s)}{\partial n_s} = \frac{\partial c_s(n_s)}{\partial n_s}$), followed by another period in which the function is decreasing because the costs continue to rise with n_s as the EVSI plateaus (for example, see Figure 10.1).

Secondly, the EVSI function is not necessarily smooth and continuously differentiable. For example, if the parameter uncertainty is Beta-distributed and the information-generating process is Beta-Binomial-distributed, the EVSI function has a piecewise-concave, scallop-like shape (for example, see Figure 5.11 in Raiffa and Schlaifer (1961)).

Thirdly, the $\text{ENBS}_s(n_s)$ function may contain discontinuities, if, for example, the $c(n)$ function contains steps because increasing study size above certain levels would require opening additional study sites which imply incurring additional fixed costs.

10.2.1.1 Insights about Optimal Solutions

Fairley, Cipriano and Goldhaber-Fiebert (2020) reformulate OPT1 into a mixed-integer optimisation problem (OPT2) that more readily yields insights about the properties of its optimal solutions.

The transformed optimisation problem involves two decisions: (a) whether to fund the study at all; (b) at what level to fund the study (e.g., at what sample size). This transformation removes sample sizes for which $\text{ENBS}_s(n_s) < 0$ (e.g., due to fixed study costs) and appropriately redefines the cost function so that the fixed cost is equal to the total cost of the study at the sample size above ENBS is no longer negative.

This mixed-integer program's optimal solution properties are well known – under assumptions that the cost function and ENBS are continuously differentiable or well-approximated by functions that are continuously differentiable. Numerous applied examples suggest this (McKenna et al. 2009; Stevenson et al. 2009; Cipriano and Weber 2018; Kunst et al. 2019). Given this, the reformulation provides insights into the research portfolio optimal investment problem (even though finding optimal solutions to large-scale mixed-integer non-linear programs can be extremely computationally intensive).

Foremost, the optimal sample size(s) of at least some of the studies in the research portfolio problem with a research budget constraint will be smaller than in the standard (unconstrained) one-decision-problem-at-a-time sample size optimisation.

In the unconstrained problem, the solution is that for any $n_s^* > 0$, the optimal sample size will satisfy

$$\frac{\partial\big(N_s v_s(n_s^*) - c_s(n_s^*)\big)}{\partial n_s} = 0 \quad \forall s \in \mathcal{S}. \tag{10.1}$$

This means that at the optimal sample size, the marginal value of information equals the marginal cost of information (i.e., $N_s \frac{\partial v_s(n_s^*)}{\partial n_s} = \frac{\partial c_s(n_s^*)}{\partial n_s}$).

However, with the research budget constraint, the solution is that for any $n_s^{**} > 0$, the optimal sample size will satisfy

$$\frac{\partial\big(N_s v_s(n_s^{**}) - c_s(n_s^{**})\big)}{\partial n_s} = \lambda^R \frac{\partial c_s(n_s^{**})}{\partial n_s} \quad \forall s \in \mathcal{S} \tag{10.2}$$

where $\lambda^R \geq 0$ is the Lagrange multiplier used in constrained optimisation problems that represents the shadow price for the research budget constraint. This implies that $N_s \frac{\partial v_s(n_s^{**})}{\partial n_s} = (\lambda^R + 1)\frac{\partial c_s(n_s^{**})}{\partial n_s}$.

Instead of increasing the sample size until the marginal value of information is equal to the marginal cost of information for each study, the sample size in each study is increased until the marginal ENBS from one more unit of sample size divided by the cost of one more unit of sample size is the same for all studies in which the research funder invests.

Comparing the two optimal solutions, one can see that $\frac{\partial v_s(n_s^{**})}{\partial n_s} \geq \frac{\partial v_s(n_s^{*})}{\partial n_s}$ when the budget constraint binds (i.e., $\lambda^R > 0$) which implies that $n_s^{**} \leq n_s^{*}$.

Thus, λ^R identifies the minimum required return on investment per dollar spent on research, and, consequently, $(\lambda^R)^{-1}$ implies the willingness to pay for one additional expected dollar of value gained from research. At lower budgets, a higher return on research investment tends to be required, because investing in studies with relatively lower returns represents a relatively larger opportunity cost, as there is then less available budget to invest in studies with higher rates of return.

Of note, because studies can enter and leave the optimal solution as the budget increases, the value of λ^R does not always monotonically decrease with increasing budget.

10.2.1.2 Optimal Sample Sizes for a Portfolio of Studies: Numerical Methods

Fairley, Cipriano and Goldhaber-Fiebert (2020) provide a second reformulation of OPT1 into an integer programming problem (OPT3) that is amenable to existing computational solution methods that yield globally optimal solutions with only small approximation error even when the assumptions of OPT2 are not satisfied. The efficient solution method relies on the assumption that the EVSI function and the cost function are additively separable (Keha, De Farias and Nemhauser 2006).

OPT3 involves approximating $v_s(n_s)$ and $c_s(n_s)$ as 1-dimensional piecewise linear functions to formulate a problem that can be solved using integer programming. Let n_s^k for $k \in \{0, \ldots, T\}$, where T is the number of piecewise segments, be a grid of points at which to evaluate $v_s(n_s)$ and $c_s(n_s)$ with $n_s^0 = 0$ and $n_s^T = u_s$, where u_s is the maximum value of n_s such that $c_s(n_s) = B$. The formulation of the sample size allocation problem is then:

$$\max_{\gamma} \sum_{s \in \mathcal{S}} \sum_{k=0}^{T} N_s\big(v_s(n_s^k) - c_s(n_s^k)\big)\gamma_s^k \qquad \text{(OPT3)}$$

$$\text{s.t.} \sum_{s \in \mathcal{S}} \sum_{k=0}^{T} c_s(n_s^k)\gamma_s^k \leq B$$

$$\sum_{k=0}^{T} \gamma_s^k = 1 \quad \forall s \in \mathcal{S}$$

$$\gamma_s^k \geq 0 \quad \forall s \in \mathcal{S}, \quad \forall k \in \{0, \ldots, T\}$$

$$\{\gamma_s^k : k \in \{0, \ldots, T\}\} \text{ is SOS2} \quad \forall s \in \mathcal{S}$$

where the constraint that $\{\gamma_s^k : k \in \{0, \dots, T\}\}$ is SOS2 means that the set of decision variables, $\{\gamma_s^k : k \in \{0, \dots, T\}\}$, is a Special Ordered Set of type 2 (SOS2). For a Special Ordered Set of type 2, there can be at most two variables that are positive and if two are positive, they must be adjacent in the ordered set. The intuition for this condition is that it enables the integer program to be solved efficiently and with a good approximation of the optimum – the program involves evaluating a grid of points n_s^k where each γ_s^k corresponds to one of the points. The SOS2 condition means that the two adjacent γ_s^ks that are positive identify the two n_s^ks that correspond to the segment of the piecewise linear function. The optimal sample size is either one of the identified grid points itself or a convex combination of two adjacent sample sizes on the grid where those sample sizes are close enough that a linear approximation of EVSI and cost between those points is accurate. SOS2 enforces that we can only select one point, and if it is between points on the grid then those points must be adjacent so the linear approximation is accurate. If the points were far apart, a linear approximation of cost and EVSI across that range would likely have insufficient accuracy.

10.3 Example of Optimal Sample Size Allocation for a Portfolio of Research Studies

Consider a research funder contemplating funding three potential studies who must decide which of these studies to fund and at what level given a total research budget constraint. Each study relates to a different clinical decision problem and hence has separate uncertain parameters θ_s and linear cost function $c_s(n_s)$ such that the marginal cost for each unit of additional sample size is different across studies.

Figure 10.2 shows the ENBS as a function of sample size for each of the three example studies. Without a budget constraint (i.e., if one found the optimal sample size for each study separately), approximately 1,065 people would be enrolled in total. For each study, the marginal gains from an additional unit of sample would equal its marginal cost – the optima (shown with stars in the figure) would occur at the top of the ENBS curves.

With a budget constraint (shown as the points labelled B1 in the figure) applied to the portfolio decision, the optimal sample sizes for the three example studies are smaller than the unconstrained optimal sample sizes. If the cost of each unit of sample were the same for each of the studies, the tangent at the constrained optimal sample sizes would be equal; if the x-axes were mapped to the dollar cost of each unit of sample size they would be even in this example where the marginal costs for each study are not equal.

In this example, without a budget constraint, the total population ENBS is \$9.7 million, and the total cost is \$5.6 million. Budget constraints at or above

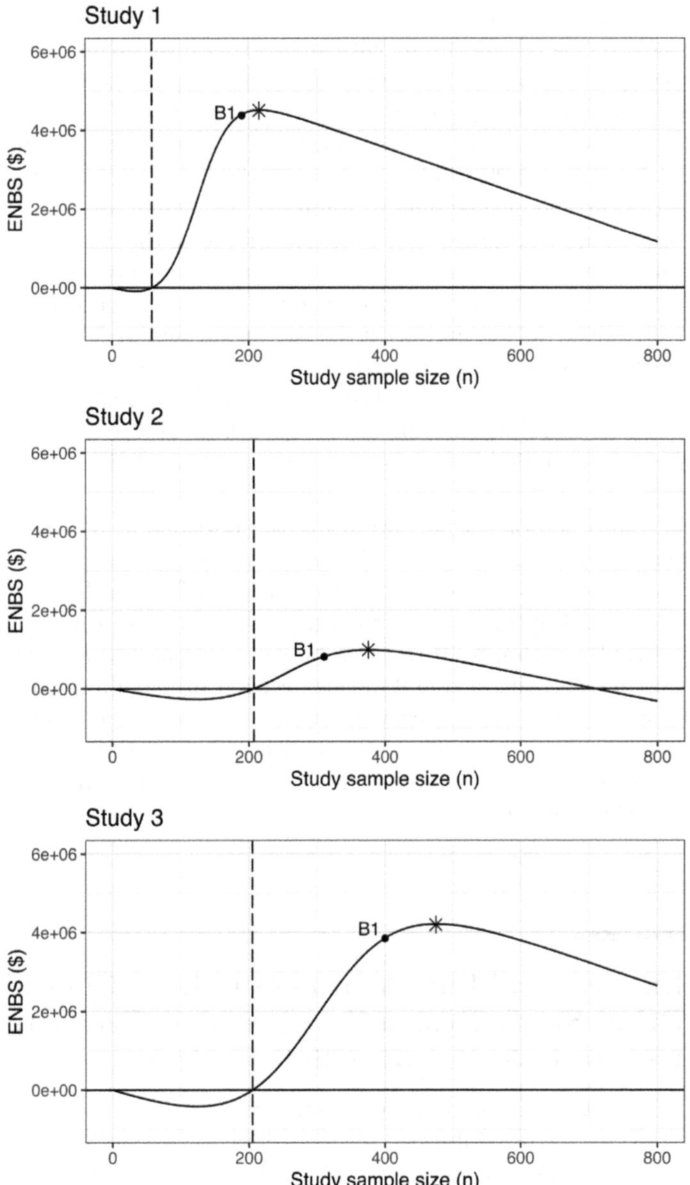

FIGURE 10.2
ENBS curves for studies 1–3 depicting the optimal sample size for each study without a budget constraint (star) and the optimal sample size for each study with a total research budget constraint (B1) (dark dot).

this level will not be binding, since it is not optimal to spend more in this example, as ENBS will be declining with any further investment. The optimum with the budget constraint occurs when one spends up to the constraint ($4.7 million) to yield a total population ENBS of $9.1 million.

Tighter budgets lead to a lower total achievable ENBS. Study sample sizes below 58, 207 and 205 yield negative ENBS for studies 1–3 respectively (shown as the vertical dashed lines in Figure 10.2). Therefore, if the budget is not sufficient to allow a sample size at least as large as the minimum required for a given study, it is better not to invest in that study at all. One can imagine that as budgets tighten, the optimal points shift down and to the left. However, since an optimal point for one study may reach the minimum threshold before the others do, that study will suddenly be defunded entirely, and the remaining amount now in the budget would be reallocated to other studies, pushing their optimal sample size(s) back up and to the right. Hence, sample sizes of all studies are not necessarily strictly declining as budgets are tightened.

It should also be noted that optimising the portfolio as a whole to invest in a greater number of studies at their unconstrained optimal levels can yield greater total ENBS at the same cost, compared to funding fewer studies at their unconstrained optimal levels. Hence, a one-study-at-a-time approach to maximising ENBS with a fixed research budget can lead to inefficiencies that can be overcome by using a simultaneous decision problem approach.

For further examples see Fairley, Cipriano and Goldhaber-Fiebert (2020) who provide numerical examples and a real-world case study that only uses the outputs of probabilistic analyses from cost-effectiveness analyses (CEA) published in the literature.

10.4 Extensions: Multiple Studies in Each Clinical Area, Other Study Design Features, and Correlated Uncertainties Across Clinical Areas

The previous sections focused on optimisations with the simplifying assumptions that: (a) each study informed a single clinical decision and each clinical decision had only one study; and (b) the only feature being optimised was study sample size and not other design features like study duration or frequency and intensity of monitoring study subjects. Such assumptions can be relaxed, yielding a richer and more general formulation of the research portfolio investment optimisation problem.

The more general formulation drops the subscript, s, from parameters, data and sample sizes. In it, $\boldsymbol{\theta}$ is a vector of the union of all parameters across all decision problems; $\mathbf{X}$ is the union of data from all possible data-generating studies, and $\mathbf{n}$ is a vector of possible research study designs amongst which to choose (e.g., sample sizes, duration, intensity, etc.) for different data collection exercises. The number of elements in $\mathbf{n}$ is not necessarily equal to the number of clinical decision problems the studies seek to inform, $|\mathcal{S}|$. The data-generating studies collect information relevant to some subset of the parameters which may influence decision certainty in multiple decision problems. Each decision problem, $s \in \mathcal{S}$, has a decision-problem-specific net monetary benefit function, $u_{sd}(\boldsymbol{\theta})$, and a per-person EVSI:

$$v_s(\mathbf{n}) = \mathrm{E}_{\mathbf{X}}[\max_d \mathrm{E}_{\boldsymbol{\theta}|\mathbf{X}}[u_{sd}(\boldsymbol{\theta}) \mid \mathbf{X}]] - \max_d \mathrm{E}_{\boldsymbol{\theta}}[u_{sd}(\boldsymbol{\theta})]. \tag{10.3}$$

However, the cost function, $c(\mathbf{n})$ is no longer necessarily additively separable on the components of $\mathbf{n}$. The research funder's decision problem continues to be to maximise the expected net present value of research by selecting which studies will receive any funding and the design of each funded study subject to the funder's budget constraint:

$$\begin{aligned} &\max_{\mathbf{n}} \sum_{s \in \mathcal{S}} N_s v_s(\mathbf{n}) - c(\mathbf{n}) \\ &\text{s.t.} \quad c(\mathbf{n}) \leq B \end{aligned} \tag{OPT1$'$}$$

where e indexes each component of the vector $\mathbf{n}$.

EVSI for each decision problem can depend on all parameters ($\boldsymbol{\theta}$), and the decision to collect data for any subset of those parameters can affect potentially multiple decision problems. Furthermore, due to correlation between parameters, information collection specifically focused on one parameter may provide information about a larger set of parameters that are correlated with it. Such correlation substantially complicates the calculation of EVSI for a single decision problem. Fairley, Cipriano and Goldhaber-Fiebert (2020) derive the rather complex closed-form expression for EVSI with correlation between parameters across multiple decision problems even in a relatively simple example. Extending Jalal and Alarid-Escudero (2018), they also derive an expression for the preposterior mean in the presence of correlation as a convex combination of a sample from the prior and the prior mean, where the prior sample is weighted by ν_s^{ρ}, and the prior mean is weighted by $1 - \nu_s^{\rho}$ where $\nu_s^{\rho} \in [0, 1]$.

When correlation is greater in magnitude, a study that informs the parameter for one decision problem also informs other decision problems. Hence, all else equal, when such correlation exists, there is more value to conducting such a study at a given sample size compared to the case of zero correlation.

In general, quantifying the additional value that accompanies correlation so as to optimise sample size allocation can be extremely challenging because the EVSI function and cost function are no longer necessarily additively separable. General non-linear solvers would be required, and the optimal solution might be estimated using Monte Carlo methods.

Even so, some insights from the optimisation problem without correlation are still applicable. When a single study collects information relevant to multiple decision problems, optimal level of funding occurs when the marginal population ENBS (summed over all influenced decision problems) expected from one more unit of budget will be the same for all funded studies. However, the optimal sample size for an individual study may be lower or higher than in the case when there is no correlation, because the optimal investment in each study is influenced by the contribution to population EVSI from others as well.

10.5 Conclusions and Discussion

Under a research budget constraint, the optimal level of investment in research is not generally identified as the maximum of the ENBS(n) curve for individual decision problem studies, but rather requires considering considering the optimal portfolios of studies to fund and the investment levels for each funded study.

The optimal allocation will occur when the marginal ENBS divided by the marginal cost is equal for all studies with non-zero sample size allocations, when ENBS(n) and $c(n)$ are continuously differentiable in the decision-relevant region. Generally, the sample size allocation of at least some studies will be lower than in the one-study-at-a-time (without budget constraint) optimisation since the marginal return to investing in more sample size for a study approaches zero as it nears its unconstrained optimum sample size; instead of collecting these small returns, the research funder can achieve larger gains by investing more in another study that is further from its optimal sample size, such that its marginal return to increasing sample size is bigger.

Real-world applications of the research portfolio optimisation are feasible. Computationally efficient solutions of the research portfolio optimal investment problem are facilitated by Fairley, Cipriano and Goldhaber-Fiebert (2020) through a reformulation of the optimisation problem as an integer programming problem, using piecewise linear approximations of ENBS(n) and $c(n)$ that do not require them to be continuously differentiable. Existing computational methods that yield the globally optimal solution with small approximation error are available for this problem (Keha, De Farias and Nemhauser 2006). Since the optimisation requires computing EVSI as a function of sample size, its computational tractability is further supported by published EVSI approximation methods (Strong et al. 2015; Menzies 2016; Jalal and Alarid-Escudero 2018; Heath, Manolopoulou and Baio 2018). Furthermore, a number of these methods can be used even when access to the underlying cost-effectiveness models are not available, as long as the raw outputs from probabilistic sensitivity analyses using the cost-effectiveness models are available. This further emphasizes the importance of making such outputs publicly available for published cost-effectiveness analyses through online appendices or in open web repositories. Journal or funder requirements to do so could facilitate the open-sourcing of these outputs.

The solutions to the research portfolio optimal investment problem should be seen as important inputs into the research funder's decision-making process, rather than final prescriptive answers. Real-world research funders like the US National Institutes of Health and the UK National Institute for Health Research consider a number of factors when making funding decisions for the thousands of studies they consider each year. These factors include, but are not limited to, the impact of the portfolio of studies in which they invest (Gillum et al. 2011; Wahls 2019). Quantifying the ENBS of studies individually, as described earlier in this book, and considering the ENBS of the portfolio of studies collectively, as described in this section, are thus important advances in achieving research funders' stated goals. These advances are made more feasible and practicable because the portfolio optimisation problem could be conducted ex post and without access to the underlying models for each study, as long as each study submitted to the research funder also provides its ENBS curve over a range of investment levels relevant both to study design and sample size.

Just as there is much discussion about willingness-to-pay thresholds for health and their relationship to hard or soft annual budgets (Neumann et al. 2016), the work on the optimal research portfolio highlights the willingness-to-pay threshold for expected net gains from health research. Given siloed budgets, it is not clear that these two thresholds are (or should necessarily be) equivalent. Future research building on work like that published by McKenna et al. (2010) would therefore be useful in illuminating such questions.

Extensions to this work that integrate implementation decisions and research decisions with a common budget constraint could yield important insights. Likewise, considering of multi-period allocation problems (perhaps with the ability to borrow against future budgets), that might involve path dependence of research studies funded sequentially over a period of years, would be interesting and fruitful areas for further research.

Interestingly and importantly, the health research portfolio optimisation problem also applies to value of information analyses related to personalisation and individualisation of healthcare. It can be used to consider allocation of sample size across subpopulations within a single clinical trial or research study. The question of how much sample size should be allocated to each study within a research budget constraint can be mapped to the question of how much sample size to allocate to each subpopulation within a study given a study budget constraint. As a simple example, if there is some existing data that suggest that there is heterogeneity of treatment response based on patient genotype (e.g., genotypes A and B) and one is more certain about the treatment response of patients with genotype A, then it may be more efficient to fund a further study that oversamples patients with genotype B. However, if additional information about the treatment response of patients with genotype A also informs the treatment response of those with type B and it is less costly to recruit patients with genotype A to a study, then oversampling patients with genotype A may be more efficient. As subpopulations may be defined by collections of patient features, the selection and mapping of patient features to partition patients into subpopulations, and the consideration of correlations in the uncertainty with respect to these subpopulations, are important components of such optimisation problems and should be the subject of future research. Likewise, future research could explore nested portfolio problems in which optimal resource allocation across studies is considered and each such resource allocation defines a study budget that then can be optimally allocated across populations within the study.

In conclusion, allocation of limited research funds across and within studies implies that current approaches to study design optimisation can and should be extended to include the trade-offs inherent in the decision and hence optimal research portfolio allocation.

11

Assessing Value of Information Given Non-Optimal Previous Decisions

Doug Coyle, David Glynn, Jeremy Goldhaber-Fiebert, and Edward C. F. Wilson

The objective of economic evaluation of healthcare is to identify which course of action by a decision maker would lead to the highest level of health within the population of interest: i.e., maximise the expectation of net benefit (Stinnett and Mullahy 1998; Paulden 2020a). This requires consideration of the long-term costs and benefits associated with each potential treatment option and the placing of these within the context of the treatments which would need to be forgone given the limited budget for healthcare.

The study of economics has traditionally been categorized into positive and normative economics, although the two subdisciplines are inherently linked (Hausman 2018). Positive economics relates to what occurs and provides an explanation for why this is the case. It describes and explains various economic phenomena. Normative economics focuses on what ought to occur based on specific societal objectives. Positive economics is therefore concerned with verifiable facts about the world, whereas normative economics is concerned with values. For example, a positive analysis will describe and predict the consequences of a policy change. A normative analysis will assess the relative desirability of those consequences.

Models for assessing optimal decision choices based on economic evaluation can adopt features of both a positive and a normative framework. A positive model considers how decisions about implementation or disinvestment are made, and directly considers both previous decisions and the likely impact of the different decisions that a decision maker can make. A normative model considers which decision should be made, with a focus on the determination of the optimal decision. Features of either a positive and/or normative framework can be applied to both the assessment of cost-effectiveness and the assessment of the Value of Information (VOI).

Economic evaluation is traditionally conducted from a normative framework (Karlsson and Johannesson 1996). In the traditional framework, the

DOI: 10.1201/9781003156109-11

optimal treatment choice is determined as the treatment with the highest net benefit (Stinnett and Mullahy 1998). When placing economic evaluation in the context of whether a new treatment should be reimbursed, the current normative framework assesses whether the new treatment under consideration would maximise the level of the health of the population, if it were the only treatment choice reimbursed. An alternative but complementary framework, which may be considered to contain both positive and normative features, would involve acceptance of previous decisions with respect to treatment reimbursement and its effect on treatment adoption (i.e., the market shares of the currently reimbursed treatment choices) and would focus on how these would change following an approval and thus how the level of the health of the whole population would change (Coyle et al. 2015; Kim and Basu 2017).

Value of information analysis assesses the potential value to be obtained from reducing uncertainty in parameter estimates within the economic evaluation, with respect to the underlying uncertainty over the optimal decision (Felli and Hazen 1999). That is, it translates reductions in parameter uncertainty into reductions in decision uncertainty and therefore reductions in the expected loss associated with the decision. Analyses based on an entirely normative framework would incorporate the results of an economic evaluation conducted under the traditional normative framework, and would involve assessing the information value based on determining what is the optimal treatment choice by assuming full adoption of this technology for an explicit period of time.

Value of information analysis could be based on an economic evaluation conducted from the alternative framework identified above, which incorporates the previous decisions made, focusing on the uncertainty over whether the treatment should be added to those previously reimbursed. A model incorporating positive features could also consider the implications of information (e.g., trial evidence), which may not just change the uncertainty regarding long-term costs and benefits of treatment; but could also influence the uptake of new technologies (i.e., bring further changes to the market shares of treatments after the new treatment is reimbursed).

In the remainder of this chapter, we first outline an example showing cost-effectiveness analyses under the traditional normative and the alternative positive framework, showing how they can lead to different adoption decisions. We then extend the comparison into how this affects the value of further information.

11.1 Assessing Cost-Effectiveness

The traditional normative framework for assessing cost-effectiveness compares each treatment alternative and identifies the optimal treatment choice based on the decision maker's threshold value of a QALY (λ): the optimal treatment will be that which maximises the expectation of net benefit (Stinnett and Mullahy 1998). The output is a recommendation of which product should be used in preference to the comparator(s).

The revised framework estimates the expected long-term costs and QALYs for the patient population with and without coverage of the new treatment. These values are obtained by weighting the expected costs and QALYs of each treatment alternative by the forecasted market share under both coverage scenarios. Analysis thus focuses on determining the incremental cost per QALY gained of reimbursing the new technology versus no reimbursement. The output is a statement of changes in costs and outcomes that are predicted to actually occur if the product is reimbursed.

When there are only two treatment options (the existing technology and a new technology), the value of reimbursing the new technology is simply assessed by estimating the incremental cost per QALY gained from the new treatment choice. Thus, a decision maker can determine whether to reimburse the new technology based on their threshold value of a QALY. Both the traditional framework and the revised framework would lead to the same conclusion with respect to cost-effectiveness, since for any market share, the incremental costs and effects of the new treatment would be downweighted by the same amount, leading to an identical incremental cost per QALY.

When there is more than one current treatment choice, the two approaches to decision-making can lead to different recommendations with respect to coverage of a new technology. These differences are highlighted by a hypothetical analysis detailed below.

The simulated analysis compares five treatment options (no treatment, existing Treatments A, B and C, and a new Treatment D) for a hypothetical progressive disease. Data inputs are detailed in Table 11.1. The analysis was conducted using a Markov model with 3-month cycles and a 50-year (lifetime) horizon. Expected values for long-term costs and benefits were based on probabilistic analysis with 5,000 replications (Briggs 2000).

When there are multiple treatment options, the normative framework involves identifying which treatment is optimal (producing the highest net benefit) for given threshold values for a QALY. If the decision maker's threshold value is unstated, then analysis can take the form of a sequential analysis, which allows identification of which treatment alternative will maximise net benefit given different threshold values for a QALY (Karlsson and Johannesson 1996; CADTH: Canadian Agency for Drugs and Technologies in Health 2017). The first step in a sequential analysis involves identifying which treatments cannot be optimal regardless of the threshold, based on the concepts of dominance and extended dominance. Further, the analysis requires the calculation of incremental cost-effectiveness ratios (ICER) for a less effective comparator compared to the next most effective comparator. This allows identification of the interventions that lie on the cost-effectiveness frontier (i.e., the treatments that may be optimal depending on the threshold value of a QALY). Thus, a

TABLE 11.1
Example model for analysis of non-optimal decisions: input parameters for simulation.

Variable	Expected Value	Probability Distribution
Annual rates		
Rate of progression without treatment	0.4	Beta (40, 60)
Mortality rate with moderate disease	0.04	Beta (4, 96)
Mortality rate with severe disease	0.25	Beta (25, 75)
Disutility		
Moderate disease	0.2	log [Normal (−1.611, 0.060)]
Severe disease	0.5	log [Normal (−0.694, 0.032)]
Disease Costs per Three Months		
Managing moderate disease	1000	Gamma (100, 10)
Managing severe disease	5000	Gamma (100, 50)
Treatment Costs per Three Months		
Treatment A	100	Fixed
Treatment B	3200	Fixed
Treatment C	3200	Fixed
Treatment D	4600	Fixed
Hazard Rates of Progression versus no treatment		
Treatment A	0.95	log [Normal (−0.051, 0.05)]
Treatment B	0.76	log [Normal (−0.274, 0.10)]
Treatment C	0.74	log [Normal (−0.301, 0.10)]
Treatment D	0.5	log [Normal (−0.734, 0.10)]
Market Share without Treatment D		
No Treatment	20%	Dirichlet (20, 20, 30, 30)
Treatment A	20%	
Treatment B	30%	
Treatment C	30%	
Capture Rates for Treatment D		
No Treatment	20%	Beta (20, 80)
Treatment A	20%	Beta (20, 80)
Treatment B	50%	Beta (50, 50)
Treatment C	50%	Beta (50, 50)

TABLE 11.2
Example model for analysis of non-optimal decisions: sequential cost-effectiveness analysis under traditional framework.

	QALYs	Total Costs	Sequential Cost-Effectiveness Results (Incremental Cost per QALY gained)	Net benefit ($\lambda = \$50,000$)
No treatment	3.43	$78,285		$92,974
Treatment A	3.50	$79,138	$11,654	$95,780
Treatment D	4.66	$149,237	$60,278	$83,827
Treatment B	3.84	$112,702	Subject to extended dominance	$79,453
Treatment C	3.88	$78,285	Subject to extended dominance	$80,797

decision maker can determine whether the new treatment should be funded as the optimal treatment choice based on their threshold value of a QALY. If the threshold value, λ, is stated explicitly, then the task is made much simpler mathematically by calculating the net benefit of each comparator at the given willingness-to-pay and choosing the option with the maximum (Stinnett and Mullahy 1998; Paulden 2020a).

Table 11.2 presents the results based on the traditional framework. Treatments B and C cannot be optimal as they are subject to extended dominance. If λ is less than $11,654, no treatment is optimal. Treatment A is optimal if λ is between $11,654 and $60,278. Treatment D will be optimal if λ is greater than $60,278. In our example, if we assume that λ is $50,000 per QALY, treatment B is optimal.

Under the positive framework, the analysis simply compares two states of the world: the treatment of the patient population without funding Treatment D and the treatment of the patient population with funding Treatment D. The cost-effectiveness of funding Treatment D is simply addressed by whether the funding scenario will maximise net benefits given the threshold value for a QALY. We have simplified the analysis for illustration by assuming the market share of existing products without funding the new treatment will remain constant (Table 11.1). The expected market shares are assumed to be 20% for Treatment A and 30% for Treatments B and C; with 20% untreated. Similarly, we assume that the expected capture rates for the new product are constant over time: Treatment D is expected to capture 20% of the market for Treatment A and 50% of the market for Treatments B and C; and take 20% of the market which were previously untreated. This is clearly a simplifying

TABLE 11.3
Example model for analysis of non-optimal decisions: cost-effectiveness analysis under revised framework.

	Market Share	QALYs	Total Costs	Incremental Cost per QALY Gained
Do not fund Treatment E		3.70	$99,354	
No Treatment	*20%*	*3.43*	*$78,285*	
Treatment A	*20%*	*3.50*	*$79,138*	
Treatment B	*30%*	*3.84*	*$112,702*	
Treatment C	*30%*	*3.88*	*$78,285*	
Fund Treatment E		4.04	$115,838	$49,204
No Treatment	*16%*	*3.43*	*$78,285*	
Treatment A	*16%*	*3.50*	*$79,138*	
Treatment B	*15%*	*3.84*	*$112,702*	
Treatment C	*15%*	*3.88*	*$78,285*	
Treatment D	*38%*	*4.66*	*$149,237*	

assumption, as market dynamics will likely mean the market share of a new product will take a number of years to reach its steady state. Thus, for the scenario where Treatment D is funded, the revised expected market shares will be 16% for no treatment and Treatment A, 15% for Treatments B and C and 38% for Treatment D. The market shares and the capture rates are both assumed to be uncertain and represented by probability distributions.

To estimate the long-term costs and QALYs for the no funding scenario, we simply weight the costs and QALYs for each treatment alternative from each replication of our Monte Carlo simulation, weight these by a random sample from the probability distribution relating to market share, and take the expectation across all replications of these probability-weighted sums. For the funding scenario, for each replication, we derive the market share of each of the existing treatment alternatives by taking the same random sample from the probability distribution relating to market share, and applying a random sample for the capture rate for the new product under consideration. Thus, the market shares under the funding scenario for each replication can be derived. We can then estimate the long-term costs and QALYs for the funding scenario using the same approach as above.

The alternative framework concludes that funding Treatment D leads to greater costs and greater QALYs than not funding Treatment D with an incremental cost per QALY gained of $49,204 (Table 11.3). Thus, based on a threshold value of a QALY of $50,000, analysis concludes it would be optimal to include coverage of Treatment D.

TABLE 11.4
Value of Information analysis in the example model for analysis of non-optimal decisions (the EVPPIs for each other individual variable are zero or less than $1)

	Expected Value of Perfect (Partial) Information ($) per Person	
	Traditional Framework	**Revised Framework**
Expected value of perfect information (EVPI)	1,396.00	1,544.00
Expected value of perfect partial information (EVPPI)		
Annual rates		
Rate of progression without treatment	0.00	31.65
Mortality rate with moderate disease	0.00	183.81
Disutility		
Moderate disease	0.00	85.32
Disease costs per 3 months		
Managing moderate disease	0.00	6.86
Hazard rates of progression versus no treatment		
Treatment A	379.45	8.26
Treatment B	8.60	219.85
Treatment C	9.13	182.57
Treatment D	792.23	1,427.47
Market share without Treatment D	N/A	12.79
Capture rates for Treatment D	N/A	87.59

11.2 Assessing the Value of Further Information

The first step in assessing the value of further information is to estimate the expected value of perfect information and the expected value of perfect partial information on a per patient basis. Table 11.4 depicts the results of the analysis based on the economic evaluation being conducted from the traditional normative approach and from the revised approach (Strong, Oakley and Brennan 2014). Analysis from the traditional framework suggests there is little or no potential value in obtaining further information on any parameters other than the hazard rates of progression with Treatments A and D. Results from the revised approach are contrary to this. There is little value in obtaining further information on the hazard rate of progression with Treatment A, as a limited proportion of the potential market for Treatment D will come from patients who would otherwise have been treated by Treatment A. As much of the assumed market for treatment D came from Treatments B and C, there is, however, information value from obtaining further information regarding the hazard rates of progression with Treatments B, C and D.

The above assumes that the type and results of the further information obtained will not affect the projected market shares of different treatment options. Value of information may also have to consider the impact of research results, especially clinical trials, on the market share of treatment alternatives (Fenwick, Claxton and Sculpher 2008). For example, in relation to the hypothetical study above, a further clinical trial comparing Treatment D to no treatment may reduce the uncertainty over the relative effectiveness of the new treatment and/or may find a more favourable hazard ratio than the existing study. The impact on the hazard ratio will have two effects. First, the relative cost-effectiveness of Treatment D may improve due to any improved effectiveness. Second, the uptake of Treatment D may increase due to the reduced uncertainty over the benefit from treatment and/or more favourable clinical results. In this example, if a favourable clinical trial leads to greater capture of market share from no treatment and Treatment A, potentially, and possibly paradoxically, this may reduce the cost-effectiveness of funding the new treatment. An assessment of the value of a future clinical trial, therefore, would need to allow for both the impact on costs and benefits for the different treatment alternatives and the impact on market share of potential results of the trial.

11.3 Discussion

This chapter illustrates the alternative analytical approaches to adopt with respect to determining the cost-effectiveness of reimbursing new technologies.

The above analysis illustrates the differences in the alternative frameworks in assessing the cost-effectiveness of a new technology. Both the traditional and revised frameworks focus on determining which alternative will result in the greatest health gains. Crucially, however, they differ in what they consider the alternative decisions available to a decision maker.

The traditional framework provides information on what the optimal choice of treatment is assuming the decision maker will elect to fund only one treatment within this context. Thus, the alternative decisions available to the decision maker relate solely to which one treatment alternative should be funded. Under this framework, Treatment D will not be funded as Treatment B is optimal.

The alternative framework assumes that reimbursement decision facing the decision maker is binary – should they fund the new treatment given the forecasted distribution of treatments for this patient population with and without reimbursement. Thus, although the revised framework recognizes the lack of health maximisation as a criterion for previous decisions, unlike the traditional framework, it allows decision makers to identify if funding the new treatment will increase or decrease the level of health within the population.

The normative framework is based on the assumption that decision makers will choose only to reimburse the optimal treatment option, given the assumed objective of maximising the level of health in the population. Thus, analysis focuses on whether a new treatment should be reimbursed as a replacement for the sole optimal treatment that is currently funded. However, in reality, decision makers chose to fund treatments for a variety of reasons not just health maximisation. Health maximisation is not the sole objective of decision makers with respect to reimbursement decisions, leading to the typical situation where more than one treatment option is funded within a specific context. The following is an inexhaustive list of criteria which may impact decisions: political pressure, the focus of clinicians and patients on the individual patient decision and societal impacts through net health benefit, reluctance to withdraw therapies from coverage when new therapies appear optimal, preference for older therapies with more real world experience. Thus, analysis based on the assumption that only the optimal treatment is funded does not address whether funding the new treatment will lead to health maximisation.

Arguments in favour of adopting the revised framework are based on the assumption that economic evaluations are only targeted towards agencies who assist decision makers in making reimbursement decisions. Economic evaluation may inform other users such as producers of practice guidelines. In such circumstances, a focus on the normative approach may be justifiable.

The revised framework incorporates both positive and normative features allowing for previous funding decisions which may not have been consistent with the goal of maximisation of health. The analysis compares two states of the world: the treatment of the patient population without funding the new treatment and the treatment of the patient population with funding the new treatment. Thus, although the analysis takes a positive approach by accepting previous decision-making, it provides a normative framework to addressing the actual decision problem facing the decision maker – should they fund the new intervention? This revised approach is consistent with the objective of maximising the level of health within the population.

A major limitation with the revised approach relates to the precision and potential bias with respect to the estimates of future market share (Regnier and Ridley 2015). These estimates are clearly uncertain and value of information analysis can easily identify, as above, the value in obtaining more precise estimates of these crucial parameters. The revised approach can lead to the potential for gaming by sponsors in terms of the estimates of future market share and from which products market share will be captured. This is especially the case where adding a new product to a market will likely lead to increased market size (Coyle et al. 2015).

The revised approach of considering current market share is similar to concepts introduced with respect to the value of implementation (Fenwick, Claxton and Sculpher 2008). Value of implementation relates to the current level of adoption of already reimbursed technologies and determines the value of moving towards perfect implementation – i.e., only the treatment with the highest net benefit is adopted. Within this framework, value of information analysis is assumed to be conducted by assuming perfect implementation. A noted difference between the value of implementation framework and the issues considered in this chapter is that value of implementation relates to currently adopted therapies and not to the decision facing a decision maker with respect to whether a new therapy should be reimbursed. In this chapter, we focus on the implications with respect to net health benefit of deciding whether or not to reimburse a new technology recognizing that previous decisions on reimbursement are often made based on other criteria. Thus, this chapter focuses on the aim of optimisation given previous decisions. Thus, value of information analysis relates to the value of obtaining information given these decisions, rather than assuming perfect implementation. The value of implementation will be discussed further in Chapter 12.

This chapter illustrates how the results of a value of information analysis will differ based on the framework adopted. The results will be further affected if further information also impacts the market share of different treatment options. This suggests that to have more relevance with the decisions being made, both economic evaluation and value of information analysis may wish to be rooted in a framework which incorporates a positive approach to previous decision-making and the impact of further research on treatment uptake.

12

Value of Information and Implementation

Sabine Grimm, Alan Brennan, and Anna Heath

Chapter 11 discussed the concept of VOI within the context of non-optimal decisions. In this chapter, we discuss the use of VOI with a specific type of non-optimal decision: sub-optimal implementation of the interventions. In standard VOI analyses discussed throughout this book, the individual-level VOI measures are calculated assuming that the implementation of health technologies is perfect. This means that when computing VOI, we assume that the technology with the highest expected net benefit will be used in practice as soon as the relevant information has been collected. This follows from the assumption that decision makers act in a rational manner, which is fundamental to Bayesian decision theory and thus VOI. The assumption of perfect implementation underlies both of the two terms in the calculation of EVPI, EVPPI and EVSI:

- The value of the current decision is assumed to be the value of the intervention that maximises expected net benefit, but this is only correct if the current optimal intervention in used in practice.
- The value of the decision made with updated evidence is assumed to be the value of the intervention that maximises the expected net benefit conditional on the updated information, which will only be realised if this intervention is implemented.

However, in practice, implementation is rarely perfect, meaning that the practical implementation of interventions is not strictly rational. Imperfect or irrational implementation may be caused by a variety of factors, including the belief that important factors are not considered in the health economic decision model (e.g., heterogeneity of patients and how they respond to treatments), or, simply, that diffusion of a new health technology takes time. Thus, the assumptions in VOI analysis rarely, if ever, match reality.

Analysts may be, and have been, tempted to "adjust" VOI analysis for imperfect or irrational implementation. In this section, we discuss the implications of such an adjustment and the measures that have been suggested in

DOI: 10.1201/9781003156109-12

the literature to address these concerns. This is a challenging and divisive topic, which has been incompletely discussed in the literature. Another note before diving into the topic of "adjusting" VOI for imperfect or irrational implementation: this section does not discuss methods to adjust the estimate of the population-level VOI for assumptions about the implementation of interventions. Population-level VOI adjustments can be made to suggest that only a small number of patients would be eligible for an intervention, for example, a public health intervention that rolls out over time across the country. Relevant adjustments can also be made to reflect that the full patient population may not benefit from the research as the intervention will not be used by all patients. However, these methods for adjusting for imperfect implementation do not alter the definition of the underlying individual-level VOI measure.

In the literature, different approaches can be found to calculate Value of Information (VOI) while considering imperfect implementation. However, the implications of these approaches have not been sufficiently discussed, and the links between the different approaches and measures have not been drawn. Most notably, despite being named as value of information methods (indicating that they calculate the value of reducing uncertainty) that "adjust" for implementation, many of the approaches actually compute the value of improving *implementation* only. This may be desired, but the analyst should be clear about what they are computing. Thus, this chapter will start by reviewing the current literature on "adjusted" VOI measures and their key implications. We then present a unified framework for value of information and implementation that was proposed by Heath, Grimm and Brennan (2023). We then provide recommendations and guidance on how these concepts should be used in practice. Finally, we discuss how these measures can be computed (Heath 2022).

12.1 Existing Frameworks

The initial framework that considered information and implementation (Fenwick, Claxton and Sculpher 2008) assumes that both information and implementation can have the two states "current" and "perfect." This leads to a four-state world, in which we can compute the expected value generated by:

(i) Current information and current implementation (A).
(ii) Perfect information and current implementation (B).
(iii) Current information and perfect implementation (C).
(iv) Perfect information and perfect implementation (D).

Current information reflects the existing uncertainty in the parameters of the health economic decision model, as in standard VOI analyses. Current

TABLE 12.1
The states of information and implementation considered in the Fenwick, Claxton and Sculpher (2008) value of implementation framework. A is the expected value generated given current information and current implementation. B is the expected value of obtaining perfect information but retaining the current level of implementation. C is the expected value of perfectly implementing the currently expected optimal intervention. D is the expected value of perfectly implementing the true optimal intervention after parameter uncertainty is resolved.

		Information	
		Current	**Perfect**
Implementation	Current	A	B
	Perfect	C	D

implementation is the proportion of the eligible population who would currently (or upon reimbursement recommendation) receive the different interventions. This framework then defines EVPI as the difference between states D and C in Table 12.1, i.e., the difference between the expected value of perfectly implementing the true optimal intervention after uncertainty is resolved and the expected value of perfectly implementing the currently expected optimal intervention. The Expected Value of Perfect Implementation (EVPIM) was defined in this framework as the difference between states C and A, i.e., the expected value of perfectly implementing the optimal intervention based on current evidence and the expected value of the current levels of implementation. The Expected Value of Perfection (EVP) was defined as the sum of the EVPI and EVPIM, i.e., $D - C + C - A = D - A$. This is then the value of resolving both uncertainty and imperfect implementation.

The last concept proposed within this framework is the "realisable EVPI," which describes the difference between the expected value of obtaining perfect information while maintaining the current level of implementation (B) and the expected value generated given current information and current implementation (A). These definitions for states A and B result in a "realisable EVPI" is equal to 0 as the current implementation levels are maintained with and without information. In general, as stated in Section 3.2, information has value as it prevents the decision maker from implementing a non-cost-effective intervention. However, if the implementation levels do not change, i.e., the decision maker does not change their action, then there is no value from collecting information.

Fenwick et al. (2020) suggest that the "realisable EVPI" could be investigated across different assumptions about the implementation of each intervention under perfect information through a sensitivity analysis. By naming this quantity the "realisable EVPI," it suggests that this is the value that could be generated by obtaining perfect information. However, as it is calculated

TABLE 12.2
The different states of information and implementation that are considered in the extended value of implementation framework, developed by Andronis and Barton (2016).

		Information		
		Current	**Improved (Sample)**	**Perfect**
Implementation	Current	A	K	B
	Improved (Expanded)	L	M	N
	Perfect	C	O	D

by changing the assumed level of implementation after perfect information has been collected, it simply calculates the value of improving implementation. Thus, we argue that it does not compute the value of "information" but rather the value of improved implementation. This illustrates the potential problems with a static framework, i.e., where we consider the levels of implementation fixed after data collection, to make sense of implementation and information.

An attempt to create a more dynamic framework extended the initial framework by allowing information and implementation to change to "improved" in addition to "current" and "perfect" (Andronis and Barton 2016). This resulted in a nine-state world presented in Table 12.2. This extension allows us to clarify that EVSI is defined as $(O - C)$, i.e., the value of moving from current to improved information under perfect implementation. It also defines the value of a specific implementation measure $(L - A)$, which maintains current information but improves implementation.

However, issues similar to those seen in the initial Fenwick et al. (2020) framework remain and relate to the "realisable EVPI," defined as above, and the "implementation-adjusted EVSI." The "implementation-adjusted" EVSI was defined as the value of moving from current information and implementation to improved information and implementation $(M - A)$. However, through this definition, the "implementation-adjusted EVSI" may only compute the value of improving the implementation of the current optimal intervention. For example, if a proposed study has zero EVSI where the reduction in parameter uncertainty does not change the optimal decision, it would still have a positive implementation-adjusted EVSI, as the implementation of the optimal intervention is improved. In this case, $M - A = L - A$, when the change in implementation is equal. Note furthermore, the values of the states A, K and B in Table 12.2 are equal, as implementation remains "current" irrespective of the collected information.

12.1.1 The Challenge of Considering Implementation Changes

In general, analysts and decision makers may be interested in the expected effect a study has on implementation and the value of improving implementation in this way. For example, clinical studies that achieve a statistically significant effect will often increase the implementation of an intervention. However, the nomenclature of "realisable EVPI" and "implementation-adjusted EVSI" suggests that the information directly provides value through its potential to avoid the use of non-optimal interventions, as is the case in standard VOI analysis. Thus, it is important to recognise that these measures often only compute the value of improving implementation. To clarify this, we propose that the terminology used for these measures should be updated. This is especially crucial as a lack of clarity could result in undesired policy and research decisions. For example, a research funding body could estimate the "implementation-adjusted EVSI" for a study where the optimal decision does not change following research. If this "implementation-adjusted EVSI" exceeded the cost of the study, the research funding body may fund this research even though it merely aids implementation. This study would be ethically problematic, however, as patients should not be subjected to research when the optimal intervention is already known and uncertainty about the optimal intervention will not be resolved through the proposed research.

Grimm, Dixon and Stevens (2017a) initially aimed to address this conflation between the value of information and implementation by moving away from the static framework in Table 12.2 and introducing separate concepts for value of implementation and information (Grimm, Dixon and Stevens 2017a). The value of implementation could also be estimated for research when we assume that the research could lead to a change in implementation. They named this measure the "research expected value of specific implementation measure" (EVSIM^R), represented by $M - A$ in Table 12.2. They then extended this measure to consider that the level of implementation for each intervention could directly be influenced by the *strength* of the evidence in favour of each intervention (denoted as $M_{\mathbf{X}} - A$). Grimm, Dixon and Stevens (2017a) then proposed that EVSIM^R and EVSI could be considered independent and added together to estimate the total value that can be derived from research, in a measure they named the "expected value of research" (EVR).

The concept of adjusting EVSI for imperfect implementation if implementation was related to the strength of the evidence ($M_{\mathbf{X}} - A$) was also discussed by Willan and Eckermann (2010) and Heath (2022). In these frameworks, the post-study implementation was related to the *probability* that an intervention is optimal. As an example, we could assume that the probability of an intervention being the best (in a cost-effectiveness analysis) could directly relate to the proportion of patients receiving it. So, if strategy 1 is 80% likely to be most cost-effective versus strategy 2 being 20% likely to be most cost-effective, then 80% of patients would receive strategy 1 and 20% strategy 2.

In this setting, implementation is directly related to uncertainty through an assumption. This means that under this assumption, the adjusted EVSI measure truly reflects the impact of reducing parameter uncertainty, aligning with our standard understanding of what VOI measures.

12.2 Taxonomy for Value of Information and Implementation Measures

To further clarify these concepts, Heath, Grimm and Brennan (2023) provided a new taxonomy for combining value of information and implementation and drawing on insights from all these frameworks (Table 12.3). Here, we do not consider that the value of information can be "adjusted" for imperfect implementation, avoiding the conflation that has been seen previously. These measures are not additive and are split into measures that consider

(i) A change in information when decision makers are considered rational.

(ii) A change in implementation with no further information collected.

(iii) A change in both the level of information available and the implementation of the optimal intervention.

The definitions in Table 12.3 are based on the states of the world specified in Table 12.2.

12.3 Calculation of Value of Information and Implementation Measures

As with all the key VOI measures discussed in this book, the computation of the measures in Table 12.3 usually requires simulation. Evidently EVPI, EVPPI and EVSI, included in the first section of Table 12.3, can be computed using the methods discussed in Chapters 3 and 4. When we assume that further information is not collected, the EVPIm and EVEIm can be computed directly from the PA samples as these compute the value of the each interventions under current information. To achieve this, the "current" and "improved" levels of implementation must be defined, either from the literature, expert knowledge or the diffusion of previous interventions (Grimm, Dixon and Stevens 2017a). Next, provided the "current" implementation levels have been defined, the EVPIIm can be calculated directly from the PA samples by combining the estimate of value under current implementation and the

TABLE 12.3
A complete summary of the value of information and implementation measures described in this chapter. The definitions are based on the states of the world specified in Table 12.2, and present an alternative nomenclature to improve the use of these measures.

<table>
<tr><th rowspan="2">Definition</th><th colspan="2">Previous</th><th colspan="2">Proposed</th></tr>
<tr><th>Terminology</th><th>Abbreviation</th><th>Terminology</th><th>Abbreviation</th></tr>
<tr><td colspan="5">Assumes decision makers are fully rational in their implementation</td></tr>
<tr><td>$D - C$</td><td>Expected Value of Perfect Information</td><td>EVPI</td><td>Expected Value of Perfect Information</td><td>EVPI</td></tr>
<tr><td>$O - C$</td><td>Expected Value of Sample Information</td><td>EVSI</td><td>Expected Value of Sample Information</td><td>EVSI</td></tr>
<tr><td colspan="5">Assumes no further information is collected</td></tr>
<tr><td>$C - A$</td><td>Expected Value of Perfect Implementation</td><td>EVPIM</td><td>Expected Value of Perfect Implementation</td><td>EVPIm</td></tr>
<tr><td>$L - A$</td><td>Expected Value of Specific Implementation Measure</td><td>EVSIM</td><td>Expected Value of Expanded Implementation</td><td>EVEIm</td></tr>
<tr><td colspan="5">Information and levels of implementation both change</td></tr>
<tr><td>$D - A$</td><td>Expected Value of Perfection</td><td>EVP</td><td>Expected Value of Perfect Information and Implementation</td><td>EVPIIm</td></tr>
<tr><td rowspan="2">$M - A$</td><td>Implementation Adjusted EVSI</td><td>IA-EVSI</td><td rowspan="2">Expected Value of Expanded Implementation ($M = L$, when the M is not impacted by the data)</td><td rowspan="2">EVEIm</td></tr>
<tr><td>Research Expected Value of Specific Implementation Measure</td><td>EVSIM^R</td></tr>
<tr><td rowspan="3">$M_{\mathbf{X}} - A$</td><td>EVSI with imperfect implementation</td><td>-</td><td rowspan="3">Expected Value of Sample Expanded Implementation</td><td rowspan="3">EVSEIm</td></tr>
<tr><td>Implementation-adjusted EVSI</td><td>EVSI^{IM}</td></tr>
<tr><td>Research EVSIM (Alternative Definition)</td><td>EVSIM^R</td></tr>
</table>

value of perfect decision-making required to compute EVPI. However, calculating EVSEIm is substantially more complex and requires bespoke estimation methods.

12.3.1 Nested Monte Carlo Simulation of EVSEIm

EVSEIm can be computed by nested Monte Carlo simulation using a similar procedure to the Monte Carlo estimation for EVSI (Ades, Lu and Claxton 2004) presented in Section 4.3.1. This algorithm simulates S datasets from the predictive distribution $p(\boldsymbol{X})$ and then R parameter simulations from the posterior distribution of $\boldsymbol{\theta}$ for each dataset, denoted $\boldsymbol{\theta}_{r,s}$. The net benefit for each intervention is computed for each simulated parameter set $\boldsymbol{\theta}_{r,s}$, $r = 1, \dots, R$, $s = 1, \dots, S$ and the expected net benefit for treatment $d = 1, \dots, D$, conditional on $\boldsymbol{X}_s$, is estimated by calculating the sample average net benefit. The EVSEIm assumes that the implementation level of the intervention, known as the market share, is related to the *strength of evidence* in favour of the intervention. This is commonly operationalised in terms of the probability that each intervention is cost-effective, which can be estimated as the proportion of simulations in which treatment d is optimal;

$$\widehat{p_d(\boldsymbol{X}_s)} = \frac{1}{R}\sum_{r=1}^{R} \mathbb{1}\left[\text{NB}_d\left(\boldsymbol{\theta}_{r,s}\right) = \max_d \text{NB}_d\left(\boldsymbol{\theta}_{r,s}\right)\right], \tag{12.1}$$

where $\mathbb{1}[\cdot]$ equals 1 if the condition is true and 0 otherwise. The market share for each intervention can then be estimated as a function of $\widehat{p_d(\boldsymbol{X}_s)}$; $\widehat{m_d(\boldsymbol{X}_s)} = f_d^m\left(\widehat{p_d(\boldsymbol{X}_s)}\right)$, where $f_d^m(\cdot)$ is a intervention-specific function that computes the market share. Finally, the EVSEIm can be estimated as

$$\begin{aligned}\widehat{\text{EVSEIm}} = {} & \frac{1}{S}\sum_{s=1}^{S}\sum_{d=1}^{D} \widehat{m_d(\boldsymbol{X}_s)}\left\{\frac{1}{R}\sum_{r=1}^{R}\text{NB}_d\left(\boldsymbol{\theta}_{r,s}\right)\right\} \\ & - \sum_{d=1}^{D} m_d \frac{1}{S}\sum_{s=1}^{S}\frac{1}{R}\sum_{r=1}^{R}\text{NB}_d\left(\boldsymbol{\theta}_{r,s}\right)\end{aligned} \tag{12.2}$$

However, as with the standard EVSI approximation, this algorithm is computationally intensive as it requires $S \times R$ evaluations of the D net benefit functions, making it infeasible in practice.

12.3.1.1 Adjusting the Efficient Computation Methods for Imperfect Implementation

In Chapter 4, four efficient computation methods were presented for EVSI. These methods focused on efficiently estimating the *mean* net benefit, conditional on the future data. As such, they are not able to estimate the probability of cost-effectiveness $p_d(\boldsymbol{X})$, which depends on the full distribution of the net benefit. However, the moment-matching method (Section 4.3.3) does estimate the full distribution for the net benefit for $Q << S$ nested simulations.

Thus, the following section outlines how the moment-matching method can be extended to compute $p_d(\boldsymbol{X})$ for all plausible datasets (Heath 2022).

It is recommended that an interested reader reviews Section 4.3.3 before proceeding with this description.

The standard moment-matching method produces simulations of $\mu_d(\boldsymbol{X}_s)$, $s = 1, \ldots, S$. To estimate EVSEIm, we note that $p_d(\boldsymbol{X})$ will be related to the value of $\mu_d(\boldsymbol{X})$, i.e., the larger the expected net benefit, the more likely the treatment is to be cost-effective. Therefore, to compute EVSEIm, we adjust the moment-matching method to estimate the probability of cost-effectiveness, $p_d(\boldsymbol{X})$, as a function of the expected net benefit, $\mu_d(\boldsymbol{X})$. To achieve this, we firstly estimate $p_d(\boldsymbol{X})$ for each of the $q = 1, \ldots, Q$ datasets, $\boldsymbol{X}_q$, from the nested simulations using Equation (12.1). The mean net benefit $\mu_d(\boldsymbol{X}_q)$ for each of the Q datasets must also be estimated as the average of the simulated values for the net benefits. From these estimates of $p_d(\boldsymbol{X}_q)$ and $\mu_d(\boldsymbol{X}_q)$, we can then use non-linear regression to approximate a function $h_d(\cdot)$;

$$p_d(\boldsymbol{X}_q) = h_d\left(\mu_d(\boldsymbol{X}_q)\right) + \varepsilon_q, \tag{12.3}$$

where $\varepsilon_q \sim N(0, \sigma^2)$ is the error due to estimating $p_d(\boldsymbol{X}_q)$ by simulation.

The functional form for $h_d(\cdot)$ is chosen to ensure that $p_d(\boldsymbol{X})$

(i) Is constrained between 0 and 1, as it is a probability.

(ii) Increases as $\mu_d(\boldsymbol{X})$ increases, i.e., the probability of cost effectiveness increases as the treatment has a higher net monetary benefit.

(iii) Increases smoothly as $\mu_d(\boldsymbol{X})$ increases, i.e., if the expected net benefits are similar than the probability of cost-effectiveness will also be similar.

The generalised logistic function is a flexible function that exhibits these three features (Richards 1959), so

$$h_d\left(\mu_d(\boldsymbol{X})\right) = \left(A + e^{-B\mu_d(\boldsymbol{X})}\right)^{-v}, \tag{12.4}$$

where A, B and v are estimated from the data $p_d(\boldsymbol{X}_q)$ and $\mu_d(\boldsymbol{X}_q)$ for $q = 1, \ldots, Q$. Heath (2022) found that Bayesian methods with weakly informative priors for estimating A, B and v improve convergence and allow the proposed non-linear model to be estimated. Once estimates have been obtained for A, B and v, the regression model can calculate the fitted values of $p_d(\boldsymbol{X})$ for each of the simulations for $\mu_d(\boldsymbol{X}_s)$, $s = 1, \ldots, S$, denoted $p_d(\boldsymbol{X}_s)$. The market share of each treatment can then be computed using $m_d(\boldsymbol{X}_s) = f_d^m(p_d(\boldsymbol{X}_s))$ and the EVSEIm estimated from these values;

$$\widehat{\text{EVSEIm}} = \frac{1}{S}\sum_{s=1}^{S}\sum_{d=1}^{D} m_d(\boldsymbol{X}_s)\mu_d(\boldsymbol{X}_s) - \sum_{d=1}^{D} m_d \frac{1}{S}\sum_{s=1}^{S}\mu_d(\boldsymbol{X}_s) \tag{12.5}$$

12.3.1.2 Estimating EVSEIm across Different Sample Sizes

The standard moment-matching method can be extended to estimate EVSI across a range of sample sizes of the proposed study $(N_{\min}, N_{\max})$ (Heath, Manolopoulou and Baio 2019) (Section 4.3.3.4) by creating a sequence of sample sizes, $N_q, q = 1, \ldots, Q$, between $N_{\min}$ and $N_{\max}$. Each simulated dataset is then generated with a different sample size, i.e., the dataset $\boldsymbol{X}_q$ contains data from N_q simulated individuals, and the variance reduction factor for the moment-matching method for a given sample size n is found using non-linear regression (Heath, Manolopoulou and Baio 2019). Thus, to estimate EVSEIm across sample sizes, we adjust the non-linear model in Equation (12.4) such that it is also a function of the sample size N_q;

$$p_d(\boldsymbol{X}_q) = h_d^N\left(\mu_d(\boldsymbol{X}_q), N_q\right) + \varepsilon_q, \tag{12.6}$$

where $\varepsilon_q \sim N(0, \sigma^2)$. In this setting, $h_d^N(\cdot)$ is still represented by a generalised logistic function, but is adjusted to account for the fact that the larger the sample size, the *faster* the probability of cost-effectiveness will increase from 0 and 1,

$$h_N(\mu_d(\boldsymbol{X}_q), N_q) = \left(A + e^{-BN_q^u \mu_d(\boldsymbol{X}_q)}\right)^{-v}, \tag{12.7}$$

where u is an additional parameter defining the rate at which the probability of cost-effectiveness increases due to the sample size. The four parameters in this model, A, B, u and v are most accurately estimated in a Bayesian framework with weakly informative priors (Heath 2022).

Once A, B, u and v have been estimated, EVSEIm can be estimated for a specific sample size n by (1) estimating the values of $\mu_d(\boldsymbol{X})$ for the sample size n using the moment-matching method, (2) estimating the probability of cost-effectiveness by computing $h_d^N(\mu_d(\boldsymbol{X}), n)$, (3) estimating the market share for each intervention and (4) calculating EVSEIm from Equation (12.5).

12.3.2 Dynamics of Implementation

A challenge of calculating EVESIm is that the function $f_d^m(p_d(\boldsymbol{X}))$ that calculates the market share based on the probability of cost-effectiveness must be specified. In practice, $f_d^m(p_d(\boldsymbol{X}))$ is likely to be complex and may be challenging to determine. Grimm, Dixon and Stevens (2017a) used diffusion models to make realistic assumptions about the implementation changes over time. However, these currently do not estimate how the strength of evidence impacts diffusion. An alternative suggestion would be to undertake a sensitivity analysis to the functional form of $f_d^m(p_d(\boldsymbol{X}))$. As both EVESIm estimation methods estimate $p_d(\boldsymbol{X})$, the sensitivity analysis to the functional form of $f_d^m(p_d(\boldsymbol{X}))$ would be relatively inexpensive as the probability of cost-effectiveness would not need to be recomputed. However, it may be challenging to determine the appropriate functional forms that should be considered in this sensitivity analyses. This is especially true if we consider changing implementation levels over time.

A potential limitation of the calculation methods proposed is that the implementation levels are assumed to be related to the outcome of a cost-effectiveness analysis, i.e., the probability of cost-effectiveness. In some settings, implementation may be more closely related to other outcomes such as the primary clinical outcome or safety concerns. The calculation methods could be adapted to estimate the probability that a given treatment is *effective*, i.e., the primary clinical outcome is largest for a specific treatment, or *safe*, i.e., adverse events are lower. The market share could then be estimated based on this probability of effectiveness or safety. This analysis would jointly consider potential complementary aspects of clinical decision-making, i.e., cost-effectiveness and clinical efficacy or safety, in study design. However, it would move away from considering the EVESIm as a measure of the value of reducing uncertainty to improve decision-making and would be closer to improving implementation through data collection.

12.4 Final Reflections on Value of Information and Implementation

One of the difficulties with considering implementation in VOI calculations relates to the fundamental assumptions of value of information; they are defined assuming rational decision-making. It has to be emphasized that the assumption of rational decision-making, i.e., that the decision maker acts rationally to minimise expected loss, is at the very core of VOI and Bayesian decision theory (Raiffa and Schlaifer 1961). When the decisions are based on outside factors, i.e., implementation is not directly related to evidence, then we break the key assumption of VOI measure, particularly when considering perfect information.

The joint presentation of value of information and implementation measures has caused confusion and made it challenging to draft a new taxonomy. We therefore propose that value of information and implementation measures are best considered separately as the generation of value derives from different sources. However, it is crucial to highlight that the value of implementation measures are highly relevant for decision makers. Furthermore, it is important to consider the relevance of VOI measures when we know that decision-making is not rational and that implementation does not follow patterns that would be expected based on the evidence. Thus, we consider the following questions: is uncertainty reduction valuable if it does not lead to improved implementation? And can VOI and value of implementation measures be considered independently and in an additive way?

There is no straightforward answer to these questions. Based on the above considerations, we suggest that both value of information and implementation measures should be considered separately. In addition to this, EVSEIm can help to estimate the value that can be derived in practice when lower decision uncertainty will improve implementation. It is important to note, however, that individual-level EVSEIm could be larger or smaller than EVSI. Furthermore, EVSI calculations are important as a positive ENBS would in most settings be a prerequisite for conducting further research with patients. Conversely, there may be situations where the reduction of uncertainty in a given parameter can significantly impact implementation but does not significantly influence decision uncertainty. For example, such a situation could be the small incidence of a very severe adverse event that does not significantly affect the health economic decision model but does affect physician or patient behaviour. In this case, EVSI may be insignificant and not warrant further research, but the proposed research could reassure patients and increase their willingness to adopt the intervention. However, a study randomising patients to receive this intervention may be ethically questionable. It is therefore important to investigate both EVSI and EVSEIm in order to understand how value is derived from a research study. However, EVSI and EVSEIm are likely not independent and therefore not additive. This is because these measures make different assumptions about whether the decision maker is rational or irrational.

We conclude this chapter by summarising our recommendations about how the effect of additional information on implementation should be considered. Firstly, as we discussed, making structural assumptions about the relationship between post-study uncertainty and implementation can be challenging, and there is currently minimal research on how to accurately derive these relationships. It has been suggested that this relationship could be estimated using structured expert elicitation (Grimm, Dixon and Stevens 2017a). However, this is not easy, as one would have to elicit a functional relationship between strength of evidence and implementation, which is unlikely to be within the comfort zone of clinical or health technology experts. Based on the above considerations, we make the following recommendations:

- Value of information and implementation measures are best considered separately, as conflating them in static frameworks can be misleading.
- VOI measures that are "adjusted" by implementation are no longer VOI measures, but instead measures of the value of changing implementation.
- If the value of information of a research study is to be assessed while accounting for implementation challenges, it is important to link the post-study implementation to the strength of evidence (as was denoted by $M_{\mathbf{X}} - A$).
- Further research is needed on how the link between implementation and strength of evidence can be established.
- EVSI or ENBS can be used:

    (a) As a minimum requirement to decide whether research involving patients should be commissioned.

   (b) To assess the value of reducing the chance of recommending a suboptimal intervention.
   (c) To understand how value is accrued when implementation is improved by reducing uncertainty in parameters that do not have a significant impact on decision uncertainty.

- EVSEIm can assess the realistic value of the research study if implementation is changed by reducing health economic decision uncertainty. This does not consider the potential value of uncertainty reduction itself.
- Value of information and implementation measures are likely not additive, as they make different assumptions about the decision maker.
- Decision makers should consider improving implementation with measures other than collecting data, and estimate the value of these measures using the EVEIm.

12.5 Conclusion

This chapter has clarified the relationship between value of information and value of implementation and noted that characterising the relationship between the strength of the evidence and implementation levels is challenging but crucial to ensure accurate research prioritisation. Future research must develop and test methods to understand and mathematically describe the drivers of implementation following the collection of evidence and how this influences health policy and implementation decisions. The clarifications provided here, combined with further methodological research, will improve the prioritisation of research and implementation strategies in an imperfect world.

Bibliography

Aas, T., A.-L. Børresen, S. Geisler, B. Smith-Sørensen, H. Johnsen, J. E. Varhaug, L. A. Akslen and P. E. Lønning. 1996. "Specific P53 mutations are associated with de novo resistance to doxorubicin in breast cancer patients". *Nature Medicine* 2 (7): 811–814.

Ades, A., G. Lu and K. Claxton. 2004. "Expected value of sample information calculations in medical decision modeling". *Medical Decision Making* 24: 207–227.

Ades, A., M. Sculpher, A. Sutton, K. Abrams, N. Cooper, N. Welton and G. Lu. 2006. "Bayesian methods for evidence synthesis in cost-effectiveness analysis". *PharmacoEconomics* 24 (1): 1–19.

Akobeng, A. K. 2005. "Understanding randomised controlled trials". *Archives of Disease in Childhood* 90 (8): 840–844.

Alarid-Escudero, F., E. A. Enns, K. M. Kuntz, T. L. Michaud and H. Jalal. 2019a. "'Time traveling is just too dangerous' but some methods are worth revisiting: the advantages of expected loss curves over cost-effectiveness acceptability curves and frontier". *Value in Health* 22 (5): 611–618.

Alarid-Escudero, F., G. Knowlton, C. Easterly and E.A. Enns. 2021. "Decision Analytic Modeling Package (dampack)". *R package version 1.0.0,* `https://github.com/DARTH-git/dampack`.

Alarid-Escudero, F., E. M. Krijkamp, P. Pechlivanoglou, H. Jalal, S.-Y. Z. Kao, A. Yang and E. A. Enns. 2019b. "A need for change! A coding framework for improving transparency in decision modeling". *PharmacoEconomics* 37 (11): 1329–1339.

Alderfer, M. A., K. Zelley, R. B. Lindell, A. Novokmet, P. L. Mai, J. E. Garber, D. Nathan, S. Scollon, N. M. Chun, A. F. Patenaude et al. 2015. "Parent decision-making around the genetic testing of children for germline TP53 mutations". *Cancer* 121 (2): 286–293.

Andronis, L. and P. Barton. 2016. "Adjusting estimates of the expected value of information for implementation: theoretical framework and practical application". *Medical Decision Making* 36 (3): 296–307.

Arrow, K. J. and R. C. Lind. 1970. "Uncertainty and the Evaluation of Public Investment Decisions". *The American Economic Review* 60 (3): 364–378.

Baio, G. 2012. *Bayesian Methods in Health Economics.* CRC Press, Boca Raton, FL.

Baio, G., A. Berardi and A. Heath. 2017. *Bayesian Cost Effectiveness Analysis with the R package BCEA.* Springer, Cham, Switzerland.

Baio, G. and A. Heath. 2017. "When Simple Becomes Complicated: Why Excel Should Lose its Place at the Top Table". *Global & Regional Health Technology Assessment* 4 (1): grhta-5000247.

Bardeesy, N., D. Falkoff, M.-J. Petruzzi, N. Nowak, B. Zabel, M. Adam, M. C. Aguiar, P. Grundy, T. Shows and J. Pelletier. 1994. "Anaplastic Wilms' tumour, a subtype displaying poor prognosis, harbours p53 gene mutations". *Nature Genetics* 7 (1): 91–97.

Barton, G., A. Briggs and E. Fenwick. 2008. "Optimal Cost-Effectiveness Decisions: The Role of the Cost-Effectiveness Acceptability Curve (CEAC), the Cost-Effectiveness Acceptability Frontier (CEAF), and the Expected Value of Perfection Information (EVPI)". *Value in Health* 11 (5): 886–897. https://doi.org/10.1111/j.1524-4733.2008.00358.x.

Basu, A. 2020. "A welfare-theoretic model consistent with the practice of cost-effectiveness analysis and its implications". *Journal of Health Economics* 70: 102287.

Bennette, C. S., D. L. Veenstra, A. Basu, L. H. Baker, S. D. Ramsey and J. J. Carlson. 2016. "Development and evaluation of an approach to using value of information analyses for real-time prioritization decisions within SWOG, a large cancer clinical trials cooperative group". *Medical Decision Making* 36 (5): 641–651.

Berger, J. O. 2013. *Statistical Decision Theory and Bayesian Analysis.* Springer Science & Business Media, New York, NY.

Berger, M. L., M. Mamdani, D. Atkins and M. L. Johnson. 2009. "Good research practices for comparative effectiveness research: defining, reporting and interpreting nonrandomized studies of treatment effects using secondary data sources: the ISPOR Good Research Practices for Retrospective Database Analysis Task Force Report–Part I". *Value in Health* 12 (8): 1044–1052.

Bernard, E., Y. Nannya, R. P. Hasserjian, S. M. Devlin, H. Tuechler, J. S. Medina-Martinez, T. Yoshizato, Y. Shiozawa, R. Saiki, L. Malcovati et al. 2020. "Implications of TP53 allelic state for genome stability, clinical presentation and outcomes in myelodysplastic syndromes". *Nature Medicine* 26 (10): 1549–1556.

Bernardo, J. M. and A. F. Smith. 2009. *Bayesian Theory.* Vol. 405. John Wiley & Sons, New York, NY.

Birch, J. M., V. Blair, A. M. Kelsey, D. G. Evans, M. Harris, K. J. Tricker and J. M. Varley. 1998. "Cancer phenotype correlates with constitutional TP53 genotype in families with the Li–Fraumeni syndrome". *Oncogene* 17 (9): 1061–1068.

Blazeby, J. 2020. "Bluebelle pilot randomised controlled trial of three wound dressing strategies to reduce surgical site infection in primary surgical wounds". (on behalf of the Bluebelle Study Group), *BMJ Open* 10 (1): e030615.

Bojke, L., M. Soares, K. Claxton, A. Colson, A. Fox, C. Jackson, D. Jankovic, A. Morton, L. Sharples and A. Taylor. 2021. "Developing a reference protocol for structured expert elicitation in health-care decision-making: a mixed-methods study". *Health Technology Assessment (Winchester, England)* 25 (37): 1.

Borgonovo, E., G. B. Hazen and E. Plischke. 2016. "A common rationale for global sensitivity measures and their estimation". *Risk Analysis* 36 (10): 1871–1895.

Borgonovo, E. and E. Plischke. 2016. "Sensitivity analysis: a review of recent advances". *European Journal of Operational Research* 248 (3): 869–887.

Børresen, A.-L., T. I. Andersen, J. E. Eyfjörd, R. S. Cornelis, S. Thorlacius, Å. Borg, U. Johansson, C. Theillet, S. Scherneck, S. Hartman et al. 1995. "TP53 mutations and breast cancer prognosis: particularly poor survival rates for cases with mutations in the zinc-binding domains". *Genes, Chromosomes and Cancer* 14 (1): 71–75.

Bowles, E. J. A., R. Wellman, H. S. Feigelson, A. A. Onitilo, A. N. Freedman, T. Delate, L. A. Allen, L. Nekhlyudov, K. A. Goddard, R. L. Davis et al. 2012. "Risk of heart failure in breast cancer patients after anthracycline and trastuzumab treatment: a retrospective cohort study". *Journal of the National Cancer Institute* 104 (17): 1293–1305.

Bratvold, R. B., J. E. Bickel and H. P. Lohne. 2009. "Value of information in the oil and gas industry: past, present, and future". *SPE Reservoir Evaluation & Engineering* 12 (04): 630–638.

Brazier, J., J. Ratcliffe, J. Saloman and A. Tsuchiya. 2017. *Measuring and Valuing Health Benefits for Economic Evaluation.* Oxford University Press, Oxford, UK.

Brazier, J., J. Roberts and M. Deverill. 2002. "The estimation of a preference-based measure of health from the SF-36". *Journal of Health Economics* 21 (2): 271–292.

Brazier, J. E., Y. Yang, A. Tsuchiya and D. L. Rowen. 2010. "A review of studies mapping (or cross walking) non-preference based measures of health to generic preference-based measures". *The European Journal of Health Economics* 11: 215–225.

Brennan, A. and S. Kharroubi. 2007a. "Efficient computation of partial expected value of sample information using Bayesian approximation". *Journal of Health Economics* 26 (1): 122–148.

———. 2007b. "Expected value of sample information for Weibull survival data". *Health Economics* 16 (11): 1205–1225.

Brennan, A., S. Kharroubi, A. O'Hagan and J. Chilcott. 2007. "Calculating partial expected value of perfect information via Monte Carlo sampling algorithms". *Medical Decision Making* 27: 448–470.

Briggs, A., M. Sculpher and K. Claxton. 2006. *Decision Modelling for Health Economic Evaluation.* Oxford University Press, Oxford, UK.

Briggs, A. H. 2000. "Handling uncertainty in cost-effectiveness models". *PharmacoEconomics* 17 (5): 479–500.

Briggs, A. H., M. C. Weinstein, E. A. L. Fenwick, J. Karnon, M. J. Sculpher, A. D. Paltiel and ISPOR-SMDM Modeling Good Research Practices Task Force. 2012. "Model parameter estimation and uncertainty: a report of the ISPOR-SMDM Modeling Good Research Practices Task Force-6". *Value in Health* 15 (6): 835–842.

Buchanan, J. M. 2008. "Opportunity Cost". In *The New Plalgrave Dictionary of Economics Online.* Palgrave Macmillan, London, United Kingdom.

Burstein, H., D. Hayes and S. Vora. 2018. *Adjuvant chemotherapy for HER2-negative breast cancer.* `https://www.uptodate.com/contents/adjuvant-chemotherapy-for-her2-negative-breast-cancer`. Accessed July 10, 2018.

CADTH: Canadian Agency for Drugs and Technologies in Health. 2017. *Guidelines for the Economic Evaluation of Health Technologies (4th Edition).* Canada.

Calip, G. S., J. A. Malmgren, W.-J. Lee, S. M. Schwartz and H. G. Kaplan. 2015. "Myelodysplastic syndrome and acute myeloid leukemia following adjuvant chemotherapy with and without granulocyte colony-stimulating factors for breast cancer". *Breast Cancer Research and Treatment* 154: 133–143.

Caro, J. J., A. H. Briggs, U. Siebert and K. M. Kuntz. 2012. "Modeling good research practices-overview: a report of the ISPOR-SMDM Modeling Good Research Practices Task Force–1". *Medical Decision Making* 32 (5): 667–677.

Carpenter, B., A. Gelman, M. Hoffman, D. Lee, B. Goodrich, M. Betancourt, M. Brubaker, J. Guo, P. Li and A. Riddell. 2016. "Stan: A probabilistic programming language". *Journal of Statistical Software* 20: 1–37.

Carvalho, T. M. de, J. van Rosmalen, H. B. Wolff, H. Koffijberg and V. M. Coupé. 2022. "Choosing a metamodel of a simulation model for uncertainty quantification". *Medical Decision Making* 42 (1): 28–42.

Chen, M. and A. Willan. 2014. "Value of information methods for assessing a new diagnostic test". *Statistics in Medicine* 33 (11): 1801–1815.

Cipriano, L. E. and T. A. Weber. 2018. "Population-level intervention and information collection in dynamic healthcare policy". *Health Care Management Science* 21 (4): 604–631. ISSN: 13869620. https://doi.org/10.1007/s10729-017-9415-5.

Claxton, K. 1999a. "Bayesian approaches to the value of information: implications for the regulation of new pharmaceuticals". *Health Economics* 8: 269–274.

———. 1999b. "The irrelevance of inference: a decision-making approach to stochastic evaluation of health care technologies". *Journal of Health Economics* 18: 342–364.

Claxton, K., S. Martin, M. Soares, N. Rice, E. Spackman, S. Hinde, N. Devlin, P. C. Smith, M. Sculpher. 2015. "Methods for the Estimation of the National Institute for Health and Care Excellence Cost-Effectiveness Threshold". *Health Technology Assessment* 19 (14): 1–503, v-vi. https://doi.org/10.3310/hta19140.

Claxton, K., S. Griffin, H. Koffijberg and C. McKenna. 2015a. "How to estimate the health benefits of additional research and changing clinical practice". *BMJ* 351.

Claxton, K., S. Martin, M. Soares, N. Rice, E. Spackman, S. Hinde, N. Devlin, P. C. Smith and M. Sculpher. 2015b. "Methods for the estimation of the National Institute for Health and Care Excellence cost-effectiveness threshold". *Health Technology Assessment (Winchester, England)* 19 (14): 1.

Claxton, K. and J. Posnett. 1996. "An economic approach to clinical trial design and research priority-setting". *Health Economics* 5 (6): 513–524. ISSN: 1057-9230. https://doi.org/10.1002/(sici)1099-1050(199611)5:6<513::aid-hec237>3.0.co;2-9.

Claxton, K. and K. M. Thompson. 2001. "A dynamic programming approach to the efficient design of clinical trials". *Journal of Health Economics* 20 (5): 797–822. ISSN: 01676296. https://doi.org/10.1016/S0167-6296(01)00093-5.

Connecticut Tumor Registry. 2018. *The Connecticut Tumor Registry (CTR): Part of the National Cancer Institute's Surveillance, Epidemiology, and End Results (SEER) program.* `https://portal.ct.gov/DPH/Tumor-Registry/CTR-Home`. Accessed Sep. 7, 2018.

Conti, S. and K. Claxton. 2009. "Dimensions of design space: a decision-theoretic approach to optimal research design". *Medical Decision Making* 29 (6): 643–660. ISSN: 0272989X. `https://doi.org/10.1177/0272989X09336142`.

Coyle, D. 2004. "The development and application of a normative framework for considering uncertainty and variability in economic evaluation". PhD diss.

Coyle, D., M. J. Buxton and B. J. O'Brien. 2003. "Measures of importance for economic analysis based on decision modeling". *Journal of Clinical Epidemiology* 56 (10): 989–997.

Coyle, D., K. M. Lee, M. Mamdani, K.-A. Sabarre and K. Tingley. 2015. "Reimbursement-based economics–what is it and how can we use it to inform drug policy reform?" *Headache: The Journal of Head and Face Pain* 55: 236–247.

Coyle, D. and J. Oakley. 2008. "Estimating the expected value of partial perfect information: a review of methods". *The European Journal of Health Economics* 9 (3): 251–259.

Crump, M., D. Tu, L. Shepherd, M. Levine, V. Bramwell and K. Pritchard. 2003. "Risk of acute leukemia following epirubicin-based adjuvant chemotherapy: a report from the National Cancer Institute of Canada Clinical Trials Group". *Journal of Clinical Oncology* 21 (16): 3066–3071.

Cunningham, R. M., M. A. Walton and P. M. Carter. 2018. "The major causes of death in children and adolescents in the United States". *New England Journal of Medicine* 379 (25): 2468–2475.

Dabbous, M., L. Chachoua, A. Caban and M. Toumi. 2020. "Managed entry agreements: policy analysis from the European perspective". *Value in Health* 23 (4): 425–433.

Davis, D. R., C. C. Kisiel and L. Duckstein. 1972. "Bayesian decision theory applied to design in hydrology". *Water Resources Research* 8 (1): 33–41.

Davis, P. and P. Rabinowitz. 2007. *Methods of numerical integration.* Courier Corporation.

De Angelis, D., A. M. Presanis, S. Conti and A. E. Ades. 2014. "Estimation of HIV burden through Bayesian evidence synthesis". *Statistical Science* 29 (1): 9–17. ISSN: 0883-4237. `https://doi.org/10.1214/13-STS428`.

Dewis, R. and J. Gribbin. 2009. *Breast Cancer: Diagnosis and Treatment: An Assessment of Need. (NICE Clinical Guidelines, No. 80-81S.)* Available from: https://www.ncbi.nlm.nih.gov/books/NBK61914/.

Dias, S., A. E. Ades, N. J. Welton, J. P. Jansen and A. J. Sutton. 2018. *Network Meta-Analysis for Decision-Making.* John Wiley & Sons, New York, NY.

Dimairo, M., P. Pallmann, J. Wason, S. Todd, T. Jaki, S. A. Julious, A. P. Mander, C. J. Weir, F. Koenig, M. K. Walton et al. 2020. "The adaptive designs CONSORT extension (ACE) statement: a checklist with explanation and elaboration guideline for reporting randomised trials that use an adaptive design". *BMJ* 369.

Donlan, P. 1997. "Modeling valuations for EuroQol health states". *Medical Care* 35 (11): 1095–1108.

Drummond, M., M. Sculpher, K. Claxton, G. Stoddart and G. Torrance. 2015. *Methods for the Economic Evaluation of Health Care Programmes.* Oxford University Press, Oxford, UK.

Dumville, J. C., T. A. Gray, C. J. Walter, C. A. Sharp, T. Page, R. Macefield, N. Blencowe, T. K. Milne, B. C. Reeves and J. Blazeby. 2016. "Dressings for the prevention of surgical site infection". *Cochrane Database of Systematic Reviews,* no. 12.

Durrleman, S. and R. Simon. 1989. "Flexible regression models with cubic splines". *Statistics in Medicine* 8 (5): 551–561.

Early Breast Cancer Trialists' Collaborative Group. 2005. "Effects of chemotherapy and hormonal therapy for early breast cancer on recurrence and 15-year survival: an overview of the randomised trials". *The Lancet* 365 (9472): 1687–1717.

Eckermann, S., A. Briggs and A. Willan. 2008. "Health technology assessment in the cost-disutility plane". *Medical Decision Making* 28 (2): 172–181.

Eckermann, S., J. Karnon and A. Willan. 2010. "The value of value of information". *PharmacoEconomics* 28 (9): 699–709.

Eckermann, S. and A. R. Willan. 2008. "Time and expected value of sample information wait for no patient". *Value in Health* 11 (3): 522–526. ISSN: 15244733. https://doi.org/10.1111/j.1524-4733.2007.00296.x. http://dx.doi.org/10.1111/j.1524-4733.2007.00296.x.

Eidsvik, J., T. Mukerji and D. Bhattacharjya. 2015. *Value of Information in the Earth Sciences: Integrating Spatial Modeling and Decision Analysis.* Cambridge University Press, Cambridge, UK.

Enns, E. A., L. E. Cipriano, C. T. Simons and C. Y. Kong. 2015. “Identifying best-fitting inputs in health-economic model calibration: a Pareto frontier approach”. *Medical Decision Making* 35 (2): 170–182.

Erber, R., C. Conradt, N. Homann, C. Enders, M. Finckh, A. Dietz, H. Weidauer and F. X. Bosch. 1998. “TP53 DNA contact mutations are selectively associated with allelic loss and have a strong clinical impact in head and neck cancer”. *Oncogene* 16 (13): 1671–1679.

Fairley, M., L. E. Cipriano and J. D. Goldhaber-Fiebert. 2020. “Optimal allocation of research funds under a budget constraint”. *Medical Decision Making* 40 (6): 797–814.

Fang, W., Z. Wang, M. B. Giles, C. H. Jackson, N. J. Welton, C. Andrieu and H. Thom. 2022. “Multilevel and quasi Monte Carlo methods for the calculation of the expected value of partial perfect information”. *Medical Decision Making* 42 (2): 168–181.

FDA. accessed 2023. *Guidance for Industry and FDA Staff: Best Practices for Conducting and Reporting Pharmacoepidemiologic Safety Studies Using Electronic Healthcare Data.* https://www.fda.gov/files/drugs/published/Best-Practices-for-Conducting-and-Reporting-Pharmacoepidemiologic-Safety-Studies-Using-Electronic-Healthcare-Data-Sets.pdf.

Felli, J. and G. Hazen. 1998. “Sensitivity analysis and the expected value of perfect information”. *Medical Decision Making* 18: 95–109.

———. 1999. “A Bayesian approach to sensitivity analysis”. *Health Economics* 8: 263–268.

Fenwick, E., K. Claxton and M. Sculpher. 2008. “The value of implementation and the value of information: combined and uneven development”. *Medical Decision Making* 28 (1): 21–32. ISSN: 0272989X. https://doi.org/10.1177/0272989X07308751.

Fenwick, E., L. Steuten, S. Knies, S. Ghabri, A. Basu, J. F. Murray, H. E. Koffijberg, M. Strong, G. D. S. Schmidler and C. Rothery. 2020. “Value of information analysis for research decisions—an introduction: report 1 of the ISPOR Value of Information Analysis Emerging Good Practices Task Force”. *Value in Health* 23 (2): 139–150.

Frazier, P. I. and W. B. Powell. 2010. “Paradoxes in learning and the marginal value of information”. *Decision Analysis* 7 (4): 378–403. ISSN: 15458490. https://doi.org/10.1287/deca.1100.0190.

Friedman, J. H. 1991. “Multivariate adaptive regression splines”. *The Annals of Statistics,* 1–67.

Garrison Jr, L. P., P. J. Neumann, P. Erickson, D. Marshall and C. D. Mullins. 2007. "Using real-world data for coverage and payment decisions: the ISPOR real-world data task force report". *Value in Health* 10 (5): 326–335.

Genetic Alliance; District of Columbia Department of Health. 2010. *Understanding Genetics: a District of Columbia guide for patients and health professionals.* Genetic Alliance, Washington, DC.

Gillum, L. A., C. Gouveia, E. R. Dorsey, M. Pletcher, C. D. Mathers, C. E. McCulloch and S. C. Johnston. 2011. "NIH disease funding levels and burden of disease". *PLoS ONE* 6 (2). ISSN: 19326203. `https://doi.org/10.1371/journal.pone.0016837`.

Glover, J., D. Izzo, K. Odato and W. Lei. 2006. *Evidence-Based Medicine Pyramid.* `http://www.dartmouth.edu/~biomed/resources.htmld/guides/ebm\_psych\_resources.html`.

Gold, M. R., J. E. Siegel and L. B. Russell. 1996. *Cost-Effectiveness in Health and Medicine.* Oxford University Press, Oxford, UK.

Greenberg, D., C. Earle, C.-H. Fang, A. Eldar-Lissai and P. J. Neumann. 2010. "When is cancer care cost-effective? A systematic overview of cost–utility analyses in oncology". *Journal of the National Cancer Institute* 102 (2): 82–88.

Greenhalgh, T. 1997. "How to read a paper: getting your bearings (deciding what the paper is about)". *BMJ* 315 (7102): 243–246.

Griffin, S., N. J. Welton and K. Claxton. 2010. "Exploring the research decision space: the expected value of information for sequential research designs". *Medical Decision Making* 30 (2): 155–162. ISSN: 0272989X. `https://doi.org/10.1177/0272989X09344746`.

Grimes, D. A. and K. F. Schulz. 2002a. "An overview of clinical research: the lay of the land". *The Lancet* 359 (9300): 57–61.

———. 2002b. "Descriptive studies: what they can and cannot do". *The Lancet* 359 (9301): 145–149.

Grimm, S., S. Dixon and J. Stevens. 2017a. "Assessing the expected value of research studies in reducing uncertainty and improving implementation dynamics". *Medical Decision Making* 37 (5): 523–533.

Grimm, S., X. Pouwels, B. Ramaekers, N. van Ravesteyn, V. Sankatsing, J. Grutters and M. Joore. 2021a. "Implementation barriers to Value of Information analysis in health technology decision making: results from a process evaluation". *Value in Health* 24 (8): 1126–1136.

Grimm, S., X. Pouwels, B. Ramaekers, B. Wijnen, T. Otten, J. Grutters and M. Joore. 2021b. "State of the ART? Two new tools for risk communication in health technology assessments". *PharmacoEconomics* 39 (10): 1185–1196.

Grimm, S. E., S. Dixon and J. W. Stevens. 2017b. "Assessing the expected value of research studies in reducing uncertainty and improving implementation dynamics". *Medical Decision Making* 37 (5): 523–533.

Grimm, S. E., M. Strong, A. Brennan and A. J. Wailoo. 2017. "The HTA risk analysis chart: visualising the need for and potential value of managed entry agreements in health technology assessment". *PharmacoEconomics* 35 (12): 1287–1296.

Guyatt, G. H., D. L. Sackett, J. C. Sinclair, R. Hayward, D. J. Cook and R. J. Cook. 1995. "Users' guides to the medical literature. IX. A method for grading health care recommendations. Evidence-Based Medicine Working Group". *JAMA* 274 (22): 1800–1804.

Guyot, P., A. Ades, M. J. Ouwens and N. J. Welton. 2012. "Enhanced secondary analysis of survival data: reconstructing the data from published Kaplan-Meier survival curves". *BMC Medical Research Methodology* 12: 1–13.

Hall, P. S., R. Edlin, S. Kharroubi, W. Gregory and C. McCabe. 2012. "Expected net present value of sample information: from burden to investment". *Medical Decision Making* 32 (3): 11–21. ISSN: 0272989X. https://doi.org/10.1177/0272989X12443010.

Hassett, M. J., S. M. Silver, M. E. Hughes, D. W. Blayney, S. B. Edge, J. G. Herman, C. A. Hudis, P. K. Marcom, J. E. Pettinga, D. Share et al. 2012. "Adoption of gene expression profile testing and association with use of chemotherapy among women with breast cancer". *Journal of Clinical Oncology* 30 (18): 2218–2226.

Hastie, T. and R. Tibshirani. 1990. *Generalized Additive Models.* Vol. 43. CRC Press, Boca Raton, FL.

Hausman, D. M. 2018. "The bond between positive and normative economics". *Revue d'economie Politique* 128 (2): 191–208.

Hayes, D. F. 2012. "Targeting adjuvant chemotherapy: a good idea that needs to be proven!" *Journal of Clinical Oncology: Official Journal of the American Society of Clinical Oncology* 30 (12): 1264–1267.

Heath, A. 2022. "Calculating expected value of sample information adjusting for imperfect implementation". *Medical Decision Making* 42 (5): 626–636.

Heath, A. and G. Baio. 2018. "Calculating the expected value of sample information using efficient nested Monte Carlo: a tutorial". *Value in Health* 21 (11): 1299–1304.

Heath, A., S. Grimm and A. Brennan. 2023. "Considering implementation levels alongside value of information calculations: challenges and considerations". *Under Review: Medical Decision Making.*

Heath, A., I. Manolopoulou and G. Baio. 2016. "Estimating the expected value of partial perfect information in health economic evaluations using integrated nested Laplace approximation". *Statistics in Medicine* 35 (23): 4264–4280.

———. 2017. "A review of methods for analysis of the expected value of information". *Medical Decision Making* 37 (7): 747–758.

———. 2018. "Efficient Monte Carlo estimation of the expected value of sample information using moment matching". *Medical Decision Making* 38 (2): 163–173.

Heath, A., N. Kunst, C. Jackson, M. Strong, F. Alarid-Escudero, J. D. Goldhaber-Fiebert, G. Baio, N. A. Menzies and H. Jalal. 2020. "Calculating the expected value of sample information in practice: considerations from 3 case studies". *Medical Decision Making* 40 (3): 314–326.

Heath, A., I. Manolopoulou and G. Baio. 2019. "Estimating the expected value of sample information across different sample sizes using moment matching and nonlinear regression". *Medical Decision Making* 39 (4): 347–359.

Heath, A., M. Strong, D. Glynn, N. Kunst, N. J. Welton and J. D. Goldhaber-Fiebert. 2022. "Simulating study data to support expected value of sample information calculations: a tutorial". *Medical Decision Making* 42 (2): 143–155.

Hernan, M. and J. Robins. 2020. *Causal Inference: What If.* Boca Raton: Chapman & Hall/CRC.

Hornberger, J., M. D. Alvarado, C. Rebecca, H. R. Gutierrez, T. M. Yu and W. J. Gradishar. 2012. "Clinical validity/utility, change in practice patterns, and economic implications of risk stratifiers to predict outcomes for early-stage breast cancer: a systematic review". *Journal of the National Cancer Institute* 104 (14): 1068–1079.

Hornberger, J. C., B. W. Brown and J. Halpern. 1995. "Designing a cost-effective clinical trial". *Statistics in Medicine* 14 (20): 2249–2259. ISSN: 10970258. https://doi.org/10.1002/sim.4780142008.

Hsu, C.-C. and B. A. Sandford. 2007. "The Delphi technique: making sense of consensus". *Practical Assessment, Research, and Evaluation* 12 (1): 10.

Hubbard, R., TE. Paradis. accessed 2023. *Real World Evidence: A New Era for Health Care Innovation.* `https://www.nehi-us.org/writable/publication\_files/file/rwe\_issue\_brief\_final.pdf`.

Hunink, M. M., M. C. Weinstein, E. Wittenberg, M. F. Drummond, J. S. Pliskin, J. B. Wong and P. P. Glasziou. 2014. *Decision Making in Health and Medicine: Integrating Evidence and Values.* Cambridge University Press, Cambridge, UK.

Husereau, D., M. Drummond, F. Augustovski, E. de Bekker-Grob, A. H. Briggs, C. Carswell, L. Caulley, N. Chaiyakunapruk, D. Greenberg, E. Loder et al. 2022. "Consolidated health economic evaluation reporting standards (CHEERS) 2022 explanation and elaboration: a report of the ISPOR CHEERS II good practices task force". *Value in Health* 25 (1): 10–31.

Incerti, D., H. Thom, G. Baio and J. P. Jansen. 2019. "R you still using Excel? The advantages of modern software tools for health technology assessment". *Value in Health* 22 (5): 575–579.

Jackson, C. H., G. Baio, A. Heath, M. Strong, N. J. Welton and E. C. F. Wilson. 2022. "Value of information analysis in models to inform health policy". *Annual Review of Statistics and Its Application* 9: 95–118.

Jackson, C. H., L. Bojke, S. G. Thompson, K. Claxton and L. D. Sharples. 2011. "A framework for addressing structural uncertainty in decision models". *Medical Decision Making* 31 (4): 662–674.

Jackson, C. H., R. Johnson, A. de Nazelle, R. Goel, T. H. de Sá, M. Tainio and J. Woodcock. 2021. "A guide to value of information methods for prioritising research in health impact modelling". *Epidemiologic Methods* 10 (1).

Jackson, C. H., A. M. Presanis, S. Conti and D. De Angelis. 2019. "Value of information: sensitivity analysis and research prioritisation in Bayesian evidence synthesis". *Journal of the American Statistical Association* 114: 1436–1449.

Jalal, H. and F. Alarid-Escudero. 2018. "A Gaussian approximation approach for value of information analysis". *Medical Decision Making* 38 (2): 174–188.

Jalal, H., P. Pechlivanoglou, E. Krijkamp, F. Alarid-Escudero, E. Enns and M. M. Hunink. 2017. "An overview of R in health decision sciences". *Medical Decision Making* 37 (7): 735–746.

Jemal, A., E. M. Ward, C. J. Johnson, K. A. Cronin, J. Ma, A. B. Ryerson, A. Mariotto, A. J. Lake, R. Wilson, R. L. Sherman et al. 2017. "Annual report to the nation on the status of cancer, 1975–2014, featuring survival". *JNCI: Journal of the National Cancer Institute* 109 (9): djx030.

Jutkowitz, E., F. Alarid-Escudero, H. K. Choi, K. M. Kuntz and H. Jalal. 2017. "Prioritizing future research on allopurinol and febuxostat for the management of gout: value of information analysis". *Pharmacoeconomics* 35:1073–1085.

Jutkowitz, E., F. Alarid-Escudero, K. M. Kuntz and H. Jalal. 2019. "The curve of optimal sample size (COSS): a graphical representation of the optimal sample size from a value of information analysis". *Pharmacoeconomics* 37 (7): 871–877.

Karlsson, G. and M. Johannesson. 1996. "The decision rules of cost-effectiveness analysis". *PharmacoEconomics* 9 (2): 113–120.

Karnon, J., J. Stahl, A. Brennan, J. J. Caro, J. Mar and J. Möller. 2012. "Modeling using discrete event simulation: a report of the ISPOR-SMDM Modeling Good Research Practices Task Force–4". *Medical Decision Making* 32 (5): 701–711.

Keha, A. B., I. R. De Farias and G. L. Nemhauser. 2006. "A branch-and-cut algorithm without binary variables for nonconvex piecewise linear optimization". *Operations Research* 54 (5): 847–858. ISSN: 0030364X. `https://doi.org/10.1287/opre.1060.0277`.

Kharroubi, S., A. Brennan and M. Strong. 2011. "Estimating expected value of sample information for incomplete data models using Bayesian approximation". *Medical Decision Making* 31 (6): 839–852.

Kim, D. D. and A. Basu. 2017. "New metrics for economic evaluation in the presence of heterogeneity: focusing on evaluating policy alternatives rather than treatment alternatives". *Medical Decision Making* 37 (8): 930–941.

Kim, D. D., M. C. Silver, N. Kunst, J. T. Cohen, D. A. Ollendorf and P. J. Neumann. 2020. "Perspective and costing in cost-effectiveness analysis, 1974–2018". *PharmacoEconomics* 38: 1135–1145.

Koffijberg, H., C. Rothery, K. Chalkidou and J. Grutters. 2018. "Value of information choices that influence estimates: a systematic review of prevailing considerations". *Medical Decision Making* 38 (7): 888–900.

Kratz, C. P., M. I. Achatz, L. Brugieres, T. Frebourg, J. E. Garber, M.-L. C. Greer, J. R. Hansford, K. A. Janeway, W. K. Kohlmann, R. McGee et al. 2017. "Cancer screening recommendations for individuals with Li-Fraumeni syndrome". *Clinical Cancer Research* 23 (11): e38–e45.

Kruschke, J.K. 2015. *Doing Bayesian Data Analysis: A Tutorial with R, JAGS and Stan.* 2nd ed., Academic Press, Burlington, MA.

Kuhn, M. 2022. *Caret: Classification and Regression Training.* R package version 6.0-92. `https://CRAN.R-project.org/package=caret`.

Kunst, N., A. Siu, M. Drummond, S. Grimm, J. Grutters, D. Husereau, H. Koffijberg, C. Rothery, E. Wilson and A. Heath. 2023a. "CHEERS Value of Information (CHEERS-VOI) Reporting Standards – Explanation and Elaboration". *Value in Health* 26 (10): 1461–1473.

Kunst, N., E. A. Burger, V. M. H. Coupe, K. M. Kuntz and E. Aas. 2023b. "A guide to an iterative approach to model-based decision making in health and medicine: An iterative decision-making framework". *Pharmaco Economics.* In press.

Kunst, N., N. K. Stout, G. O'Brien, K. D. Christensen, P. M. McMahon, A. C. Wu, L. R. Diller and J. M. Yeh. 2022. "Population-based newborn screening for germline TP53 variants: clinical benefits, cost-effectiveness, and value of further research". *JNCI: Journal of the National Cancer Institute* 114 (5): 722–731.

Kunst, N., E. C. Wilson, D. Glynn, F. Alarid-Escudero, G. Baio, A. Brennan, M. Fairley, J. D. Goldhaber-Fiebert, C. Jackson, H. Jalal et al. 2020. "Computing the expected value of sample information efficiently: practical guidance and recommendations for four model-based methods". *Value in Health* 23 (6): 734–742.

Kunst, N. R., F. Alarid-Escudero, A. D. Paltiel and S. Y. Wang. 2019. "A value of information analysis of research on the 21-gene assay for breast cancer management". *Value in Health* 22 (10): 1102–1110. `https://doi.org/10.1016/j.jval.2019.05.004`.

Lindley, D. V. 1956. "On a measure of the information provided by an experiment". *The Annals of Mathematical Statistics,* 986–1005.

Lumley, T. 2002. "Network meta-analysis for indirect treatment comparisons". *Statistics in Medicine* 21 (16): 2313–2324.

Lunn, D., A. Thomas, N. Best and D. Spiegelhalter. 2000. "WinBUGS-a Bayesian modelling framework: concepts, structure, and extensibility". *Statistics and Computing* 10 (4): 325–337.

Madan, J., A. Ades, M. Price, K. Maitland, J. Jemutai, P. Revill and N. Welton. 2014. "Strategies for efficient computation of the expected value of partial perfect information". *Medical Decision Making,* 0272989X13514774.

Makady, A., H. Stegenga, A. Ciaglia, T. P. Debray, M. Lees, M. Happich, B. Ryll, K. Abrams, R. Thwaites, S. Garner et al. 2017. "Practical implications of using real-world evidence (RWE) in comparative effectiveness research: learnings from IMI-GetReal". *Journal of Comparative Effectiveness Research* 6 (6): 485–490.

Marchetti, A., F. Buttitta, G. Merlo, F. Diella, S. Pellegrini, S. Pepe, P. Macchiarini, A. Chella, C. A. Angeletti, R. Callahan et al. 1993. "P53 alterations in non-small cell lung cancers correlate with metastatic involvement of hilar and mediastinal lymph nodes". *Cancer research* 53 (12): 2846–2851.

Marrone, M., A. Stewart and W. D. Dotson. 2015. "Clinical utility of gene-expression profiling in women with early breast cancer: an overview of systematic reviews". *Genetics in Medicine* 17 (7): 519–532.

McCabe, C., K. Claxton and A. Culyer. 2008. "The NICE cost-effectiveness threshold". *PharmacoEconomics* 26 (9): 733–744.

McCabe, C., M. Paulden, I. Awotwe, A. Sutton and P. Hall. 2020. "One-way sensitivity analysis for probabilistic cost-effectiveness analysis: conditional expected incremental net benefit". *PharmacoEconomics* 38 (2): 135–141.

McCabe, C. J., T. Stafinski, R. Edlin and D. Menon. 2010. "Access with evidence development schemes: a framework for description and evaluation". *Pharmacoeconomics* 28: 143–152.

McKenna, C., Z. Chalabi, D. Epstein and K. Claxton. 2010. "Budgetary policies and available actions: a generalisation of decision rules for allocation and research decisions". *Journal of Health Economics* 29 (1): 170–181. ISSN: 01676296. `https://doi.org/10.1016/j.jhealeco.2009.11.005`.

McKenna, C. and K. Claxton. 2011. "Addressing adoption and research design decisions simultaneously: the role of value of sample information analysis". *Medical Decision Making* 31 (6): 853–865.

McKenna, C., S. Griffin, H. Koffijberg and K. Claxton. 2016. "Methods to place a value on additional evidence are illustrated using a case study of corticosteroids after traumatic brain injury". *Journal of Clinical Epidemiology* 70: 183–190.

McKenna, C., C. McDaid, S. Suekarran, N. Hawkins, K. Claxton, K. Light, M. Chester, J. Cleland, N. Woolacott and M. Sculpher. 2009. "Enhanced external counterpulsation for the treatment of stable angina and heart failure: a systematic review and economic analysis". *Health Technology Assessment* 13 (24). ISSN: 13665278. `https://doi.org/10.3310/hta13240`.

McKenna, C., M. Soares, K. Claxton, L. Bojke, S. Griffin, S. Palmer and E. Spackman. 2015. "Unifying research and reimbursement decisions: case studies demonstrating the sequence of assessment and judgments required". *Value in Health* 18 (6): 865–875.

Meltzer, D. O., T. Hoomans, J. W. Chung and A. Basu. 2011. "Minimal modeling approaches to value of information analysis for health research". *Medical Decision Making* 31 (6): E1–E22.

Menzies, N. 2016. "An efficient estimator for the expected value of sample information". *Medical Decision Making* 36 (3): 308–320.

Michel, Y. A., L. Engel, K. Rand-Hendriksen, L. A. Augestad and D. G. Whitehurst. 2016. "'When I saw walking I just kind of took it as wheeling': interpretations of mobility-related items in generic, preference-based health state instruments in the context of spinal cord injury". *Health and Quality of Life Outcomes* 14: 1–11.

Milborrow, S. 2011. *Earth: Multivariate Adaptive Regression Splines.* R package. Derived from mda:mars by T. Hastie and R. Tibshirani. `http://CRAN.R-project.org/package=earth`.

Moher, D., K. F. Schulz, I. Simera and D. G. Altman. 2010. "Guidance for developers of health research reporting guidelines". *PLoS Medicine* 7 (2): e1000217.

Morita, S., P. F. Thall and P. Müller. 2008. "Determining the effective sample size of a parametric prior". *Biometrics* 64 (2): 595–602.

Moureau, N. and V. Chopra. 2016. "Indications for peripheral, midline and central catheters: summary of the MAGIC recommendations". *British Journal of Nursing* 25 (8): S15–S24.

Nabhan, C., A. Klink and V. Prasad. 2019. "Real-world evidence—what does it really mean?" *JAMA Oncology* 5 (6): 781–783.

National Cancer Institute. 2021. *Cancer in Children and Adolescents.* Accessed January 6, 2021. `https://www.cancer.gov/types/childhood-cancers/child-adolescent-cancers-fact-sheet#r3`.

———. accessed 2020. *NCI Dictionary of Cancer Terms.* `https://www.cancer.gov/publications/dictionaries/cancer-terms/def/levels-of-evidence?redirect=true`.

National Health Service. 2018. *PREDICT tool version 2.1: Breast cancer overall survival.* `http://www.predict.nhs.uk/predict\_v2.1/`. Accessed Jul. 7, 2018.

National Pressure Ulcer Advisory Panel, European Pressure Ulcer Advisory Panel, Pan Pacific Pressure Injury Alliance. 2019. *Prevention and Treatment of Pressure Ulcers: Clinical Practice Guideline. The International Guideline.* Edited by E. Haesler.

National Statistics, O. for. 2019. *Births in England and Wales: summary tables, release 31st July 2019.* `https://www.ons.gov.uk/peoplepopulationandcommunity/birthsdeathsandmarriages/livebirths/datasets/birthsummarytables`.

Neumann, P. J., J. T. Cohen, M. C. Weinstein. 2014. "Updating cost-effectiveness—the curious resilience of the $50,000-per-QALY threshold". *New England Journal of Medicine* 371 (9): 796–797.

Neumann, P. J., G. D. Sanders, L. B. Russell, J. E. Siegel and T. G. Ganiats. 2016. *Cost-Effectiveness in Health and Medicine.* Oxford University Press, Oxford, UK.

Oakley, J. 2002. "Value of information for complex cost-effectiveness models". *Research Report.*

Oakley, J. E. and A. O'Hagan. 2004. "Probabilistic sensitivity analysis of complex models: a Bayesian approach". *Journal of the Royal Statistical Society, Series B* 66 (3): 751–769.

O'Brien, B. J., M. F. Drummond, R. J. Labelle and A. Willan. 1994. "In search of power and significance: issues in the design and analysis of stochastic cost-effectiveness studies in health care". *Medical Care,* 150–163.

O'Hagan, A., C. E. Buck, A. Daneshkhah, J. R. Eiser, P. H. Garthwaite, D. J. Jenkinson, J. E. Oakley and T. Rakow. 2006. *Uncertain Judgements: Eliciting Experts' Probabilities.* John Wiley & Sons, New York, NY.

Orucevic, A., R. E. Heidel and J. L. Bell. 2016. "Utilization and impact of 21-gene recurrence score assay for breast cancer in clinical practice across the United States: lessons learned from the 2010 to 2012 National Cancer Data Base analysis". *Breast Cancer Research and Treatment* 157: 427–435.

Page, M. J., J. E. McKenzie, P. M. Bossuyt, I. Boutron, T. C. Hoffmann, C. D. Mulrow, L. Shamseer, J. M. Tetzlaff, E. A. Akl, S. E. Brennan et al. 2021. "The PRISMA 2020 statement: an updated guideline for reporting systematic reviews". *International Journal of Surgery* 88: 105906.

Parmigiani, G. and L. Inoue. 2009. *Decision Theory: Principles and Approaches.* John Wiley & Sons, New York, NY.

Paulden, M. 2020a. "Calculating and interpreting ICERs and net benefit". *PharmacoEconomics* 38 (8): 785–807.

———. 2020b. "Why it's time to abandon the ICER". *PharmacoEconomics* 38 (8): 781–784.

Petitjean, A., M. Achatz, A. Borresen-Dale, P. Hainaut and M. Olivier. 2007. "TP53 mutations in human cancers: functional selection and impact on cancer prognosis and outcomes". *Oncogene* 26 (15): 2157–2165.

Philips, Z., K. Claxton and S. Palmer. 2008. "The half-life of truth: what are appropriate time horizons for research decisions?" *Medical Decision Making* 28 (3): 287–299.

Plummer, M. 2003. "JAGS: A program for analysis of Bayesian graphical models using Gibbs sampling". In *Proceedings of the 3rd International Workshop on Distributed Statistical Computing,* 124: 125. Vienna, Austria.

Plummer, M. 2015. *rjags: Bayesian Graphical Models using MCMC.* R package version 4-4. `http://CRAN.R-project.org/package=rjags`.

Presanis, A. M., O. N. Gill, T. R. Chadborn, C. Hill, V. Hope, L. Logan, B. D. Rice, V. C. Delpech, A. E. Ades and D. De Angelis. 2010. "Insights into the rise in HIV infections, 2001 to 2008: a Bayesian synthesis of prevalence evidence". *AIDS (London, England)* 24 (18): 2849–2858.

R Core Team. 2014. *R: A Language and Environment for Statistical Computing.* Vienna, Austria: R Foundation for Statistical Computing. `http://www.R-project.org/`.

Raiffa, H. 1968. *Decision Analysis: Introductory Lectures on Choices Under Uncertainty.* Addison-Wesley, Boston, MA.

Raiffa, H. and H. Schlaifer. 1961. *Applied Statistical Decision Theory.* Harvard University Press, Boston, MA.

Rasmussen, C. and C. Williams. 2006. *Gaussian Processes for Machine Learning.*

Ravdin, P. M., L. A. Siminoff, G. J. Davis, M. B. Mercer, J. Hewlett, N. Gerson and H. L. Parker. 2001. "Computer program to assist in making decisions about adjuvant therapy for women with early breast cancer". *Journal of Clinical Oncology* 19 (4): 980–991.

Reeves, B. C., L. Rooshenas, R. C. Macefield, M. Woodward, N. J. Welton, B. R. Waterhouse, A. D. Torrance, S. Strong, D. Siassakos, W. Seligman et al. 2019. "Three wound-dressing strategies to reduce surgical site infection after abdominal surgery: the Bluebelle feasibility study and pilot RCT". *Health Technology Assessment* 23 (39): 1.

Regnier, S. and D. B. Ridley. 2015. "Forecasting market share in the US pharmaceutical market". *Nature Reviews Drug Discovery* 14 (9): 594–595.

Repo, A. J. 1989. "The value of information: approaches in economics, accounting, and management science". *Journal of the American Society for Information Science* 40 (2): 68–85.

Richards, F. 1959. "A flexible growth function for empirical use". *Journal of Experimental Botany* 10 (2): 290–301.

Rickard, C. M., N. Marsh, J. Webster, N. Runnegar, E. Larsen, M. R. McGrail, F. Fullerton, E. Bettington, J. A. Whitty, M. A. Choudhury et al. 2018. "Dressings and securements for the prevention of peripheral intravenous catheter failure in adults (SAVE): a pragmatic, randomised controlled, superiority trial". *The Lancet* 392 (10145): 419–430.

Roberts, M., L. B. Russell, A. D. Paltiel, M. Chambers, P. McEwan and M. Krahn. 2012. "Conceptualizing a model: a report of the ISPOR-SMDM modeling good research practices task force–2". *Medical Decision Making* 32 (5): 678–689.

Robson, M. E., C. D. Storm, J. Weitzel, D. S. Wollins, K. Offit et al. 2010. "American Society of Clinical Oncology policy statement update: genetic and genomic testing for cancer susceptibility". *Journal of Clinical Oncology* 28 (5): 893–901.

Rothery, C., M. Strong, H. E. Koffijberg, A. Basu, S. Ghabri, S. Knies, J. F. Murray, G. D. S. Schmidler, L. Steuten and E. Fenwick. 2020. "Value of information analytical methods: report 2 of the ISPOR value of information analysis emerging good practices task force". *Value in Health* 23 (3): 277–286.

Rouzier, R., P. Pronzato, E. Chéreau, J. Carlson, B. Hunt and W. J. Valentine. 2013. "Multigene assays and molecular markers in breast cancer: systematic review of health economic analyses". *Breast Cancer Research and Treatment* 139: 621–637.

Rubin, D. B. 1988. "Using the SIR algorithm to simulate posterior distributions". *Bayesian Statistics* 3: 395–402.

Rutter, C. E., X. Yao, B. R. Mancini, J. A. Aminawung, A. B. Chagpar, O. Saglam, E. W. Hofstatter, M. Abu-Khalaf, C. P. Gross and S. B. Evans. 2016. "Influence of a 21-gene recurrence score assay on chemotherapy delivery in breast cancer". *Clinical Breast Cancer* 16 (1): 59–62.

Sackett, D. L., W. M. Rosenberg, J. M. Gray, R. B. Haynes and W. S. Richardson. 1996. *Evidence Based Medicine: What It Is and What It Isn't,* 7023.

Sadatsafavi, M., N. Bansback, Z. Zafari, M. Najafzadeh and C. Marra. 2013. "Need for speed: an efficient algorithm for calculation of single-parameter expected value of partial perfect information". *Value in Health* 16 (2): 438–448.

Sadatsafavi, M., C. Marra and S. Bryan. 2013. "Two-level resampling as a novel method for the calculation of the expected value of sample information in economic trials". *Health Economics* 22 (7): 877–882.

Saltelli, A., K. Chan and E. M. Scott, eds. 2000. *Sensitivity Analysis.* Wiley, New York, NY.

Sapsford, R. 2009. *Deputy Minister's response to the investigation into the Ministry of Health and Long-Term Care's decision-making concerning the funding of Avastin for colorectal cancer patients.* Letter dated August 26, 2009.

Savage, L. J. 1954. *The Foundations of Statistics.* Dover, Mineola, NY.

Schulz, K. F., D. G. Altman and D. Moher. 2010. "CONSORT 2010 statement: updated guidelines for reporting parallel group randomised trials". *Journal of Pharmacology and Pharmacotherapeutics* 1 (2): 100–107.

Sculpher, M. 2008. "Subgroups and heterogeneity in cost-effectiveness analysis". *PharmacoEconomics* 26: 799–806.

Sculpher, M., M. Drummond and M. Buxton. 1997. "The iterative use of economic evaluation as part of the process of health technology assessment". *Journal of Health Services Research and Policy* 2 (1): 26–30.

Sculpher, M. J., K. Claxton, M. Drummond and C. McCabe. 2006. "Whither trial-based economic evaluation for health care decision making?" *Health Economics* 15 (7): 677–687.

Seidinger, A. L., M. J. Mastellaro, F. Paschoal Fortes, J. Godoy Assumpção, I. Aparecida Cardinalli, M. Aparecida Ganazza, R. Correa Ribeiro, S. R. Brandalise, S. dos Santos Aguiar and J. A. Yunes. 2011. "Association of the highly prevalent TP53 R337H mutation with pediatric choroid plexus carcinoma and osteosarcoma in southeast Brazil". *Cancer* 117 (10): 2228–2235.

Sherman, R. E., S. A. Anderson, G. J. Dal Pan, G. W. Gray, T. Gross, N. L. Hunter, L. LaVange, D. Marinac-Dabic, P. W. Marks, M. A. Robb et al. 2016. "Real-world evidence—what is it and what can it tell us". *New England Journal of Medicine* 375 (23): 2293–2297.

Shewmaker, P., S. A. Chrysanthopoulou, R. Iskandar, D. Lake and E. Jutkowitz. 2022. "Microsimulation model calibration with approximate Bayesian computation in R: A tutorial". *Medical Decision Making* 42 (5): 557–570.

Siebert, U., O. Alagoz, A. M. Bayoumi, B. Jahn, D. K. Owens, D. J. Cohen, K. M. Kuntz, ISPOR-SMDM Modeling Good Research Practices Task Force. 2012. "State-transition modeling: a report of the ISPOR-SMDM modeling good research practices task force-3". *Value in Health* 15 (6): 812–820.

Siegel, R. L., K. D. Miller and A. Jemal. 2018. "Cancer statistics, 2018". *CA: A Cancer Journal for Clinicians* 68 (1): 7–30.

Signorovitch, J. E., E. Q. Wu, A. P. Yu, C. M. Gerrits, E. Kantor, Y. Bao, S. R. Gupta and P. M. Mulani. 2010. "Comparative effectiveness without head-to-head trials: a method for matching-adjusted indirect comparisons applied to psoriasis treatment with adalimumab or etanercept". *Pharmacoeconomics* 28: 935–945.

Simera, I., D. Moher, A. Hirst, J. Hoey, K. F. Schulz and D. G. Altman. 2010. "Transparent and accurate reporting increases reliability, utility, and impact of your research: reporting guidelines and the EQUATOR Network". *BMC Medicine* 8 (1): 1–6.

Smith, G. C., A. A. Moraitis, D. Wastlund, J. G. Thornton, A. Papageorghiou, J. Sanders, A. E. Heazell, S. C. Robson, U. Sovio, P. Brocklehurst et al. 2021. "Universal late pregnancy ultrasound screening to predict adverse outcomes in nulliparous women: a systematic review and cost-effectiveness analysis". *Health Technology Assessment* 25 (15): 1.

Speich, B., B. von Niederhäusern, N. Schur, L. G. Hemkens, T. Fürst, N. Bhatnagar, R. Alturki, A. Agarwal, B. Kasenda, C. Pauli-Magnus et al. 2018. "Systematic review on costs and resource use of randomized clinical trials shows a lack of transparent and comprehensive data". *Journal of Clinical Epidemiology* 96: 1–11.

Spiegelhalter, D., K. Abrams and J. Myles. 2004. *Bayesian Approaches to Clinical Trials and Health-Care Evaluation.* John Wiley / Sons, Chichester, UK.

Stanfill, B., H. Mielenz, D. Clifford and P. Thorburn. 2015. "Simple approach to emulating complex computer models for global sensitivity analysis". *Environmental Modelling & Software* 74: 140–155.

Stevenson, M. D., J. E. Oakley, M. Lloyd Jones, A. Brennan, J. E. Compston, E. V. McCloskey and P. L. Selby. 2009. "The cost-effectiveness of an RCT to establish whether 5 or 10 years of bisphosphonate treatment is the better duration for women with a prior fracture". *Medical Decision Making* 29 (6): 678–689. ISSN: 0272989X. https://doi.org/10.1177/0272989X09336077.

Stinnett, A. and J. Mullahy. 1998. "Net health benefits: a new framework for the analysis of uncertainty in cost-effectiveness analysis". *Medical Decision Making* 18 (2): S68–S80.

Strong, M., A. Brennan and J. Oakley. 2015. "How to calculate value of information in seconds using 'SAVI', the Sheffield Accelerated Value of Information web app". *Value in Health* 18 (7): A725–A726.

Strong, M. and J. Oakley. 2013. "An efficient method for computing single-parameter partial expected value of perfect information". *Medical Decision Making* 33 (6): 755–766.

Strong, M. and J. E. Oakley. 2014. "When is a model good enough? Deriving the expected value of model improvement via specifying internal model discrepancies". *SIAM/ASA Journal on Uncertainty Quantification* 2 (1): 106–125.

Strong, M., J. Oakley and A. Brennan. 2014. "Estimating multiparameter partial expected value of perfect information from a probabilistic sensitivity analysis sample: a nonparametric regression approach". *Medical Decision Making* 34 (3): 311–326.

Strong, M., J. Oakley, A. Brennan and P. Breeze. 2015. "Estimating the expected value of sample information using the probabilistic sensitivity analysis sample a fast nonparametric regression-based method". *Medical Decision Making* 35 (5): 570–583.

Strong, M., J. E. Oakley and J. Chilcott. 2012. "Managing structural uncertainty in health economic decision models: a discrepancy approach". *Journal of the Royal Statistical Society: Series C (Applied Statistics)* 61 (1): 25–45.

Suhrcke, M., K. Claxton, J. Lomas and J. Ochalek. accessed 2023. *Cost-effectiveness threshold estimation.* `https://www.york.ac.uk/che/research/global-health/cost-effectiveness-thresholds/`.

Sweeting, T. and S. Kharroubi. 2003. "Some new formulae for posterior expectations and Bartlett corrections". *Test* 12 (2): 497–521.

Tabori, U., A. Shlien, B. Baskin, S. Levitt, P. Ray, N. Alon, C. Hawkins, E. Bouffet, M. Pienkowska, L. Lafay-Cousin et al. 2010. "TP53 alterations determine clinical subgroups and survival of patients with choroid plexus tumors". *Journal of Clinical Oncology* 28 (12): 1995–2001.

Tang, G., S. Shak, S. Paik, S. J. Anderson, J. P. Costantino, C. E. Geyer, E. P. Mamounas, D. L. Wickerham and N. Wolmark. 2011. "Comparison of the prognostic and predictive utilities of the 21-gene Recurrence Score assay and Adjuvant! for women with node-negative, ER-positive breast cancer: results from NSABP B-14 and NSABP B-20". *Breast cancer research and treatment* 127: 133–142.

Therrell Jr, B. L. 2001. "US newborn screening policy dilemmas for the twenty-first century". *Molecular Genetics and Metabolism* 74 (1–2): 64–74.

Thompson, M. S. 1981. "Decision-analytic determination of study size: the case of electronic fetal monitoring". *Medical Decision Making* 1 (2): 165–179. ISSN: 1552681x. `https://doi.org/10.1177/0272989X8100100206`.

Tuffaha, H., L. Gordon and P. Scuffham. 2016. "Value of Information analysis informing adoption and research decisions in a portfolio of health care interventions". *MDM Policy & Practice* 1 (1).

Tuffaha, H., C. Rothery, N. Kunst, C. Jackson, M. Strong and S. Birch. 2021. "A review of web-based tools for Value of Information analysis". *Applied Health Economics and Health Policy* 19 (5): 645–651.

Tuffaha, H. W., B. M. Gillespie, W. Chaboyer, L. G. Gordon and P. A. Scuffham. 2015a. "Cost-utility analysis of negative pressure wound therapy in high-risk cesarean section wounds". *Journal of Surgical Research* 195 (2): 612–622.

Tuffaha, H. W., H. Reynolds, L. G. Gordon, C. M. Rickard and P. A. Scuffham. 2014a. "Value of information analysis optimizing future trial design from a pilot study on catheter securement devices". *Clinical Trials* 11 (6): 648–656.

Tuffaha, H. W., C. M. Rickard, J. Webster, N. Marsh, L. Gordon, M. Wallis and P. A. Scuffham. 2014b. "Cost-effectiveness analysis of clinically indicated versus routine replacement of peripheral intravenous catheters". *Applied Health Economics and Health Policy* 12: 51–58.

Tuffaha, H. W., S. Roberts, W. Chaboyer, L. G. Gordon and P. A. Scuffham. 2015. "Cost-effectiveness and value of information analysis of nutritional support for preventing pressure ulcers in high-risk patients: implement now, research later". *Applied Health Economics and Health Policy* 13: 167–179.

Tuffaha, H. W., M. Strong, L. G. Gordon and P. A. Scuffham. 2016. "Efficient value of information calculation using a nonparametric regression approach: an applied perspective". *Value in Health* 19 (4): 505–509.

Tuuli, M. G., J. Liu, A. T. Tita, S. Longo, A. Trudell, E. B. Carter, A. Shanks, C. Woolfolk, A. B. Caughey, D. K. Warren et al. 2020. "Effect of prophylactic negative pressure wound therapy vs standard wound dressing on surgical-site infection in obese women after cesarean delivery: a randomized clinical trial". *JAMA* 324 (12): 1180–1189.

Valentelyte, G., C. Keegan and J. Sorensen. 2022. "A comparison of four quasi-experimental methods: an analysis of the introduction of activity-based funding in Ireland". *BMC Health Services Research* 22 (1): 1311.

Vanni, T., J. Karnon, J. Madan, R. G. White, W. J. Edmunds, A. M. Foss and R. Legood. 2011. "Calibrating models in economic evaluation: a seven-step approach". *Pharmaco Economics* 29: 35–49.

Vervaart, M., E. Aas, K. P. Claxton, M. Strong, N. J. Welton, T. Wisløff and A. Heath. 2023. "General Purpose Methods for Simulating Survival Data for Expected Value of Sample Information Calculations". *Medical Decision Making,* 0272989X231162069.

Vervaart, M., M. Strong, K. P. Claxton, N. J. Welton, T. Wisløff and E. Aas. 2022. "An efficient method for computing expected value of sample information for survival data from an ongoing trial". *Medical Decision Making* 42 (5): 612–625.

Von Elm, E., D. G. Altman, M. Egger, S. J. Pocock, P. C. Gøtzsche and J. P. Vandenbroucke. 2007. "The Strengthening the Reporting of Observational Studies in Epidemiology (STROBE) statement: guidelines for reporting observational studies". *The Lancet* 370 (9596): 1453–1457.

Wahls, W. P. 2019. "Opinion: The National Institutes of Health needs to better balance funding distributions among US institutions". *Proceedings of the National Academy of Sciences* 116 (27): 13150–13154. ISSN: 0027-8424. https://doi.org/10.1073/pnas.1909217116. http://www.pnas.org/lookup/doi/10.1073/pnas.1909217116.

Wakker, P. and M. P. Klaassen. 1995. "Confidence intervals for cost/effectiveness ratios". *Health Economics* 4 (5): 373–381.

Walden University Library. accessed 2020. *Evidence-Based Research: Overview.* https://academicguides.waldenu.edu/library/healthevidence/evidencepyramid.

Wang, S.-Y., W. Dang, I. Richman, S. S. Mougalian, S. B. Evans and C. P. Gross. 2018. "Cost-effectiveness analyses of the 21-gene assay in breast cancer: systematic review and critical appraisal". *Journal of Clinical Oncology* 36 (16): 1619.

Ward, S., A. Scope, R. Rafia, A. Pandor, S. Harnan, P. Evans and L Wyld. 2013. "Gene expression profiling and expanded immunohistochemistry tests to guide the use of adjuvant chemotherapy in breast cancer management: a systematic review and cost-effectiveness analysis". *Health Technology Assessment* 17 (44).

Wasserman, J. D., A. Novokmet, C. Eichler-Jonsson, R. C. Ribeiro, C. Rodriguez-Galindo, G. P. Zambetti and D. Malkin. 2015. "Prevalence and functional consequence of TP53 mutations in pediatric adrenocortical carcinoma: a children's oncology group study". *Journal of Clinical Oncology* 33 (6): 602.

Weinstein, M. C. 1983. "Cost-effective priorities for cancer prevention". *Science* 221 (4605): 17–23. ISSN: 00368075. https://doi.org/10.1126/science.6407110.

Weiss, N. 2006. *A Course in Probability.* Addison-Wesley, Boston, MA.

Welch, J. S., A. A. Petti, C. A. Miller, C. C. Fronick, M. O'Laughlin, R. S. Fulton, R. K. Wilson, J. D. Baty, E. J. Duncavage, B. Tandon et al. 2016. "TP53 and decitabine in acute myeloid leukemia and myelodysplastic syndromes". *New England Journal of Medicine* 375 (21): 2023–2036.

Welton, N., J. Madan, D. Caldwell, T. Peters and A. Ades. 2014. "Expected value of sample information for multi-arm cluster randomized trials with binary outcomes". *Medical Decision Making* 34 (3): 352–365.

Welton, N. J., A. J. Sutton, N. Cooper, K. R. Abrams and A. Ades. 2012. *Evidence Synthesis for Decision Making in Healthcare.* Vol. 132. John Wiley & Sons, New York, NY.

Willan, A. and A. Briggs. 2006. *The Statistical Analysis of Cost-Effectiveness Data.* John Wiley / Sons, Chichester, UK.

Willan, A. and S. Eckermann. 2010. "Optimal clinical trial design using value of information methods with imperfect implementation". *Health Economics* 19 (5): 549–561. `https://doi.org/10.1002/hec.1493`.

Willan, A., R. Goeree and K. Boutis. 2012. "Value of information methods for planning and analyzing clinical studies optimize decision making and research planning". *Journal of Clinical Epidemiology* 65 (8): 870–876.

Wilson, E. 2015. "A practical guide to value of information analysis". *PharmacoEconomics* 33 (2): 105–121. ISSN: 1170-7690. `https://doi.org/10.1007/s40273-014-0219-x`. `http://dx.doi.org/10.1007/s40273-014-0219-x`.

Wilson, E. C., D. Wastlund, A. A. Moraitis and G. C. Smith. 2021. "Late pregnancy ultrasound to screen for and manage potential birth complications in nulliparous women: a cost-effectiveness and value of information analysis". *Value in Health* 24 (4): 513–521.

Wishart, G. C., E. M. Azzato, D. C. Greenberg, J. Rashbass, O. Kearins, G. Lawrence, C. Caldas and P. D. Pharoah. 2010. "PREDICT: a new UK prognostic model that predicts survival following surgery for invasive breast cancer". *Breast Cancer Research* 12 (1): 1–10.

Wood, S. 2011. "Fast stable restricted maximum likelihood and marginal likelihood estimation of semiparametric generalized linear models". *Journal of the Royal Statistical Society: Series B (Statistical Methodology)* 73 (1): 3–36.

———. 2003. "Thin plate regression splines". *Journal of the Royal Statistical Society: Series B (Statistical Methodology)* 65 (1): 95–114.

Wood, S. 2017. *Generalized Additive Models: An Introduction with R.* 2nd ed. Chapman / Hall/CRC.

Wunder, J. S., N. Gokgoz, R. Parkes, S. B. Bull, S. Eskandarian, A. M. Davis, C. P. Beauchamp, E. U. Conrad, R. J. Grimer, J. H. Healey et al. 2005. "TP53 mutations and outcome in osteosarcoma: a prospective, multicenter study". *Journal of Clinical Oncology* 23 (7): 1483–1490.

Yeh, J. M., N. K. Stout, A. Chaudhry, K. D. Christensen, M. Gooch, P. M. McMahon, G. O'Brien, N. Rehman, C. L. B. Zawatsky, R. C. Green et al. 2021. "Universal newborn genetic screening for pediatric cancer predisposition syndromes: model-based insights". *Genetics in Medicine* 23 (7): 1366–1371.

Yokota, F. and K. M. Thompson. 2004. "Value of information literature analysis: a review of applications in health risk management". *Medical Decision Making* 24 (3): 287–298.

Index